T0199656

Marine Navigation and Safety of Sea Transportation

STCW, Maritime Education and Training (MET), Human Resources and Crew Manning, Maritime Policy, Logistics and Economic Matters

Editors

Adam Weintrit & Tomasz Neumann
Gdynia Maritime University, Gdynia, Poland

CRC Press
Taylor & Francis Group
Boca Raton   London   New York   Leiden

CRC Press is an imprint of the
Taylor & Francis Group, an **informa** business

A BALKEMA BOOK

Published by:
CRC Press/Balkema
P.O. Box 447, 2300 AK Leiden, The Netherlands
e-mail: Pub.NL@taylorandfrancis.com
www.crcpress.com – www.taylorandfrancis.com

First issued in paperback 2020

© 2013 Taylor & Francis Group, London, UK
CRC Press/Balkema is an imprint of the Taylor & Francis Group, an informa business

No claim to original U.S. Government works

ISBN 13: 978-0-367-57640-0 (pbk)
ISBN 13: 978-1-138-00104-6 (hbk)

Visit the Taylor & Francis Web site at
http://www.taylorandfrancis.com

and the CRC Press Web site at
http://www.crcpress.com

Typeset by V Publishing Solutions Pvt Ltd., Chennai, India

MARINE NAVIGATION AND SAFETY OF SEA TRANSPORTATION

List of reviewers

Prof. Roland **Akselsson**, Lund University, Sweden
Prof. Anatoli **Alop**, Estonian Maritime Academy, Tallin, Estonia
Prof. Yasuo **Arai**, Independent Administrative Institution Marine Technical Education Agency,
Prof. Terje **Aven**, University of Stavanger (UiS), Stavanger, Norway
Prof. Andrzej **Banachowicz**, West Pomeranian University of Technology, Szczecin, Poland
Prof. Michael **Barnett**, Southampton Solent University, United Kingdom
Prof. Eugen **Barsan**, Constanta Maritime University, Romania
Prof. Milan **Batista**, University of Ljubljana, Ljubljana, Slovenia
Prof. Angelica **Baylon**, Maritime Academy of Asia & the Pacific, Philippines
Prof. Christophe **Berenguer**, Grenoble Institute of Technology, Saint Martin d'Heres, France
Prof. Heinz Peter **Berg**, Bundesamt für Strahlenschutz, Salzgitter, Germany
Prof. Tor Einar **Berg**, Norwegian Marine Technology Research Institute, Trondheim, Norway
Prof. Alfred **Brandowski**, Gdynia Maritime University, Poland
Sr. Jesus **Carbajosa Menendez**, President of Spanish Institute of Navigation, Spain
Prof. Pierre **Cariou**, Word Maritime University, Malmö, Sweden
Prof. A. Güldem **Cerit**, Dokuz Eylül University, Izmir, Turkey
Prof. Shyy Woei **Chang**, National Kaohsiung Marine University, Taiwan
Prof. Kevin **Cullinane**, University of Newcastle upon Tyne, UK
Prof. German **de Melo Rodriguez**, Polytechnical University of Catalonia, Barcelona, Spain
Prof. Daniel **Duda**, Naval University of Gdynia, Polish Nautological Society, Poland
Prof. Włodzimierz **Filipowicz**, Gdynia Maritime University, Poland
Prof. Jens **Froese**, Jacobs University Bremen, Germany
Prof. Masao **Furusho**, Kobe University, Japan
Prof. Wiesław **Galor**, Maritime University of Szczecin, Poland
Prof. Jerzy **Gaździcki**, President of the Polish Association for Spatial Information; Warsaw, Poland
Prof. Avtandil **Gegenava**, Georgian Maritime Transport Agency, Head of Maritime Rescue Coordination Center, Georgia
Prof. Witold **Gierusz**, Gdynia Maritime University, Poland
Prof. Stanislaw **Górski**, Gdynia Maritime University, Poland
Prof. Andrzej **Grzelakowski**, Gdynia Maritime University, Poland
Prof. Jerzy **Hajduk**, Maritime University of Szczecin, Poland
Prof. Marianna **Jacyna**, Warsaw University of Technology, Poland
Prof. Ales **Janota**, University of Žilina, Slovakia
Prof. Jacek **Januszewski**, Gdynia Maritime University, Poland
Prof. Jung Sik **Jeong**, Mokpo National Maritime University, South Korea
Prof. Tae-Gweon **Jeong**, Korean Maritime University, Pusan, Korea
Prof. Mirosław **Jurdziński**, Gdynia Maritime University, Poland
Prof. John **Kemp**, Royal Institute of Navigation, London, UK
Prof. Serdjo **Kos**, University of Rijeka, Croatia
Prof. Andrzej **Królikowski**, Maritime Office in Gdynia; Gdynia Maritime University, Poland
Prof. Shashi **Kumar**, U.S. Merchant Marine Academy, New York
Prof. Bogumił **Łączyński**, Gdynia Maritime University, Poland
Prof. Andrzej **Lewiński**, University of Technology and Humanities in Radom, Poland
Prof. Józef **Lisowski**, Gdynia Maritime University, Poland
Prof. Vladimir **Loginovsky**, Admiral Makarov State Maritime Academy, St. Petersburg, Russia
Prof. Evgeniy **Lushnikov**, Maritime University of Szczecin, Poland
Prof. Melchor M. **Magramo**, John B. Lacson Foundation Maritime University, Iloilo City, Philippines
Prof. Marek **Malarski**, Warsaw University of Technology, Poland
Prof. Francesc Xavier **Martinez de Oses**, Polytechnical University of Catalonia, Barcelona, Spain
Prof. Boyan **Mednikarov**, Nikola Y. Vaptsarov Naval Academy,Varna, Bulgaria
Prof. Jerzy **Mikulski**, Silesian University of Technology, Katowice, Poland
Prof. Janusz **Mindykowski**, Gdynia Maritime University, Poland
Prof. Daniel Seong-Hyeok **Moon**, World Maritime University, Malmoe, Sweden
Prof. Wacław **Morgaś**, Polish Naval Academy, Gdynia, Poland

Prof. Takeshi **Nakazawa**, World Maritime University, Malmoe, Sweden
Prof. Rudy R. **Negenborn**, Delft University of Technology, Delft, The Netherlands
Prof. Gabriel **Nowacki**, Military University of Technology, Warsaw
Mr. David **Patraiko**, The Nautical Institute, UK
Prof. Vytautas **Paulauskas**, Maritime Institute College, Klaipeda University, Lithuania
Prof. Jan **Pawelski**, Gdynia Maritime University, Poland
Prof. Thomas **Pawlik**, Bremen University of Applied Sciences, Germany
Prof. Francisco **Piniella**, University of Cadiz, Spain
Prof. Malek **Pourzanjani**, Australian Maritime College, Australia
Prof. Boris **Pritchard**, University of Rijeka, Croatia
Prof. Michael **Roe**, University of Plymouth, Plymouth, United Kingdom
Prof. Władysław **Rymarz**, Gdynia Maritime University, Poland
Prof. Abdul Hamid **Saharuddin**, Universiti Malaysia Terengganu (UMT), Terengganu, Malaysia
Prof. Aydin **Salci**, Istanbul Technical University, Maritime Faculty, ITUMF, Istanbul, Turkey
Prof. Viktoras **Sencila**, Lithuanian Maritime Academy, Klaipeda, Lithuania
Prof. Jacek **Skorupski**, Warsaw University of Technology, Poland
Prof. Wojciech **Ślączka**, Maritime University of Szczecin, Poland
Prof. Roman **Śmierzchalski**, Gdańsk University of Technology, Poland
Prof. Jac **Spaans**, Netherlands Institute of Navigation, The Netherlands
Cmdr. Bengt **Stahl**, Nordic Institute of Navigation, Sweden
Prof. Elżbieta **Szychta**, University of Technology and Humanities in Radom, Poland
Prof. Leszek **Szychta**, University of Technology and Humanities in Radom, Poland
Prof. El **Thalassinos**, University of Piraeus, Greece
Prof. Gert F. **Trommer**, Karlsruhe University, Karlsruhe, Germany
Prof. Mykola **Tsymbal**, Odessa National Maritime Academy, Ukraine
Prof. Elen **Twrdy**, University of Ljubljana, Slovenia
Capt. Rein van **Gooswilligen**, Netherlands Institute of Navigation
Prof. Nguyen **Van Thu**, Ho Chi Minh City University of Transport, Ho Chi Minh City, Vietnam
Prof. George Yesu Vedha **Victor**, International Seaport Dredging Limited, Chennai, India
Prof. Vladimir A. **Volkogon**, Baltic Fishing Fleet State Academy, Kaliningrad, Russian Federation
Prof. Ryszard **Wawruch**, Gdynia Maritime University, Poland
Prof.. Ruan **Wei**, Shanghai Maritime University, Shanghai, China
Prof. Adam **Weintrit**, Gdynia Maritime University, Poland
Prof. Krystyna **Wojewódzka-Król**, University of Gdańsk, Poland
Prof. Adam **Wolski**, Maritime University of Szczecin, Poland
Prof. Jia-Jang **Wu**, National Kaohsiung Marine University, Kaohsiung, Taiwan (ROC)
Prof. Hideo **Yabuki**, Tokyo University of Marine Science and Technology, Tokyo, Japan
Prof. Lu **Yilong**, Nanyang Technological University, Singapore
Prof. Homayoun **Yousefi**, Chabahar Maritime University, Iran

TABLE OF CONTENTS

STCW, Maritime Education and Training (MET), Human Resources and Crew Manning, Maritime Policy, Logistics and Economic Matters. Introduction

A. Weintrit & T. Neumann
Gdynia Maritime University, Gdynia, Poland

The monograph is addressed to scientists and professionals in order to share their expert knowledge, experience and research results concerning all aspects of navigation, safety at sea and marine transportation.

The contents of the book are partitioned into seven separate subchapters: Maritime Education and Training (covering the subchapters 1.1 through 1.11), On board drills (covering the subchapters 2.1 through 2.2), Human resources and crew manning (covering the subchapters 3.1 through 3.13), Terrorism and piracy (covering the subchapters 4.1 through 4.3), Maritime policy and global excellence (covering the subchapters 5.1 through 5.3), Baltic Sea logistic and transportation problems (covering the subchapters 6.1 through 6.8), and Financial indices, freight markets and other economic matters (covering the subchapters 7.1 through 7.3).

In each of them readers can find a few subchapters. Subchapters collected in the first chapter, titled „Maritime Education and Training (MET)", concerning the using GIS as tool for met, research, and campus management enhancement, development of English for bridge watchkeeping, analyzing the perceptions of students who take the navigation course for the first time, the influence of student self-evaluation on the effectiveness of maritime studies, marine e-learning evaluation: a neuroscience approach, administrative model for quality education and training in maritime institution, a concept of examinations for seafarers in Poland, implementation of the 2010 Manila amendments to the STCW Convention and Code in Ukrainian MET system, the marine transportation program of a maritime university in the Philippines, new laboratories in the Department of Marine Electronics in Gdynia Maritime University, and Polish activities in IMO on electro-technical officers (ETO) requirements. Certainly, this subject may be seen from different perspectives.

In the second chapter there are described problems related to on board drills: lived experiences of deck cadets on board, and the research of self-evaluation on fire-fighting drill on board.

Third chapter is about human resources and crew manning. The readers can find some information about the problem of the human resource supply chain on the social network at maritime education and training, a code of conduct for shipmasters. intermediate report on a research project in progress, human factors and safety culture in maritime safety, the role of intellectual organizations in the formation of an employee of the marine industry, wasted time and flag state worries: impediments for recruitment and retention of Swedish seafarers, analysis of compensation management system for seafarers in ship management companies: an application of Turkish ship management companies, analysis of parameters and processes of Latvian seafarers' pool, the meaning and making of the first Filipino female Master Mariner: the story of Capt. Ramilie Ortega, hiring practices of shipping companies and manning agencies in the Philippines, development of maritime students' professional career planning skills: needs, issues and perspectives, the role of human organizational factors on occupational safety; a scale development through Tuzla region dockyards, Ilonggo seafarers' edge among other nationalities of the world, and influence of emotional intelligence on the work performance of seafarers.

The fourth chapter deals with terrorism and piracy. The contents of the fourth chapter are partitioned into three subchapters: characteristics of piracy in the Gulf of Guinea and its influence on international maritime transport in the region, transport infrastructure as a potential target of terrorist attacks, and Maritime Piracy Humanitarian Response Programme (MPHRP).

The fifth chapter deals with maritime policy and global excellence. The contents of the fifth chapter are partitioned into three subchapters: problem of vessels "escaping" from under flag by case of estonia and some possible measures for rising of attractiveness of them for ship-owners, analysis of safety inspections of recreational craft in the European Union, and stakeholders' satisfaction: response to global excellence.

In the sixth chapter there are described problems related to the Baltic Sea logistic and transportation problems: "ground effect" transport on the Baltic Sea, methods and models to optimize functioning of transport and industrial cluster in the Kaliningrad region, safety culture in the Baltic Sea, development of logistics functions in the Baltic Sea region ports, redefining the Baltic Sea maritime transport geography as a result of a new environmental regulation for the sulphur emission control areas, sustainable transportation development prerequisites at the example of the Polish coastal regions, and chosen problems of financing of the logistic centres in Polish seaports.

Seventh chapter is about financial indices, freight markets and other economic matters. The readers can find some information about the impact of freight markets and international regulatory mechanism on global maritime transport sector, forecasting financial indices: the Baltic Dry Indices, and analysis of the toll collection system in poland in the context of economy.

Each subchapter was reviewed at least by three independent reviewers. The Editors would like to express their gratitude to distinguished authors and reviewers of subchapters for their great contribution for expected success of the publication. They congratulate the authors for their excellent work.

Chapter 1

Maritime Education and Training (MET)

Maritime Education and Training (MET)
STCW, Maritime Education and Training (MET), Human Resources and Crew Manning, Maritime Policy,
Logistics and Economic Matters – Marine Navigation and Safety of Sea Transportation – Weintrit & Neumann (Eds)

GIS as an Innovative Tool for MET, Research, and Campus Management Enhancement: MAAP initiatives

A.M. Baylon & E.M.R. Santos
Maritime Academy of Asia & the Pacific (MAAP), Bataan, Philippines

ABSTRACT: In reference to 54 TransNav papers, it can be surmised that geospatial technologies (GST) in the global maritime industry is widely used and advanced. The 2010 STCW amendments particularly on the use of technology like ECDIS had certainly put tremendous pressure on Philippine Maritime Education and Training Institutions (METIs) to update their educational content to meet the demands of the international seafaring industry. This basic qualitative research paper presents one of the best practices spearheaded by MAAP in introducing GIS to its community (internal and external) who share similar passion for MET innovation. In addition to maritime GST uses, various MAAP initiatives on how to integrate GIS to enhance instruction, research, extension services and campus management of MET schools are discussed. There are 78 proposed GIS-based research project workshop outputs currently being implemented by 20 higher educational institutions (HEIs) in the Philippines, 17 of which are applicable to any METIs.

1 INTRODUCTION

The Philippines, known as the manning capital of the world, is one of the biggest producers and suppliers of 28% Filipino seafarers worldwide. To maintain and sustain the Philippines manning leadership position, it is necessary that the country's maritime labor force be empowered with evolving technologies that are increasingly used in the shipping industry. This is also in consonance with the implementation of the STCW 2010 known as 2010 Manila Amendment that requires the use of technology like ECDIS, as one of the emergent developments, needed to enhance the seafarers' skills on board for safety of life at sea. These skills encompass the so-called geospatial technology (GST) which is especially valuable to marine officers, having substantial responsibility of manning people and maintaining smooth maritime operations such as: deck and engine officers, port planners, administrators, marine transportation systems managers, ship managers, and maritime company personnel.

2 GEOSPATIAL TECHNOLOGY (GST) AND ITS APPLICATIONS IN MARITIME

GST, also known as geomatics or geomatics engineering, refers to technologies involved in capturing, processing, manipulating, storing, managing, and displaying information that are referenced to a particular geographic location. These include a group of disciplines such as geographic information systems (GIS), global navigation satellite systems (GNSS), and others. The term GIS, however, is commonly used to denote GST.

GST (or GIS) has become pervasive nowadays in a wide variety of applications and industries, ranging from agriculture to local governance to utilities (Cimons, 2011). The maritime industry has increasingly applied geospatial technologies such as GNSS, remote sensing (RS), hydrographic surveying and coastal mapping, ports planning and management, and charting, as well as development of a marine spatial data infrastructure (MSDI). A sample application of GST in ports management and cargo handling is a revolutionary, cost-saving technology that uses robotic balloon cranes, currently being developed by Jeremy Wiley of Tethered Air and Harvard University (Hsu, 2012; Katdare, 2012).With balloon cranes, any seaport or coastal area can be used to load and unload ship containers and heavy cargo, in lieu of conventional cranes that are seen in seaports.

Further, GST issues related to the maritime industry are being shared through conferences, such as the one organized by Geo Maritime (Geo Maritime, 2012) and by Transnav (http://transnav.eu) with published refereed papers on International Journal on Marine Navigation and Safety of Sea Transportation.

Some of these GIS-based papers demonstrate the benefits of integrated use of satellite & GIS technologies on ships to various options of using other electronic navigation systems (Boykov 2012); the GNSS for an Aviation Analysis based on EUPOS and GNSS/EGNOS Collocated Stations in PWSZ CHELM (Fellner, et. al., 2008); the digital chart application in the field of maritime traffic with the purpose of resolving *"Information Isolation"* thru development of Web chart systems using raster data instead of vector chart based , which is accessed with higher speed than most vector chart based web rendering system in wide bandwidth network with least errors for many kinds of application (Hu et al , 2007). On the other hand, the paper of Prof. Weintrit comprehensively explains the problems connected with the utilization of GIS technology and more sensitively speaking, its waterborne implementation, i.e. ECDIS (Electronic Chart Display and Information System) technology and the electronic navigational charts (ENC) in the widely comprehended maritime (open sea, coastal and harbour), inland navigation and sea-river navigation areas. (Weintrit 2010).

3 PURPOSE AND SCOPE

MAAP is always on the search for innovative means to enhance its various MET programs. Organized by MAAP's empowered Department of Research and Extension Services, the very first CHED-endorsed national conference on GIS with theme *"Enhancing Research, Extension Services and Institutional Capacity of HEIS through GIS"* was successfully accomplished in Mach 2012.

This paper presents the output of the latest MAAP initiatives in introducing GIS as an innovative tool in enhancing research, extension and campus management with insights, and list of workshop GIS-based research project outputs being implemented by 78 seminar–workshop participants represented by 20 universities and colleges nationwide.

This is a basic qualitative research paper that investigates the quality of relationships, activities, situations or materials aimed to expand man's knowledge, not to create or invent something.

4 ABOUT MAAP

The Maritime Academy of Asia and the Pacific (MAAP), is situated on a 103 hectare campus in the Bataan Peninsula about 50 km. west-southwest of the Philippine capital city, Manila, at 14°26'42.04"N and 120°32'58.79"E .

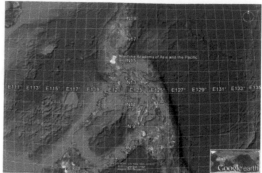

Figure 1. MAAP @ Google Earth, 12 Aug. 2012.

Figure 2 Map courtesy of Google Earth. MAAP Imagery on 19 April 2005.

Founded in January 1998, MAAP is a non-stock, non-profit academic institution established by the Associated Marine Officers' and Seamen's Union of the Philippines (AMOSUP), to ensure continues supply of competent marine deck and engine officers thru quality and full scholarship education (www.maap.edu.ph). To date MAAP has produced a total of 1, 498 competent maritime graduates (646 BSMT, 7021 BSMarE and 131 Dual course who are all officers. For the past 15 years, MAAP graduates have registered an average passing rate of more than 98% in their PRC licensure examination as operational level deck and engine officers. Similar to other Philippine maritime schools, MAAP students' curriculum consists of three -year classroom and one year on board ship. It also conducts various post-graduate maritime courses and continuing education training. As full scholars, MAAP students enjoy state of the art facilities, free board and lodging and sure employment from their sponsoring company that has collective bargaining agreement with the AMOSUP. MAAP being at the forefront of innovative MET through outcome-based program enhancement, is positively affected by the rapid rise in global demand for maritime manpower. Since its establishment on January 14, 1998, with an initial intake of only 150 scholars in June 1999, and with currently 500 fresh men scholars admitted annually,

MAAP has continued to grow with 2,015 scholars this year.

MAAP addresses the global demand for highly qualified maritime officers due to its various innovative value- added MET programs, and one of which is the introduction of GIS as one of an innovative tool for MET programs enhancement.

5 MAAP MET CURRICULUM

The Commission on Higher Education (CHED) is the government entity responsible in regulating and supervising higher educational institutions (HEIs), offering formal education programs like Bachelor degrees in Marine Transportation (BSMT) and Marine Engineering (BSMARE). To date, there are 95 maritime HEIs offering BSMT and BSMarE, 67 of which are offering both courses, while 28 offer only either one of the courses The BSMT and BSMarE curricula of MAAP consist of 155 and 158 units of professional and general education courses, respectively; plus 40 units which is equivalent to one-year shipboard training during the third year. The education core courses for BSMT and for BSMarE cover a diverse array of topics including: information and communication technology; ships and ship routines; meteorology and oceanography; world geography; maritime pollution and prevention; and merchant ships search and rescue. These courses can potentially be enhanced with the integration of GIS topics relevant for future deck officers, faculty, and researchers. Starting 2011, MAAP is the only school in the country authorized by CHED to offer vertically articulated Master degree programs in Marine Transportation (MSMT) and Marine Engineering (MSMarE). GST topics can be integrated more in-depth in its graduate curriculum.

6 EXAMPLES OF MARITIME GST USES

In reference to the 2007-2012 TRANSNAV scientific papers specifically the 54 refereed papers, it can be surmised that the maritime industry employs GIS technologies in areas such as: 1.) GPS and GPS-enabled communication systems (Januszewski 2007; Vejrazka, et al, 2007; Bosy, et al, 2007; Lemanczyk & Demkowicz, 2007; Aguila, et. al., 2008; Bober, et. al., 2008; Grzegorzewski, et. al., 2008; Fellner et al , 2008; Dziewicki, 2009; Yoo et al, 2009; Kujala, et. al., 2009; Im & Seo, 2010; Vejrazka, et. al., 2010; Janota & Koncelík, 2010; Bober, et. al., 2010; Fukuda & Hayashi, 2010; Fellner, et. al., 2010; Ilcev.2011; Januszewski, 2011; Arai, et. al., 2011; and Ambroziak, et. al., 2011); 2.) ECDIS (Rudolph, 2007; Hu, et. al., 2007); 3.) Automatic Identification System (AIS) (Aarsather &

Moan , 2007; Yousefi, 2007; Weizhang, et. al., 2007; Naus et. al., 2007; Harati-Mokhtari, et. al., 2007; Drozd et. al., 2007; Banachowicz & Wolejsza 2008; Plata & Wawruch 2009: Hu et al 2010; Wolejsza, 2010; Aarsather & Moan, 2010; Bukaty & Morozova, 2010; Park & Kim, 2011; Miyusov et al 2011; Stupak & Zurkiewicz, 2011; Ni Ni HlaingYin, et. al., 2011; Krol, et. al. 2011; Yang et. al., 2012 Xiang et. al., 2012; Kwiatkowski et al, 2012; Mazaheri et. al., 2012; Gucma & Marcjan, 2012; Goerlandt, et. al., 2012; and Krata, et. al., 2012) and 4.) Remote sensing imagery (Stateczny & Kazimierski 2009).

The fast development of the Satellite Navigation Systems (SNS) like GPS, GLONASS, Galileo and Compass, and Satellite Based Navigation Systems (SBAS) like EGNOS, WAAS, MSAS and GAGAN would even look better 10 years from now. These systems, called GNSS (Global Satellite Navigation System), are undergoing construction or modernization (new satellites, new frequencies, new signals, new monitoring stations, etc.) and continuous improvement to increase its accuracy, availability, integrity, and resistance to interference (Januszewski, 2011).

Indeed, the use of GIS has been increasingly seen in marine transportation and ports administration (Global Administrative Areas, 2012).

It can be concluded from the aforementioned TRANSNAV scientific papers , that the applications of GIS in marine transportation are certainly useful in varied areas namely: routing of vessels and the type of vessel; knowing the positions of vessels in real time; mapping and analyzing incidence; selecting new sites and analyzing marine aids such as buoys, signal lights, and other man-made coastal and offshore structures; hydrographic and bathymetric mapping of harbors, approaches, and channels; delineating shipping channels, maritime zones, and marine protected areas; producing, managing and upgrading IMO-compliant navigation charts; designing and analyzing transportation networks; and monitoring and analyzing climate patterns and ocean currents. Maritime mapping can best be accomplished by GIS software.

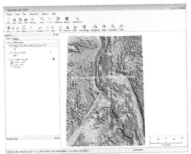

Figure 3.MAAP geographic location (red square) and Philippine map using Mapwindows software.

GIS tools can also be used in management and operations of ports namely: design of ports infrastructure; future expansions; environmentally compliant storm water systems; port asset and facility management; security planning and operations; berthing assignments and vessel tracking and routing; cargo multimodal operations; disaster response planning; and delineating restricted areas and danger zones.

Furthermore, GIS techniques can be used to help design and implement a sustainable marine environment through the following: mitigating oil spills; restoration of wetlands; coastal zone management that includes coral reefs and mangroves restoration and estuary maintenance; and assessing environmental impacts of man-made coastal structures and coastal activities such as dredging. Some examples of software tools used in maritime GIS are ESRI suite of software, CARIS, and SevenCs.

Physically, the GIS system processes spatial and non-spatial data using computer hardware, software, manpower, and methods and procedures. The system records, manipulates, analyze, stores, and display the data to produce useful information. The output information may be further processed to arrive at desired output.

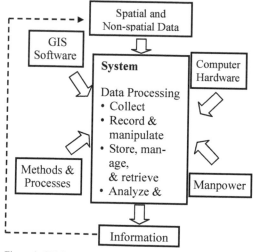

Figure 4. GIS System Diagram

7 ACADEMIC CAPABILITY ENHANCEMENT THROUGH GIS

MAAP believes that human resources are the key to institutional enhancement made possible thru capability building. Invited by MAAP as part of its research and extension services initiatives, Dr Alejandro F Tongco, GIS Specialist from Oklahoma State University, had served as technical adviser,

facilitator and trainer on GIS, organized by MAAP at various venues: Philippine Navy Education and Training Compound (March 7-10,201), Bataan Provincial Hall (March 13-16, 2012), MAAP and BPSU community (March 12, 17, 21-24, 2012) and other HEIs (March 26-29, 2012). MAAP-DRES on the other hand trained the participants on how to write a GIS-based research project proposal.

From the series of lectures that Dr. Tongco facilitated, he emphasized that the real power of GIS is in analyzing the relationships of people, places, things, and events using several spatially referenced data layers representing the earth.

Analyzing Dr Tongco's statement, the maritime or shipping industry is certainly spatial in nature, since it possesses many of these components: people (e.g., mariners of all kinds, maritime and shipping industry personnel, and MET students and faculty); places (e.g., ship destination and departure points, docks, ports, approaches, harbors, shipping lanes, mariner's and student's addresses, and MET schools); things (e.g. ships and watercraft, marine infrastructures and installations, protected marine sanctuaries, and artificial reefs); and events (e.g. transport routes and networks, hazards and incidents including fire, shipwrecks, groundings, oil and chemical spills, and piracy).

From his aforementioned lectures and thru various readings, it is opined, that the maritime industry including MET schools can leverage the power of GIS through analysis and finding innovative ways, means and solutions to the shipping industry's multifaceted and interlinked global challenges.

The increasingly widespread application of GIS in the maritime and related industries necessitates the integration of GIS techniques in METIs. GIS can be used in improving the teaching methodologies and curriculum content such as those of MAAP's. Maritime graduates that are equipped with GIS skills are not only honed with spatial analytical competence, but are more at ease as well with equipment and activities in the workplace that use GIS. It behooves MET schools to not limit itself to producing maritime graduates to simply man vessels but to give graduates the opportunity and encouragement to be analytical thinkers and to train them to be future researchers in the wide ocean of maritime research.

GIS techniques can be used as an aid to teaching and as a skill to be taught to students. For example, instructors can integrate GIS techniques in field laboratory exercises, e.g. to plan for, perform, and analyze results of fieldwork. At the same time, GIS software skills can be taught to students to enhance the student's analytical thinking abilities. These would require instructors with required GIS skills that maybe developed through training and experience. A potent method to build GIS knowledge capability in MET schools is for instructors and students to engage in GIS-based research and development activities.

GIS techniques can be integrated in the MET curriculum. For example, the enclosure of GIS topics can enrich the curricular content of several courses. Existing courses may include: information and communication technology (ICT); ships and ship routines; meteorology and oceanography; world geography; maritime pollution and prevention; and merchant ships search and rescue.

Ideally, GIS may be offered as a distinct course or subject within the BSMT degree program, e.g., Geospatial Technology in Maritime. This would be an in-depth treatment of the subject matter, taking into account the various GIS applications in the maritime industry. A similar course or subject could also be offered in the graduate school within the Master of Science in Marine Transportation (MSMT) or Shipping Management (MSSM) program which is being offered by the newly established MAAP Center for Advanced Maritime Studies (CAMS) accredited by the Commission on Higher Education (CHED) in 2011. Some examples of GIS curriculum content are those of the GST competency model developed by the GeoTech Center (2012).

To address the challenges in maritime mapping and charting, the United Kingdom Hydrographic Office (UKHO) has proposed competencies in several spatial related skills such as cartography, geodesy, GPS, International Maritime Organization (IMO)-compliant electronic navigational chart (ENC) and digital nautical chart (DNC) production based on new International Hydrographic Office (IHO) S-100 and S-101 standards, spatial database management system (SDBMS), GIS software such as ArcGIS and SevenCs, and electronic chart display and information system (ECDIS) (UK Hydrographic Office, 2007). The integration of GIS in maritime curricular programs has already started. For example, a maritime graduate program called Master of Science in International Maritime Studies - Marine Spatial Planning is offered by Southampton Solent University (www.shippingedu.com). The Faculty of Navigation of the Gdynia Maritime University and the Nautical Institute recognizes the importance of Geomatics and GIS in maritime applications, as shown in the list of the 87 Transnav conference main research topics or subjects (http://transnav2013.am.gdynia.pl/Symposium/confe rence-main-topics.html) for 10th International Navigational Conference on Maritime Navigation and Safety of Sea Transportation they organized on 19-21 June 2013. The conference gathers global scientists and professionals to meet and share their respective expertise, knowledge, experience and research results, concerning all aspects of navigation and sea transportation.

Areas for research in maritime GIS are vast. This is where any MET schools can empower and prepare itself and its students for new emerging GIS developments for the maritime industry. This requires an innovative interested GIS-skilled faculty and a supportive MET management. On the other hand, MET graduates students can enhance the research environment of their schools through discipline-based thesis as part of their study program. There is dire need for maritime research in the Philippines that is focused on technical disciplines, i.e., science, technology, engineering, mathematics, and design (STEMD). GIS-based maritime researches can contribute to these disciplines. Having the most number of maritime schools and graduates in the country, it is expected that the Philippines would be able to produce a proportionate number of maritime technical research outputs in international forums.

An equally important GIS initiative that a maritime school can create for a better impact in maritime research for the maritime industry is the establishment of a maritime spatial data infrastructure (MSDI) or similarly an integrated shipping spatial information system (ISSIS). An MSDI or ISSIS is a framework of geospatial data, software tools, and metadata for users to use the information efficiently. It is GIS in itself, which involves the collection, processing, management, storage, distribution, and display of geospatial information for the maritime industry, maritime researchers, and MET schools. An MSDI of ISSIS can assist for the planning, implementation, and assessment of spatially related content of maritime studies. The comprehensive database may include spatial information about port facilities, maritime security, marine hazards and incidents, MET students and alumni, greenhouse gas emissions, and environmental and meteorological data. Examples of web-based MSDI or ISSIS are the Narragansett Bay data portal, US National Oceanic and Atmospheric Administration (NOAA) marine web GIS, and MarineCadastre.gov. GIS-skilled maritime graduates are offered wider employment opportunities, not just sea-based, but in land-based jobs as well. They may venture into jobs such as development and production of ENCs based on IHO S-100 and S-101 standards (MarineCadastre.gov; International Hydrographic Office (IHO) and Hydrographic and Oceanographic Service of the Chilean Navy (SHOA), 2009); design, development, installation, operation, and training of ECDIS hardware and software; development of marine spatial data; establishment and maintenance of MSDI and ISIS; marine research, e.g. marine surveying, bathymetry, marine geology, and coastal projects; employment in shipping and maritime companies and agencies; and teaching and research in MET schools.

8 IMPROVING MET CAMPUS MANAGEMENT USING GIS TECHNIQUES

Campus facilities and resource management is another area, wherein the application of GIS techniques can potentially improve management processes, and thus realize cost savings. GIS is an information system that takes advantage and produces valuable information about the unique geographic locations and inter-relationships among people, places, things, and events in the campus. The application of GIS to manage campus facilities and assets is not unique to non-maritime campuses. It is applicable to any school campus or facility as well. Campus GIS involves the development and management of a central spatial database of all buildings and assets within. The spatial database allows for a fast and easy query of the campus' physical resources relative to each other. GIS provides for the visualization, query, and reporting of specific information. Additionally, a GIS data portal can provide data access for faculty, staff, researchers, and students in campus. A central spatial database provides an enhanced communication and cooperation between departments, offices, faculty, staff, students, and administration. As a result, workflows become more efficient and resources are efficiently allocated. Planning, monitoring, and assessment of campus resources are better informed through GIS. Information is more accurate as a result of GIS analysis. GIS data can be visualized and thus minimizes guesswork on factors involved in campus processes. The overall result is increased campus productivity and economic efficiency.

9 INSTITUTION-WIDE GIS IMPLEMENTATION

As part of MAAP Research & Extension Program, the Department of Research & Extension Services (DRES) with more than one year coordination (2011-2012) with GIS specialist from Oklahoma State University as regards GIS (www.philgis.org), DRES has been fascinated with the wide application of GIS in any discipline. The first MAAP GIS paper and power point presentation thru its DRES Director was entitled *"Human Security and GIS: An Introduction"* during the CHED-endorsed 20th IFFSO General Conference by the International Federation of Social Science Organizations (IFFSO) with theme *"Social Science Perspectives on Human Security"*, hosted by Lyceum of the Philippines, Batangas City on November 18-20, 2011. The second paper was entitled *"Introducing GIS as Catalyst for Research, Extension Services and Development of Catanduanes Island"* in support to the 2012 National Conference on Water and

Biodiversity (BIOME3) with thematic Scope *"Water plus Diversity Equals Food plus Life"* at Virac Catanduanes Island on Oct 21-23, 2013. Hence, MAAP thru its Department of Research and Extension Services (DRES) has embarked on GIS-based project proposals within MAAP and with other institutions who share the same passion for innovation. As early as November 7, 2011, a 46-page GIS-based project proposal entitled *"University–wide Integration of geographic information systems (GIS): enhancing campus management & academic capability building of Bataan Peninsula State University (BPSU)"* has been prepared which was also shared with Lyceum International Maritime Academy (LIMA) of the Lyceum of the Philippines University (LPU) (41-page GIS-based proposal) & the Catanduanes State University (CSU) (39-page GIS-based proposal) for their respective institution-wide GIS implementation.

The two major target areas for GIS engagements are: academic capacity building through GIS techniques and technologies, & enhancing campus management through GIS.

Academic capacity building includes faculty and staff development in GIS, GIS curriculum and instruction, and GIS-based research. GIS-based campus management includes GIS-based campus physical facilities, assets, and resource management and GIS-based student admissions and alumni management. Academic capacity building and campus management can mutually benefit and grow from each other through GIS. For example, faculty and staff can enhance their GIS skills through involvement in GIS-engaged campus facilities and asset management as well as in GIS-based student admissions and alumni management studies. On the other hand, the campus management can benefit from the involvement of faculty and staff. Involving both students and campus managers in campus wide projects can likewise mutually enrich GIS-engaged curriculum and instruction and campus management. GIS-based research in the maritime is a new wide-open field. GIS-skilled graduates are expected to be more prepared to face the new challenge, whether the research is sea or land based.

Within the campus, these activities could well be facilitated by a physical entity such as a GIS center or a Research and Extension Services center with GIS-skill technical staff, to act as the central coordinating body and knowledge base, as well as to set the direction and sustainability of GIS-engaged activities in MET schools. Furthermore, in order to institutionalize GIS, this may need to be integrated in the MET's strategic vision, mission, and goals, specifically in departments and offices where GIS activities are to be conducted.

As a summary, an institution-wide implementation of GIS for a MET school may be illustrated in the diagram below.

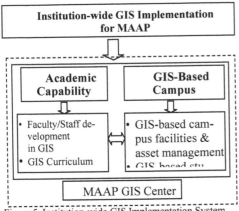

Figure 5. Institution-wide GIS Implementation System

The diagram contains:

Institution-wide GIS Implementation for MAAP

- **Academic Capability**
  - Faculty/Staff development in GIS
  - GIS Curriculum
- **GIS-Based Campus**
  - GIS-based campus facilities & asset management
  - GIS based stu...

MAAP GIS Center

MAAP has already embarked on integrating GIS in the campus in March 2012. An initial introduction of GIS for its faculty, staff, and students was conducted. Two separate four-day GIS hands-on workshops were held, one group for faculty and staff (20-23 March 2012 and 26-29 March 2012) at MAAP campus, which were also offered to interested HEIs nationwide. For upper-class students, GIS hands-on workshops were conducted every Saturdays (March 12, 17 and 24, 2012). Student-participants had presented their GIS-Based research project entitled *"Piracy Attack Analysis and Monitoring through GIS "*by 1/Cl Sun, Patrick John Austine and *"Integration of GIS to Automatic Integration System (AIS)"* by 1/Cl Espago, Marville Cullen.

Furthermore, a 3-day CHED and DOST-endorsed GIS National Conference was organized and hosted by MAAP on March 29-31, 2012 at its campus in Mariveles Bataan Philippines. The activity was conducted in cooperation with partner agencies as shown on photo from left to right: Bataan Provincial Government (*PMO Engineer Ric Yuzon*); CHED Zonal Research Regional Center/ Angeles University Foundation (*Director Dr. Roberto Pagulayan*); Commission on Higher Education (*ARMM CHED Dr. Carmencita Aquino*); Philippine Association of Extension Program Implementers with SEC CN-10059 (*PAEPI President/MAAP Director/Dr. Angelica Baylon*); Association of Universities of Asia and the Pacific (*AUCP Secretary General Dr Ruben Umaly*); Department of Science and Technology (*DOST Regional Director Dr. Victor Mariano as GOH and Speaker*); Maritime Academy of Asia and the Pacific (*MAAP President Vadm Eduardo Ma. R. Santos, AFP (Ret))* ; National Research Council of the Philippines / Philippine Normal University (*NRCP VP/Chair Division/PNU President Dr. Ester Ogena*), Oklahoma State University, USA (represented by the *OSU GIS Specialist Dr. Alejandro Tongco* as

trainer and resource speaker) ; Provincial DOST (*Director Ms. Rosalina Ona*); Asian Educational Research Association (*AERA Chair Dr. Gismo Agulan*). Other GIS partners not on photo are Philippine Navy (*PN Education and Training Head (N8) Capt. Anthony Sean Villa*), Commission on Higher Education (*CHED Regional Director Dr Virginia Akiate*); Philippine Association of Research Managers (*PhilARM President Dr. Ricardo Castro)* and Philippine Association of Institutional Rescarchers (*PAIR President Dr. Genaro Japos)*.

Figure 6. National GIS Conference Photo Souvenir

This GIS national activity was participated in by 20 Higher Education Institutions (HEIs) and has come up with a list of proposed GIS-based research projects to be implemented in their respective institutions. There are 78 GIS-based projects proposed by 20 participating institutions nationwide (*MAAP; Philippine Merchant Marine Academy; Holy Cross College of Davao; SPAMAST; Asian Institute of Maritime Studies; University of the Cordilleras; San Beda College; New Era University; Baliuag University; Bataan Peninsula State University; Philippine Normal University; Philippine Navy; University of Luzon; CPU Outreach Center Central Philippines University; Central Luzon State University; University of Northern Philippines; Cebu Normal University; Mindanao Sanitarium Hospital; some local government units participated by Balanga City and Bataan Province*) with total of 115 seminar-workshop active participants: Some of the interesting 17 GIS-based proposed projects applicable to METIs are as follows:

1 GIS-based Campus-wide Management of Physical Facilities & Assets
2 GIS Applications in Ship Management: Database Query & Visualization of Ship's Facilities & Assets
3 Development and Production of IMO Compliant Philippine Electronic Navigational Charts(ENCs) Using Geospatial Technologies
4 Development of a National Maritime Geospatial System & Data Portal: A Research & Extension Initiative
5 Students , Faculty and Staff Profiling Using GIS

6 Enhancing Alumni Data Base Using GIS
7 A Graduate Level Subject Offering "*GIS with Maritime Applications* " at the MAAP Center for Advanced Maritime Studies
8 Identifying Vulnerability of Different Barangays to Different Types of Risks Using GIS
9 Tracer Study of MET Graduates
10 Research on the Employability of MET Graduates in the Various Programs Offered
11 Community Baseline Data on Outreach Areas
12 Monitoring of Outreach Projects Using GIS
13 Impact Evaluation of Outreach Projects Using GIS
14 Enhancing Outreach Community Planning Using GIS
15 Resource Mapping of Outreach Areas Using GIS
16 Integrating GIS in Outreach Program Implementation
17 Geo-Hazard Mapping & Risk Reduction Management by GIS
18

On the other hand, 18 GIS-based research projects have been successfully presented: three GIS-based presentations by the Philippine Navy (6-9 March 2012); four GIS-based presentations from the Province of Bataan by local government units (13-16 March 2012) ; four GIS-based presentations by Bataan Peninsula State University (20-23 March 2012) ; three GIS-based presentations by the Maritime Academy of Asia and the Pacific (12,17,24 March 2012); and four GIS-based presentations by other MET/HEIs (20-26 March 2012).

10 CONCLUSIONS

An introduction into how GIS can be integrated in the MET curriculum, in research, and in campus management has been discussed. GIS as a tool can be applied in any discipline. GIS can be introduced initially as one of the topics within existing maritime-related courses and can be used as a tool in field research work. Then, it can grow into a distinct GIS course within the undergraduate and graduate programs. With GIS-skilled faculty and staff researchers, GIS-based maritime researches can be developed. A GIS-based campus management system can grow hand in hand with that of GIS in teaching, research and extension services, constantly nourishing and building each other's knowledge base toward an institution-wide integration of GIS technologies.

REFERENCES

Aarsather K. G., Moan T. 2010. *Computer Vision and Ship Traffic Analysis: Inferring Maneuver Patterns from the Automatic Identification System*. TransNav - International Journal on Marine Navigation and Safety of Sea Transportation, Vol. 4, No. 3, pp. 303-308

Aarsather K. G. & Moan T. 2007. *Combined Maneuvering Analysis, AIS and Full-Mission Simulation*. TransNav - International Journal on Marine Navigation and Safety of Sea Transportation, Vol. 1, No. 1, pp. 31-36

Aguilar E. D., Jaworski L., Kolodziejczak M. 2008. *Accuracy Analysis of the EGNOS System during Mobile Testing*. TransNav - International Journal on Marine Navigation and Safety of Sea Transportation, Vol. 2, No. 1, pp. 37-41

Ambroziak S.J., et. al. 2011. *Ground-based, Hyperbolic Radiolocation System with Spread Spectrum Signal - AEGIR*. TransNav - International Journal on Marine Navigation and Safety of Sea Transportation, Vol. 5, No. 2, pp. 233-238

Arai Y., Pedersen E., Kouguchi N., Yamada K. 2011. *Onboard Wave Sensing with Velocity Information GPS*. TransNav- International Journal on Marine Navigation and Safety of Sea Transportation, Vol. 5, No. 2, pp. 205-211, 2011

Banachowicz A. & Wolejsza P. 2008. *The Analysis of Possibilities How the Collision Between m/v 'Gdynia' and m/v 'Fu Shan Hai' Could Have Been Avoided*. TransNav - International Journal on Marine Navigation and Safety of Sea Transportation, Vol. 2, No. 4, pp. 377-381

Bober R., Szewczuk T., Wolski A. 2008. *An Effect of Urban Development on the Accuracy of the GPS/EGNOS System*. TransNav - International Journal on Marine Navigation and Safety of Sea Transportation, Vol. 2, No. 3, pp. 235-238

Bober R., Szewczuk T., Wolski A.2010. *Effect of Measurement Duration on the Accuracy of Position Determination in GPS and GPS/EGNOS Systems*. TransNav-International Journal on Marine Navigation and Safety of Sea Transportation, Vol. 4, No. 3, pp. 295-299

Bosy J., Graszka W. & Leonczyk M.2007. *ASG-EUPOS - a Multifunctional Precise Satellite Positioning System in Poland*. TransNav - International Journal on Marine Navigation and Safety of Sea Transportation, Vol. 1, No. 4, pp. 371-374

Boykov A. & Katenin V. 2012. *About Effectiveness of Complex Using of Satellite and Geoinformation Technologies on the Ship of Compound 'River-Sea' Type*. TransNav - International Journal on Marine Navigation and Safety of Sea Transportation, Vol. 6, No. 4, pp. 475-479

Bukaty V. M. & Morozova S.U. 2010. *Possible Method of Clearing-up the Close-quarter Situation of Ships by Means of Automatic Identification System*. TransNav - International Journal on Marine Navigation and Safety of Sea Transportation, Vol. 4, No. 3, pp. 309-313

Bukaty V. M. & Morozova S.U. 2010. *On Determination of the Head-on Situation under Rule 14 of Colreg-72*. TransNav - International Journal on Marine Navigation and Safety of Sea Transportation, Vol. 4, No. 4, pp. 383-388

Cimons, M. 2011. *Geospatial Technology as a Core Tool* Content provided by National Science Foundation. *U.S. News & World Report,* 11 May 2011. Retrieved August 21, 2012 from http://www.usnews.com

Crossworld Marine Services. 2011. 400,000 Filipino Seafarer Deployed Overseas while demands continue to grow. 23 September 2011. Retrieved August 20, 2012 from http://www.crossworldmarine.com

Drozd W., Dziewicki M., Waraksa M., Bibik Ł. 2007. *Operational Status of Polish AIS Network*. TransNav - International Journal on Marine Navigation and Safety of Sea Transportation, Vol. 1, No. 3, pp. 251-253

Dziewicki M. 2009. *Modernization of Maritime DGPS in Poland*. TransNav - International Journal on Marine Navigation and Safety of Sea Transportation, Vol. 3, No. 1, pp. 39-42, 2009

Environmental Systems Research Institute. 2012. *GIS Solutions for Ports and Maritime Transport*. Retrieved

August 23, 2012 from http://www.esri.com/library/brochures/pdfs/gis-sols-for-ports.pdf

Fellner A., Cwiklak J., et. al., 2008. *GNSS for an Aviation Analysis Based on EUPOS and GNSS/EGNOS Collocated Stations in PWSZ CHELM.* TransNav - International Journal on Marine Navigation and Safety of Sea Transportation, Vol. 2, No. 4, pp. 351-356

Fellner A., Banaszek K. & Tróminski P. 2010. *Alternative for Kalman Filter – Two Dimension Self-learning Filter with Memory.* TransNav - International Journal on Marine Navigation and Safety of Sea Transportation, Vol. 4, No. 4, pp. 429-431

Fukuda G. & Hayashi S. 2010. *The Basic Research for the New Compass System Using Latest MEMS.* TransNav - International Journal on Marine Navigation and Safety of Sea Transportation, Vol. 4, No. 3, pp. 317-322, 2010

Grzegorzewski M., et. a.l, A. 2008. *GNSS for an Aviation.* TransNav - International Journal on Marine Navigation and Safety of Sea Transportation, Vol. 2, No. 4, pp. 345-350

Gucma L. & Marcjan K. 2012. *Incidents Analysis on the Basis of Traffic Monitoring Data in Pomeranian Bay.* TransNav - International Journal on Marine Navigation and Safety of Sea Transportation, Vol. 6, No. 3, pp. 377-380

Harati-Mokhtari A., et. al., 2007. *AIS Contribution in Navigation Operation- Using AIS User Satisfaction Model.* TransNav - International Journal on Marine Navigation and Safety of Sea Transportation, Vol. 1, No. 3, pp. 243-249

Hsu, J. 2010. Robot Balloon Cranes Could Revolutionize World's Seaports. *Live Science, 7 March* 2012. Retrieved August 21, 2012 from http://www.livescience.com

Hu, Q., Yu, L. & Chen, J. 2007. *A Method to Build Web Raster Chart System.* TransNav - International Journal on Marine Navigation and Safety of Sea Transportation, Vol. 1, No. 2, pp. 171-174

Hu Q., Yong, J., Shi, C. & Chen, G. 2010. *Evaluation of Main Traffic Congestion Degree for Restricted Waters with AIS Reports.* TransNav - International Journal on Marine Navigation and Safety of Sea Transportation, Vol. 4, No. 1, pp. 55-58

Ilcev S. D. 2011. *Maritime Communication, Navigation and Surveillance (CNS).* TransNav - International Journal on Marine Navigation and Safety of Sea Transportation, Vol. 5, No. 1, pp. 39-50

Im N. & Seo J.-H. 2010. *Ship Manoeuvring Performance Experiments Using a Free Running Model Ship.* TransNav- International Journal on Marine Navigation and Safety of Sea Transportation, Vol. 4, No. 1, pp. 29-33

Janota, A. & Koncelík V.2010. *GPS-based Vehicle Localisation.* TransNav - International Journal on Marine Navigation and Safety of Sea Transportation, Vol. 4, No. 3, pp. 289-294, 2010

Januszewski J. 2010. *A Look at the Development of GNSS Capabilities over the Next 10 Years.* TransNav - International Journal on Marine Navigation and Safety of Sea Transportation, Vol. 5, No. 1, pp. 73-78, 2011

Januszewski, J. 2007. *Modernization of Satellite Navigation Systems and Theirs New Maritime Applications.* TransNav - International Journal on Marine Navigation and Safety of Sea Transportation, Vol. 1, No. 1, pp. 39-45

Krata P., Jachowski J., & Montewka J. 2012. *Modeling of Accidental Bunker Oil Spills as a Result of Ship's Bunker Tanks Rupture – a Case Study.* TransNav - International Journal on Marine Navigation and Safety of Sea Transportation, Vol. 6, No. 4, pp. 495-500

Krol A., Stupak T., et. al. 2011. *Fusion of Data Received from AIS and FMCW and Pulse Radar - Results of Performance Tests Conducted Using Hydrographical Vessels "Tukana" and "Zodiak".* TransNav - International Journal on Marine Navigation and Safety of Sea Transportation, Vol. 5, No. 4, pp. 463-469

Kwiatkowski M., et. al. 2012. *Integrated Vessel Traffic Control System.* TransNav - International Journal on Marine Navigation and Safety of Sea Transportation, Vol. 6, No. 3, pp. 323-327

International Hydrographic Office (IHO) and Hydrographic and Oceanographic Service of the Chilean Navy (SHOA). *IHO S-100: The New IHO Hydrographic Geospatial Standard for Marine Data and Information.* (2009). Retrieved August 31, 2012 from http://icaci.org/files/documents/ICC_proceedings/ICC2009

International Hydrographic Office (IHO). *IHO S-101: The Next Generation ENC Product Specification.* Retrieved August 31, 2012 from http://www.iho.int/mtg_docs

GEBCO. 2012. *General Bathymetric Chart of the Oceans.* 25 June 2012. Retrieved September 1, 2012 from http://www.gebco.net

Geo Maritime. 2012. *Geo Maritime Meeting: 13-14 June, St. Pauls – Geospatial Information in the Maritime Sector,* June 2012. Retrieved August 21, 2012 from http://www.wbresearch.com/geomar/agendadownload.aspx

Geo Maritime. *Information Sharing and Interoperability in a Digital Era.* (2012, June). Retrieved August 21, 2012 from http://www.wbresearch.com/uploadedFiles/Events/UK/201 2/21090_001/Download_Center_Content/Full%20agenda%20for%20website.pdf

Geotech Center. 2012. *Geospatial Competency Model.* (2012). Retrieved August 24, 2012 from http://www.careeronestop.org/competencymodel/pyramid.aspx?geo=Y

Global Administrative Areas. 2012. *GADM Database of Administrative Areas, version 2.* January 2012. Retrieved September 1, 2012 from http://www.gadm.org/

Janota A. & Koncelík V. 201. *GPS-based Vehicle Localisation.* TransNav - International Journal on Marine Navigation and Safety of Sea Transportation, Vol. 4, No. 3, pp. 289-294

Goerlandt F., et. al. 2012. *Simplified Risk Analysis of Tanker Collisions in the Gulf of Finland.* TransNav - International Journal on Marine Navigation and Safety of Sea Transportation, Vol. 6, No. 3, pp. 381-387

Katdare, A. 2012. Tethered Air's Robotic Balloon Crane Can Make any Shore a Seaport. *Crazy Engineers.* (2012, March 9). Retrieved August 21, 2012 from http://www.crazyengineers.com/tethered-airs-robotic-balloon-crane-can-make-any-shore-a-seaport-1903

Kujala P., Rogowski J. B. & Kopanska K. 2009: *Positioning Using GPS and GLONASS Systems.* TransNav - International Journal on Marine Navigation and Safety of Sea Transportation, Vol. 3, No. 3, pp. 283-286, 2009

Lemanczyk M. & Demkowicz J. 2007. *Galileo Satellite Navigation System Receiver Concept.* TransNav - International Journal on Marine Navigation and Safety of Sea Transportation, Vol. 1, No. 4, pp. 375-378, 2007

MarineCadastre.gov. *MMC Viewer and Data Registry.* Retrieved August 29, 2012 from http://marinecadastre.gov

MAPWINDOW. *Mapwindow GIS version 4.8.6.* Retrieved September 1, 2012 from http://www.mapwindow.org

Mazaheri A., et. al. 2012. *A Decision Support Tool for VTS Centers to Detect Grounding Candidates.* TransNav - International Journal on Marine Navigation and Safety of Sea Transportation, Vol. 6, No. 3, pp. 337-343, 2012

Miyusov M.V., Koshevoy V.M. & Shishkin A.V. 2011. *Increasing Maritime Safety: Integration of the Digital Selective Calling VHF Marine Radiocommunication System and ECDIS.* TransNav - International Journal on Marine Navigation and Safety of Sea Transportation, Vol. 5, No. 2, pp. 159-161

Narragansett Bay.org. *RI Marine Data Download (Physical).* Retrieved August 21, 2012 from http://www.narrbay.org/physical_data.htm

National Oceanic and Atmospheric Administration - National Geophysical Data Center. *Marine Geology Data*. Retrieved August 29, 2012 from http://maps.ngdc.noaa.gov/viewers/marine_geology

Naus K., Makar A., Apanowicz J.: *Usage AIS Data for Analyzing Ship's Motion Intensity*. TransNav - International Journal on Marine Navigation and Safety of Sea Transportation, Vol. 1, No. 3, pp. 237-242, 2007

Ni Ni HlaingYin, Hu Q., Shi Chaojian: *Studying Probability of Ship Arrival of Yangshan Port with AIS (Automatic Identification System)*. TransNav - International Journal on Marine Navigation and Safety of Sea Transportation, Vol. 5, No. 3, pp. 291-294, 2011

Ocean Samp. *RI Ocean Samp Project*. Retrieved August 21, 2012 from http://www.narrbay.org

Park G.K. & Kim Young-Ki. 2011. *On a Data Fusion Model of the Navigation and Communication Systems of a Ship*. TransNav - International Journal on Marine Navigation and Safety of Sea Transportation, Vol. 5, No. 1, pp. 51-56

Philippine Overseas Employment Administration. 2010. *Overseas Employment Statistics 2010*. Retrieved August 20, 2012 from http://www.poea.gov.ph/stats/2010_Stats.pdf

Plata S. & Wawruch R.: 2009. *CRM-203 Type Frequency Modulated Continuous Wave (FM CW) Radar*. TransNav - International Journal on Marine Navigation and Safety of Sea Transportation, Vol. 3, No. 3, pp. 311-314

Rudolph J. 2007. *ECDIS Operator Training*. TransNav - International Journal on Marine Navigation and Safety of Sea Transportation, Vol. 1, No. 1, pp. 77-81, 2007

Shippingedu.com. *MSc International Maritime Studies – Marine Spatial Planning | Southampton Solent University*. Retrieved August 29, 2012 from http://www.shippingedu.com/our-network/southampton-solent-university/msc-ims-marine-spatial-planning.html

Stateczny A. & Kazimierski W. 2009. *Target Tracking in RIS*. TransNav - International Journal on Marine Navigation and Safety of Sea Transportation, Vol. 3, No. 4, pp. 385-390

Stupak T. & Zurkiewicz S. 2011. *Congested Area Detection and Projection – the User's Requirements*. TransNav - International Journal on Marine Navigation and Safety of Sea Transportation, Vol. 5, No. 3, pp. 285-290

Transnav. 2013, *Gdynia Maritime University and Nautical Institute Symposium Conference Main Topics*. Retrieved September 2, 2012 from http://transnav2013.am.gdynia.pl/Symposium/conference-main-topics.html

United Kingdom Hydrographic Office. 2007. *REF 026B: A1-B1 M&C Functional Competencies*. 21 June 2007. Retrieved August 24, 2012 from http://www.ukho.gov.uk/AboutUs/StaffPortal/Documents/Competences/REF026B.doc

University of Arkansas Division of Agriculture. 2010. *Geospatial Technologies Introduction*. 18 May 2010. Retrieved August 21, 2012 from http://baegrisk.ddns.uark.edu/kpweb

US Embassy- Manila.2009. Seafarer Law. *Powerpoint presentation*. 15 October 2009. Retrieved August 20, 2012 from http://photos.state.gov/libraries/manila/19452/pdfs/US%20Embassy%20Seafarer%20Law1.pdf

Vejrazka F., et. al. 2007. *Software navigation receivers for GNSS and DVB*. TransNav - International Journal on Marine Navigation and Safety of Sea Transportation, Vol. 1, No. 2, pp. 137-141

Vejrazka F., Kovar P., Eska M., Puricer P.: *Software navigation receivers for GNSS and DVB*. TransNav - International Journal on Marine Navigation and Safety of Sea Transportation, Vol. 1, No. 2, pp. 137-141, 2007

Wei Z., Liu, R. & Chang, L. 2007. *The Research of Integrated Maritime Digital Information System*. TransNav - International Journal on Marine Navigation and Safety of Sea Transportation, Vol. 1, No. 2, pp. 193-196

Weintrit, A. 2010. *Six in One or One in Six Variants. Electronic Navigational Charts for Open Sea, Coastal, Off-Shore, Harbour, Sea-River and Inland Navigation*. TransNav - International Journal on Marine Navigation and Safety of Sea Transportation, Vol. 4, No. 2, pp. 165-177

Wikipedia. 2012. *Filipino seamen*. 18 August 2012. Retrieved August 21, 2012 from http://en.wikipedia.org/wiki/Filipino_seamen

Wikipedia. 2012. *Geomatics*. 25 August 2012. Retrieved August 30, 2012 from http://en.wikipedia.org/wiki/Geomatic

Wolejsza, P. 2010. *Data Transmission in Inland AIS System*. TransNav - International Journal on Marine Navigation and Safety of Sea Transportation, Vol. 4, No. 2, pp. 179-182

Xiang Z., et. al. 2012. *Applied Research of Route Similarity Analysis Based on Association Rules*. TransNav - International Journal on Marine Navigation and Safety of Sea Transportation, Vol. 6, No. 2, pp. 181-185

Yang C., Hu Q., Tu X., & Geng J. 2012. *An Integrated Vessel Tracking System by Using AIS, Inmarsat and China Beidou Navigation Satellite System*. TransNav - International Journal on Marine Navigation and Safety of Sea Transportation, Vol. 6, No. 2, pp. 175-178

Yoo Y., et. al. 2009. *Application of 3-D Velocity Measurement of Vessel by VI-GPS for STS Lightering*. TransNav - International Journal on Marine Navigation and Safety of Sea Transportation, Vol. 3, No. 1, pp. 43-48

Vejrazka F., Kovar P. & Kacmarík P. 2010. *Galileo AltBOC E5 Signal Characteristics for Optimal Tracking Algorithms*. TransNav - International Journal on Marine Navigation and Safety of Sea Transportation, Vol. 4, No. 1, pp. 37-40

Yousefi, H. 2007. *The Role of Navigational Aids Such as Radar/ARPA, ECDIS, AIS, Autopilot, on Safe Navigation at Sea*. TransNav - International Journal on Marine Navigation and Safety of Sea Transportation, Vol. 1, No. 2, pp. 177-179

Maritime Education and Training (MET)
STCW, Maritime Education and Training (MET), Human Resources and Crew Manning, Maritime Policy,
Logistics and Economic Matters – Marine Navigation and Safety of Sea Transportation – Weintrit & Neumann (Eds)

On Development of English for Bridge Watchkeeping

Z. Bezhanovi, L. Khardina & K. Zarbazoia
Batumi State Maritime Academy, Georgia

ABSTRACT: The goal of the presented paper is to propose a sample of a teaching data compilation designed in a form of a coursebook unit (aimed at non-native English students of Sea Navigation Specialty) to meet the revised STCW 78/95 requirements regarding the English Language competence and proficiency of bridge team applied to the latest changes introduced by the International Association of Marine Aids to Navigation and Lighthouse Authorities.

1 INTRODUCTION

Under the SOLAS (Chapter V – 1/7/02, REGULATION 14 - Ships' manning, 4) on all ships, to ensure effective crew performance in safety matters, a working language shall be established and recorded in the ship's log-book. On ships (merchant ones of 500 gross tonnage or more meant) English shall be used on the bridge as the working language for bridge-to-bridge and bridge-to-shore safety communications as well as for communications on board between the pilot and bridge watchkeeping personnel, unless those directly involved in the communication speak a common language other than English.

The amendments, to be known as "The Manila amendments to the STCW Convention and Code" entered into force on 1 January 2012. The stated amendments as well as the Convention in whole principally underline the significance of the English Language competence for the seafarers. The Specification of minimum standard of competence for officers in charge of a navigational watch (STCW 78/95, Chapter II, Section A-II/2, Table A-II/2) requires the adequate knowledge of the English language to enable the officer to perform the officer's duties.

The aim of the presented paper is to present a sample of a textbook (designed for non-native English speakers) unit designed to meet the revised STCW 78/95 requirements applied to the latest changes introduced by the International Association of Marine Aids to Navigation and Lighthouse Authorities.

The goal of the unit presented below is to provide the nautical students with adequate competence and proficiency to enable them to use specific nautical terminology and to develop oral communication skills by means of pair work and individual project work motivation.

Sample Unit: English for Bridge Watchkeeping Skills – Types and Functions of AtoNs

AtoNs (Aids to Navigation) include any device or system, external to a vessel, which is provided to help a mariner to determine position and course, to warn of dangers or of obstructions, or to give advice about the location of a best or preferred route. AtoNs should not be confused with a navigational aid. A navigational aid is an instrument, device, chart, etc., carried on board a vessel for the purpose of assisting navigation.

Visual Marks can be natural or man-made conspicuous objects such as mountain-tops, rocks, churches, towers, minarets, monuments, chimneys, etc.

Purpose-built AtoNs include lighthouses, beacons, leading (range) lines, lightvessels, buoys, daymarks (dayboards) and traffic signals. Visual marks can be provided with a light or left unlit. The effectiveness of a visual AtoN depends on its type, location, distance and atmospheric conditions.

AtoN distinguishing features include the location, type (fixed structure, floating platform) and characteristics (shape, size, elevation, color, lit/unlit, light intensity, signal character, construction material, names, letters and numbers).

A lighthouse is a conspicuous structure on land, close to the shoreline or in the water which acts as a daymark and also provides a marine signaling light with a range of up to 25 nautical miles; it can be a manned or automated facility.

A beacon is usually a small fixed visual mark. Visual characteristics are often presented by daymarks, topmarks and by numbers.
– Lighthouses and beacons are usually used to perform one or more of the following navigational functions:
– to mark a landfall position;
– to mark an obstruction or a danger:

– danger (depth unknown):⇨ Obstruction

cover/uncover:⇨
– Obstruction, least depth known:

⇨

– to indicate the lateral limits of a channel or navigable waterway;
– to indicate a turning point or a junction in a waterway;
– to form part of a leading (range) line:

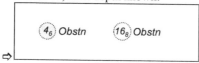

– to assist mariners to take a bearing or Line of Position (LOP);
Other purposes for which a lighthouse can be used, include:
– to indicate coastguard functions;
– to provide base for audible (fog) signals:

– to provide collection of meteorological and oceanographic data;
– to provide radio and telecommunication facilities;
– to provide Vessel Traffic Service (VTS) functions;
Lighthouses and beacons are also used to mark the entrance of a Traffic Separation Scheme (TSS) and other ship routing systems which are established in most of the major congested shipping areas of the world. Elements used in Traffic Separation/Routing Systems include a set of typical 18 elements Traffic separation scheme: 1. traffic separated by separation zone; 2. Traffic separation scheme, traffic separated by natural obstructions; 3. Traffic separation scheme

with outer separation zone (separating traffic using scheme from traffic not using it); 4. Traffic separation scheme, roundabout; 5. Traffic separation scheme, with "crossing gates"; 6. Traffic separation schemes crossing, without designated precautionary area; 7. Precautionary area; 8. Inshore traffic zone, with defined end-limits; 9. Inshore traffic zone without defined end-limits; 10. Recommended direction of traffic flow, between Traffic separation schemes; 11. Recommended direction of traffic flow, for ships not needed a deep water route; 12. Deep water route, as part of one-way traffic lane; 13. Two-way deep water route, with minimum depth stated; 14. Deep water route, centerline shown as recommended one-way or two-way track; 15. Recommended route (often marked by centerline buoys); 16. Two-way route with one-way sections; 17. Area to be avoided, around navigational aid; 18Area to be avoided, because of danger of stranding.
(see Individual work/presentation, pls).

A floating AtoN serves a similar purpose to a beacon or a lighthouse. The floating platform can be a buoy or a lightship/a lightvessel:

– BA (British Admiralty) Symbol for a Lightship:

Other forms of floating aids are spar buoys, light float and the LANBY (Large Automatic Navigation Buoy).
– BA (British Admiralty) Symbol a LANBY:

To the Examinees and Instructors – Communicative Competence Tasks:
The principles of assessment include Speech Fluency Development: pair work discussions; presentations of individual project works; presentations of the illustrations in Blind Format and the tests, based on the illustrations used in the text, e.g.:

– Danger (depth unknown)
– Obstruction cover/uncover
– Obstruction, least depth known
  Leading line
– Individual work/Presentation: be ready to comment upon the presented illustration in its blind format:

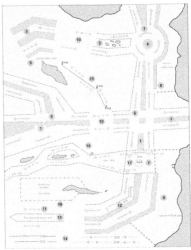

Individual work/Presentation: be ready to comment upon the presented illustrations.

Pair Work. Ask and answer the following questions:

Describe the AtoNs in general.
....
....

What is the difference between the AtoN and the navigational aid?
....
....

Describe Visual Marks in general.
....
....

What are the examples of purpose-built AtoNs?
....
....

What factors does the effectiveness of a visual AtoN depend on?
....
....

What are AtoN distinguishing features?
....
....

Describe a lighthouse in general.
....
....

Describe a beacon in general.
....
....

What elements often define the visual characteristics of a beacon?
....
....

List the typical navigational functions of a lighthouse or a beacon.
....
....

What is a typical usage of a beacon?
....
....

Where TSS and other ship routeing systems are established?
....
....

What can be a floating platform for a floating AtoN?
....
....

List possible forms of floating aids.
....
....

Individual project work:
– Skim and scan the presented text and conclude, why it is important for a future seafarer: don't simply copy the contents, put appropriate word combinations into the given sentence models, motivate your answer in the form of a short summary,:

to comply with: the COLREG rules; the STCW requirements; the SOLAS requirements; the MARPOL requirements; the IALA recommendations	to steer; to manoeuvre; to alter the course to starboard; to alter the course to port; to operate astern propulsion	to become: a rating forming part of a navigational watch; an officer in charge of a navigational watch; a Master	to perform: watchkeeping duties; the Master's orders; the pilot's advice; VTS; instructions	a vessel engaged in dredging or underwater operation; a vessel engaged in fishing	a vessel restricted in her ability to manoeuvre; a power-driven vessel engaged in a towing operation
to maintain: a proper lookout; shiphandling	to proceed at a safe speed	to prevent: the risk of collision; striking with another vessel	in narrow channel; in congested waters; in Head-on situation	to obtain: an early warning of the risk of collision; a watchkeeping license	not to run aground; not to impede the passage
on-board training; seagoing service	to provide: manoeuvrability of the vessel; safe navigation	to avoid: an accident; the immediate danger	to deem the risk of distress;	to take an early action; to allow the safe passage	to join the traffic lane; to leave the traffic lane

Of course, I naturally (obviously/evidently) agree that ....
....
....
......................................is/are important for future seafarers (mariners/seamen).

Let me present a set of arguments clearly proving my point of view:

Firstly,..is/ are necessary (compulsory/obligatory) for/to ....
....
....
....
....
....
....
....
....

Similarly (likewise), there is no doubt, (it is for sure) that ....
....
....
....
....
is/are essential (important) to/for ....
....
....
....
....
....
....
....

So, as I think (in my opinion/as to me), it is for sure, that every cadet should ....
....
....
....
....
....
....
....

In other words, it is obvious, (clear) that ....
....
....
....
....
....
....
....
....
....
....

Let's use this illustration as an example (use any of the presented ones):.................................
....
....
....
....
....
....
....
....
....

Finally, as the conclusion, if one wants to become a seafarer, he should ....
....
....
....
....

Pair work: Use the following SMCP (Traffic organization service) to build up a role play between a vessel and the VTS station:

Clearance, forward planning

Traffic clearance is required before entering ... .

Do not enter the traffic lane / ... .

Proceed to the emergency anchorage.

Keep clear of .../ avoid ... .

You have permission

~ to enter the traffic lane / route - traffic clearance granted.

~ to enter traffic lane / route in position ... at ... UTC.

Do not pass the reporting point ... until ... UTC.

Report at the next way point / way point ... / at ... UTC.

You must arrive at way point ... at ... UTC - your berth is clear.

Do not arrive in position ... before / after ... UTC.

The tide is with you / against you.

Conclusion - the expected outcome of selected data and communicative competence tasks covers the fluent use of adequate speech act modeling, self representation and critical thinking development.

REFERENCES

IALA Aids to Navigation Manual NAVGUIDE, International Association of Marine Aids to Navigation and Lighthouse Authorities, 2010.
Maritime Buoyage System and Other Aids to Navigation - International Association of Marine Aids to Navigation and Lighthouse Authorities, 2010.
The International Regulations for Preventing Collisions at Sea 1972 (COLREGs), 2006.
SOLAS Consolidated Edition 2001

Maritime Education and Training (MET)
STCW, Maritime Education and Training (MET), Human Resources and Crew Manning, Maritime Policy,
Logistics and Economic Matters – Marine Navigation and Safety of Sea Transportation – Weintrit & Neumann (Eds)

Analyzing the Perceptions of Students Who Take the Navigation Course for the First Time: The Case of Dokuz Eylul University Maritime Faculty

B. Kuleyin, B. Celik & A.Y. Kaya
Dokuz Eylul University Maritime Faculty, Izmir, Turkey

ABSTRACT: The "Navigation" course, one of the basic courses taught at the Department of Marine Transportation Engineering, is of extremely great importance for oceangoing watchkeeping officer training. Effective, efficient and safe movements of ships are greatly dependent on the quality of this training. Likewise, the desired quality in such education and training is closely related with the attitudes of prospective mariners toward the technical concept of "Navigation" The purpose of this study is to scrutinize the perceptions of the students taking this course for the first time. The analysis involves the perceptions formed at the very beginning of the term as well as those transformed at the end of the term. In this direction, a questionnaire designed by Geiger and Ogilby (2000) was conducted twice (first at the beginning and then at the end of the semester) to 61 students in the Maritime Faculty of Dokuz Eylul University. The questionnaire sought responses to the following questions: 1. Are there any correlations between former (the initial) perceptions regarding the Navigation course in terms of students' hometown (origin) and the level of success? 2. Are there any correlations between the latter perceptions regarding the Navigation course in terms of students' hometown and level of success? 3. Have students' perceptions of Navigation course changed over the period? SPSS 20 packed program and t-tests were used for the analysis of questionnaire data collected through the questionnaire. Thus such a comparative analysis is thought to bring about a thorough assessment of the development believed to be paced.

1 INTRODUCTION

International Convention on Standards of Training, Certification and Watchkeeping for Seafarers (STCW, 1978) entered into force on 28 April 1984. In STCW Code, Table A-II/1 shows the specifications of minimum standard of competence for officers in charge of navigational watch on ships of 500 gross tonnages or more. Required knowledge, understanding and proficiency for planning and conducting a passage and determining position are terrestrial and coastal navigation (IMO, 2012).

"Navigation" course is given as a lecture in the first semester at Dokuz Eylul University (DEU) Maritime Faculty. After the second semester, students are exposed to a three month ocean going training as cadets. Theoretical lecturers-especially navigation course- and this training period is the keystone of the career of the cadet in gaining or losing the position as an officer. This course provides an intensive study of marine navigation from the perspective of a surface deck officer aboard a ship. Students will understand the proper use of navigational charts, nautical publications and various aids to navigation.

In this paper, we focus on students' perception at the beginning and at the end of the term during first semester. The second part provides the literature review. The third part consists of methodology, data collecting instruments, population, sample size and limitations. In the fourth part, data analysis and findings are presented.

2 LITERATURE REWIEV

A thorough literature review reveals that several studies (Table 1) in various disciplines have analyzed the perceptions of the students who take a course for the first time.

In relation to financial management course, Civan & Cenger (2010) made a study designed to measure students' expectations from financial education. A questionnaire was conducted through eight different universities. 192 questionnaire forms, (out 248 in total) were responded by the students. In the study, a

5-point Likert scale was used. The students of the course were asked such as questions whether the course is necessary is essential in business life.

In relation to accounting course, Kaya (2007) carried out a study on the perceptions of students who take this course for the first time. At the beginning and at the end of the semester, two questionnaires were carried out with 270 freshmen. 5-point Likert scale was used and similar questions were asked relating to the course, and the results were analyzed according to gender and daytime/evening education variables.

In relation to accounting course, Geiger and Ogilby (2000) made a study on students' perceptions of this course and their effects on the decision to major in accounting. At the beginning and at the end of the semester, two questionnaires were conducted with 331 students. 11 questions were asked by using a 5-point Likert scale. The responses to each question were analyzed by coding keywords like "boring", "career", "award", "time", "course".

Table 1. Examples of Some Scientific Researches Where Problem Solving Inventory Has Been Used

Author Information	Population	Sample	Personal
Geiger & Ogilby (2000)	Accounting students in the United States at two medium-sized public universities from two different geographic areas.	331 students	1) Gender 2) Major 3) Course
Saemann & Crooker (1999)	Accounting introductory courses students taught by the same instructor.	283 students for first survey, 196 second survey students	1) Gender 2) Age 3) racial heritage 4) Major intended to pursue 5) GPA
Krishnan et al. (1999)	Finance courses in three different business schools, one each from Texas, Oklahoma and North Dakota.	386 students for first survey, 275 second survey students	1) Gender 2) Age 3) Major 4) Grade 5) GPA
Allen (2010)	Introductory accounting students at one college and two universities	421 students	1) Gender 2) Age 3) Nationality 4) Major 5) GPA
Civan & Cenger (2010)	Students in Turkey at eight public Universities who took financial management course before.	192 students	None
Kaya (2007)	All Students (freshman) of business administration faculty at one public university in Turkey	270 students	1) Gender 2) Major (daytime -night education)

In relation to accounting course, Krishnan et al. (1999) made a study designed to determine the

perceptions and expectations of students' relating to "Introduction to Finance Course" and "Finance". At the beginning of semester and at the end of the semester, questionnaires were carried out in three different business administration departments. 386 students responded to survey at the beginning of the semester and 275 students responded to the survey at the end of the semester. 35 per cent of participants are female and 84 per cent of them are in the 20-25 age brackets. Using a 5-point Likert scale, students were asked some questions. . The content of each question were categorized by coding keywords like "interesting", "useful", "compelling". The questions are categorized according to the intensity of the course content such as "in applications of mathematics more intense" or "in terms of financial theories more intense".

In relation to accounting course, the Saemann and Crooker (1999) made a study to investigate students' perceptions of profession and, its impact on students' decision to choose accounting department. The study has been done on the three initial accounting course given by the same instructor. 283 students responded to the survey for the first questionnaire, 234 students responded to the survey for the second, and 196 students responded to the survey in both. Age range of participants is 16-50 years of age and the average age is 25, of the participants 49. 5% are female. The survey asked some questions designed to measure the perceptions of students of how the course was conducted.

3 METHODOLOGY

3.1 Data Collection Instruments

The data collection instruments used in this study are two-fold: a student-based information form comprising variables whose perceptions are expected /thought to be affected by the "Navigation" course and a questionnaire developed by Geiger and Ogilby (2000) aiming to determine the perceptions of the students of the "Navigation" course. The former is a form developed by the author which includes questions aiming to reveal the profile of the students who have taken the "Navigation" course the very first time. These questions included in this form regarding the gender, the score gained on the university entrance exam, and whether the hometown where the student has grown up is at the seaside or not. The questionnaire, the second instrument, has been translated into Turkish language by Kaya (2007). The form of this questionnaire (Table 2) comprises 11 statements the first 10 of which are to be graded within a five-point Likert scale and the last one is based on a four-point Likert scale.

Table 2.Statements Used to Questionnaire

		Strongly Disagree	←——→	Strongly Agree
1	This course will help me to do well in my future business courses. (COURSES) Bu ders gelecekteki iş yaşantımda başarılı olmamı sağlayacaktır. (BAŞARI)			1 2 3 4 5
2	This course will help me do well in my career. (CAREER) Bu ders kariyerimde başarılı olmamı sağlayacaktır. (KARİYER)			1 2 3 4 5
3	Doing well in this course would be personally rewarding. (REWARDING) Bu derste başarılı olmam benim için ödüllendirici olabilir. (ÖDÜL)			1 2 3 4 5
4	I expect to spend more time on this course than my other courses. (TIME) Bu ders için diğer derslere göre daha fazla zaman harcamayı umuyorum. (ZAMAN)			1 2 3 4 5
5	I am looking forward to this course. (LOOK/ENJOY) Bu dersin ilerleyen sürecini merak ediyorum. (MERAK/ZEVKLİ)			1 2 3 4 5
6	This course will be difficult. (DIFFICULTY) Bu ders zor olacak. (ZOR)			1 2 3 4 5
7	This course will be boring. (BORING) Bu ders sıkıcı olacak. (SIKICI)			1 2 3 4 5
8	I am highly motivated to do well in this course. (MOTIVATED) Bu derste başarılı olmak için son derece motive olmuş durumdayım. (MOTİVASYON)			1 2 3 4 5
9	I expect to learn a lot in this class. (EXPLEARN) Bu dersten çok şey öğrenmeyi bekliyorum. (BİLGİ)			1 2 3 4 5
10	The instructor will affect my opinion of the usefulness of this course. (INSTRUCTOR) Eğitici benim bu dersin önemi hakkındaki düşüncelerim için etkili olacaktır. (EĞİTİCİ)			1 2 3 4 5
11	What is your expected grade in the course? (EXPGRADE) Bu dersten beklediğiniz başarı notu nedir? (PUAN) (4 üzerinden işaretleme yapınız)			1 2 3 4

Sources: Kaya, 2007, p.129; Geiger and Ogilby, 2000, p.68.

3.2 Population and Sample Size

The population of this study comprises the students studying at DEU Maritime Faculty Marine Transportation Engineering Department; and the sample included covers 61 first-year students (freshmen) of the mentioned department who has for the first time taken the "Navigation" course during 2012/2013 fall semester.

3.3 Limitations

The evaluation of the data collected concerns the sample group involved in the study and thus the overall results could not be generalized. Despite such a limitation, the findings are thought to contribute to and supplement the future similar studies concerning the effects of the "Navigation" course. The contribution expected to be made are thought to be more effective on the future studies involving the first-year students studying at the Marine Transportation Engineering Departments of many other maritime faculties.

4 DATA ANALYSIS AND FINDINGS

The statistical analysis of the data collected has been based on the SPSS 20 packet-program. The analysis has made use of the reliability analysis, frequency tables, descriptive statistics and dependent and independent samples t-tests. The reliability analysis has revealed that the Cronbach Alpha value has been 0.705, which proves the reliability to be sufficient.

4.1 Demographic Variables

The demographic specifications of the participants who have responded to the items involved in the questionnaire are detailed in Table 3.

Table 3.The Demographic Specifications of the Participants

	n	%
GENDER		
Male	57	95.0
Female	3	5.0
HOMETOWN		
At seaside/coastal	43	71.7
With no seaside/non-coastal	17	28.3
ACHIEVEMENT SCORE		
Low	30	50.0
High	30	50.0

4.2 Testing the Hypotheses

The analyses carried out on the 11 statements involved in the questionnaire have been used to test the following four hypotheses:

- H1: There exists a meaningful difference between the perceptions of the male and female students recorded at the beginning and at the end of the Navigation course.
- H2: There exists a meaningful difference between the perceptions of the students from coastal regions and of those from non-coastal regions recorded at the beginning and at the end of the Navigation course.
- H3: There exists a meaningful difference between the perceptions of the students with low scored university entrance exams and of those with high scored recorded at the beginning and at the end of the Navigation course.
- H4: There exists a meaningful difference between the perceptions of the students at the beginning of the Navigation course and their perceptions at the end of the course.

Table 4.Statements Revealing Differences at the Beginning of the Course for Independent Variables

Statements	Independent Variables	Variable Group	Mean	t	p
TIME	Gender	Male	4.07	2.240	0.029
		Female	3.00		
EXPGRADE	Gender	Male	3.65	-5.502	0.000
		Female	4.00		
DIFFICULTY	Hometown	Coastal	3.44	2.763	0.008
		Noncoastal	2.71		

The t-test results indicated in Table 4 reveal that concerning the expected duration and achievement level of the course there exists a meaningful difference between the perceptions of male and those of female in that the duration related expectations of the male are higher than those of the female and the achievement expectations of the female are higher than those of the male. In terms of the regions where the students have come from, it has been found out that there exists a meaningful difference between the perceptions of those from coastal regions and those from non-coastal regions concerning how difficult the course is expected to be. The former group thinks that the course will be difficult. This could mean that those from the coastal regions have more information about sea and maritime issues. In terms of the achievement score expected, there exists no meaningful difference between the perceptions of those with high score received in the university entrance exams and of those with low scores.

Table 5.Statements Revealing Differences at the End of the Course for Independent Variables

Statements	Independent Variables	Variable Group	Mean	t	p
CAREER	Gender	Male	4.75	-3.747	0.025
		Female	5.00		

The paired t-test results are indicated in Table 5. In terms of the gender, there exists a meaningful difference between the perceptions of male and female. Female expectations related with career are higher than those of male. Female students opine "This course will help me do well in my career".

There is not a meaningful difference between at the beginning and at end of course in terms of hometown. Students' perceptions at end of term are similar to those evolved at the beginning of term.

Table 6. Statements Revealing Differences between the End of the Course and the Beginning of Course

Statements	Dependent Variables	Mean	t	p
DIFFICULTY	At the Beginning	3.23	2.464	0.000
	At the End	2.95		
EXPGRADE	At the Beginning	3.37	10.695	0.017
	At the End	2.12		

The t-test results indicated in Table 6 reveal that, in terms of the difficulty of course, there exists a meaningful difference between the t-test results at beginning of course and at the end of course. At the beginning of the course, students' perceptions are "course will be difficult". At the end of the course, their perceptions are different; fewer students have the opinion of "course will be difficult".

The Students' expectations about the examination results are higher than the actual. The mean value of student's grade expectations at the beginning of the course is 3.37, but the mean value is in reality 2.12. It shows that in terms of student perception about course grade, there exists a meaningful difference between the result of the beginning of the course and the end of the course.

5 CONCLUSIONS

In this case, students' perceptions are examined by statements such as hometown, gender, success. There exist meaningful differences for some statements. Population consists of the students who have taken the navigation course for the first time; there is 57 male and 3 female. 1 student responded only one questionnaire. Thus, the assessment is made for 60 students. In terms of career, female expectations are quite higher than those of male. This department is rarely chosen by female students, as a matter of course, because maritime profession is hard.

In the forthcoming years, reaching the first-year students studying at the Marine Transportation Engineering Departments of many other maritime faculties might enable more effective assessment and results can be generalized.

REFERENCES

IMO (International Maritime Organization). (2012). STCW Convention. Access: 06.12.2012 http://www.imo.org.
Geiger, M. A. & Ogilby, S. M. (2000) The first course in accounting: students' perceptions and their effect on the decision to major in accounting. *Journal of Accounting Education*, 18: 63-78.
Kaya, U. (2007) İlk Defa Muhasebe Dersi Alan Öğrencilerin Derse Yönelik Algılamaları Üzerine Bir Alan Araştırması: Karadeniz Teknik Üniversitesi Örneği, *MUFAD Journal*, 36: 125-133.
Allen, C. L. (2004) Business Student's Perception of the Image of Accounting. *Managerial Auditing Journal, 19(2): 235-258.*
Civan, M. & Cenger, H. (2010) Finansal Yönetim Dersini Almış Öğrencilerin Finans Eğitimi Beklentilerini Ölçmeye Yönelik Yapılan Bir Çalışma *MUFAD Journal*, 46:84-99.
Saeman G.P. & Crooker K.J. (1999) Student Perceptions of the Profession and Its Effect on Decisions to Major in Accounting, *Journal of Accounting Education*, 17:1-22
Krishnan R.S., Bathala C.T., Bhattacharya, H.K. Ritchey R. (1999) Teaching Introductory Finance Course: What can we learned from Student Perceptions and Expectations? *Financial Practice and Education*, 70-82

Maritime Education and Training (MET)
STCW, Maritime Education and Training (MET), Human Resources and Crew Manning, Maritime Policy,
Logistics and Economic Matters – Marine Navigation and Safety of Sea Transportation – Weintrit & Neumann (Eds)

The Influence of Student Self-evaluation on the Effectiveness of Maritime Studies

I. Bartusevičiene
Lithuanian Maritime Academy, Klaipeda, Lithuania

L. Rupšiene
Klaipeda University, Klaipeda, Lithuania

ABSTRACT: The shift of the instruction/learning paradigms influences the understanding of studies and the change in the roles of the study process participants. The new paradigm highlights an active role of student in the learning process. Student self-evaluation is one of the ways to involve them in the study process. Definitions of the concept of self-evaluation describe it as learners' decisions about their progress and the level of achievement of the set standards (criteria) in order to improve studies. With the relation between student self-evaluation and the effectiveness of their studies having been chosen as the object of the research, the influence of the content, the forms, and the frequency of self-evaluation on the effectiveness of studies will be discussed in the paper. For the empirical research, a case of full-time studies of Marine Engineering was decided upon. An originally created questionnaire was used for the research. The characteristics of effective self-evaluation, such as its content, forms, and frequency, were analyzed with the aim of finding out their influence on the effectiveness of maritime studies.

1 INTRODUCTION

1.1 *Paradigm shift*

Currently, in the system of tertiary education, an intensive paradigm shift (from instruction (teaching) to learning) has been taking place. In the 20th c., constructivist theories changed the view of the learning to the effect that learners develops their own individual knowledge by means of interaction with the environment and on the basis of their previous experience (Lefrancois 1997). Given the shift of the teaching/learning paradigms, the evaluation of student achievements changes as well. A characteristic quality of the new learning paradigm is the involvement of learners in the evaluation of their achievements.

1.2 *Student self-evaluation*

Although self-evaluation is defined in rather different ways, one should note that all definitions of self-evaluation characterize it as learner's activity during which decisions are taken about their own progress and the degree of achievement of the set standard (criterion) in order to improve the process of learning (Stellwagen 1997, Garcia & Roblin 2008, Brew 1999).

Recently, self-evaluation has been analyzed in a number of different aspects. Quite a few studies provided reliable evidence of self-evaluation being beneficial in many respects. Thus, e.g., self-evaluation has been established to contribute to students' development of autonomy (Stallings & Tascione 1996, Hart 2008, Paris & Paris 2001, Cambra-Fierro & Cambra-Berdun 2007), meta-cognitive competences (Edwards 1989), as well as critical thinking and the ability to discuss (Hanrahan & Isaacs 2001); it is believed to assist in learning self-regulation (Cambra-Fierro & Cambra-Berdun 2007) and to encourage learners to accept greater responsibility for their own study outcomes (Biggs & Tang 2007).

Some researchers analyzed how self-evaluation added to the improvement of the study process. They noted that self-evaluation facilitated the study process (Stallings & Tascione 1996) and improved the effectiveness of studies (Bartuseviciene & Rupsiene 2010), and that self-evaluation with clearly defined procedures, criteria, and standards contributed to the clarity of evaluation and provided teachers with reliable information about student improvement (Ross & Starling 2008). The relationship between student and teacher self-evaluation has been extensively researched; as established in that research direction, students occasionally tended to evaluate themselves somewhat better than teachers (Boud & Falchikov 2007); the evaluations of good students and teachers tended to coincide more frequently (Ross 2006); moreover, student self-evaluation was closer to the outcomes of teacher evaluation when the evaluation criteria were clearly set and discussed in advance (Boud & Falchikov 2007, Ross & Starling 2008).

1.3 *The effectiveness of studies*

The phenomenon of the effectiveness of studies is an important issue due to its close relationship with the phenomenon of the quality of studies. The quality of maritime studies is a multidimensional and complex phenomenon; it is influenced by a wide range of factors. As revealed by some previous research (Kalvaitiene et al. 2011), the motives of choosing a maritime profession is one of the factors that influences the quality of studies; it can be consolidated by explaining to young people all the merits of a maritime profession and demonstrating an opportunity of finding emotional attractiveness and realization of their interests in the choice of maritime professions.

The concept of the *effectiveness of studies* was defined as an attribute of learning at an institution of higher education which indicates the achievement of the intended learning outcomes (Rupsiene & Bartuseviciene 2011). The assessment of students' achievements is one of the factors that influences the effectiveness of maritime studies. As witnessed by previous research, the components of the assessment of students' achievements, such as assessment frequency, assessment methods, feedback characteristics, and student involvement in the assessment process positively influence the effectiveness of maritime studies whenever properly used (Bartuseviciene & Rupsiene 2011). However, students' involvement in the assessment process in the form of self-evaluation and its influence on the effectiveness of studies needs deeper investigation.

A number of research papers analyzed the relation between student self-evaluation and the effectiveness of studies. J. Dewey (cit. Malone & Pederson 2008) noted that self-evaluation helped to achieve better study results. Currently, the amount of evidence of the said insight has been increasing (McDonald & Boud 2003, Ross & Starling 2008). However, it should be noted in the present context that, according to some researchers, student self-evaluation, unless applied appropriately, not only fails to increase the effectiveness of studies, but in come cases may reduce it (Ljungman & Sile, 2008, Andrade & Ying 2007).

Recently, scientists' attention has been focused on the search for elements, components, models, and technologies of self-evaluation which would positively influence the effectiveness of studies or some components of it. It has been discovered that self-evaluation is more effective when students' ability to evaluate themselves is taken into consideration and further developed (Andrade & Ying 2007), when students get acquainted with the standards of quality work and the criteria of its evaluation (Andrade 2000), when the assignments and the evaluation criteria of their performance are provided in the written form (Nicol & Macfariane-Dick 2006), when the examples of well-done assignments are provided (Orsmond et al. 2002), when students keep learning diaries in which they reflect upon their learning and its results (Malone & Pederson 2008), when self-evaluation methods are adjusted to self-evaluation goals, learning contexts, and learners' personal qualities (Chamorro-Premuzic & Furnham 2006), and when learners' gender is taken into account (Goodrich 1996). Despite that, quite a lot of uncertainties have been left in the interpretation of the relation between student self-evaluation and the effectiveness of studies.

Having chosen the relation between student self-evaluation and the effectiveness of the mastery of the study subject as the object of the research, the authors of the article focused on the impact of the form, content, and frequency of student self-evaluation on the effectiveness of the mastering of an academic subject in Marine Engineering study programme.

The aim of the research was to determine what influence the form, the content, and the frequency of student' self-evaluation had on the effectiveness of the mastering of an academic subject and which of the above mentioned elements was more important for the mastering of the subject in maritime studies.

The research methods included a questionnaire survey and statistical analysis of quantitative data.

2 RESEARCH METHODOLOGY

2.1 *Research sample*

For the empirical research, a case study of the programme of maritime specialty was chosen. The research population consisted of Marine Engineering students of the Lithuanian Maritime Academy in spring 2010. 132 full-time students of Marine Engineering (95 % of all population) in all the study years of the said study programme were surveyed. The research sample included merely male students.

The questionnaires were handed to each student personally. To ensure the ethical character of the research, an official permission for its conducting was obtained from the Head of the Academy. Moreover, all the participants of the research were introduced to the research objective and the specific requirements for the completing of the questionnaire. The questionnaires were compiled in such a way as to maximally ensure the participant anonymity; it was impossible to identify the respondent by means of the questionnaire data. Therefore, the essential principles of the social science research were observed, viz., voluntariness and anonymity that accounted for the situation in which the respondents were able to fully express their views of the phenomenon or the event of the research.

2.2 *Research instrument*

The research instrument was construed by the authors, given the aim of the research. The questionnaire entailed an open question about the study subject to which, in the respondents' opinion, student self-evaluation was applied the most effectively. The question was necessary to establish the limitations of the application of the research findings. 38% of the respondents pointed out such subjects: the figure proved that student self-evaluation was insufficiently practiced in the *Marine Engineering* study programme. However, the analysis of the named academic subjects revealed that student self-evaluation was effective both in the subjects of professional qualification and professional specialization.

The most frequently named subjects of professional qualification included *Chemistry, Physics, Information Technologies,* and *Physical Training.* It turned out that the majority of the teachers of the subjects of professional specialization, such as *Electrical Engineering* or *Theory of Internal Combustion Engines*, also applied student self-evaluation. Therefore, the findings of the research provided in the article are limited to the study programme (viz. *Marine Engineering*) and the subjects of the study field (subjects of professional qualification and specialization). The relevance of the research findings for other study programs could be established after additional research.

The items of the questionnaire contributed to the identification of the forms, the frequency, and the content of student self-evaluation and students' views of the influence of self-evaluation on the effectiveness of the mastering of academic subjects. All the questions were formulated in such a way as to enable students who provided answers to them to refer to their own experience in the studies of the subject where the application of student self-evaluation was the most successful. The frequency of self-evaluation was measured on the basis of the number of times that student self-evaluation was applied in the course of the studies of the subject. The form of self-evaluation was assessed in accordance with the way of its most frequent application (written, oral, or both written and oral). The content of self-evaluation was judged by the frequency that the students had to assess their

progress, to note things they had not understood, to point out learning difficulties, and to plan the prospects of further learning on the request of their teachers.

The answers to the question *Do you feel you better mastered the academic subject in which student self-evaluation was applied in the best way?* revealed the effectiveness of the mastering of the academic subject. Even though the opinions may have been subjective, in our case, the provision was observed (see Fitz-Gibbon & Kochan 2000, Petty & Green 2007) that it was possible to refer to learners' opinion when judging the effectiveness of studies. However, the second limitation of the research findings was identified there, as the teachers' opinion or other parameters of the evaluation of study effectiveness were not taken into consideration. To eliminate the said limitations, additional research would also be necessary.

3 RESEARCH OUTCOMES

3.1 *The influence of student self-evaluation on the effectiveness of studies*

Before we establish the impact of the content, the frequency, and the form of student self-evaluation on the effectiveness of their mastering of academic subjects, we have to find out how frequently, in which form(s), and in what way in terms of the content the learners evaluate their achievements during the studies. The forms of self-evaluation were determined as written, oral, and both written and oral. In accordance with the respondents' answers, the students evaluated themselves in different forms: in writing: 6 %; orally: 46%; and in writing and orally: 48%. Moreover, it turned out that self-evaluation was applied from 0 to 11 times. The respondents' answers witnessed that in 25% of the cases, self-evaluation was used only once; on average, students had an opportunity to evaluate themselves 3.32 times (sd=2.83) (Fig. 1). On the basis of the aforementioned facts, a conclusion can be drawn that self-evaluation applied in different forms and with different regularity can positively influence the effectiveness of the mastering of the subject; however, further research would be necessary to establish a reasonable number of self-evaluation events.

Figure 1. Frequency of self-evaluation.

The respondents' answers about the content of self-evaluation (Fig.2) witnessed that they most often self-evaluated the things they failed to understand (64%), planned further learning (61 %), pointed out the difficulties they encountered while studying the subject (54 %), and self-evaluated the progress made (52%). 62 % of the respondents pointed out that, in their opinion, they better mastered the subjects in which self-evaluation was applied the most successfully. Self-evaluation also helped 63 % of the respondents to better organize their individual learning.

The respondents were asked to rank the aspects of evaluation, such as the periodicity of evaluation, the ways of evaluation, the provision of feedback, and self-evaluation, in accordance with their importance for the effectiveness of studies. 23 % of respondents gave the priority to self-evaluation.

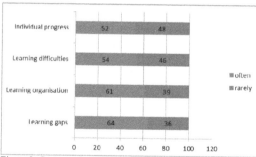

Figure 2. Content of self-evaluation.

The method of regression analysis was used to explain the meanings of the variable *The effectiveness of mastering of an academic subject* in accordance with the meanings of independent variables: self-evaluation form, self-evaluation content (which was aggregated from the four self-evaluation content variables presented in Fig.2), and self-evaluation frequency. On performing the regression analysis (Table 1), the influence of three variables on the effectiveness of the mastering of the subject was established (r=0.570). The three variables accounted for 32.5 % of dependent variable dispersion (r^2=0.325; r^2adj=0.271). The regression model was appropriate (p=0.002), and it described a statistically significant relation

(F=5.949) between the dependent variable and the three identified independent variables. T meanings showed that the dependent variable could be forecast by means of the three identified variables.

Table 1. Coefficients of the appropriateness of a regression model

Model	R	R Square	Adjusted R Square	Std. Error of the Estimate	F	Sig.
1	,570	,325	,271	,906	5,949	,002

Standardized (Beta) coefficients (Table 2) witnessed that the effectiveness of the mastering of the subject mainly depended on the content of self-evaluation.

Table 2. Coefficients of the regression model.

Model	Unstandardized Coefficients		Standardized Coefficients	t	Sig.
	B	Std. Error	Beta		
1 (Constant)	2,315	,724		3,199	,003
Self-evaluation forms	-,106	,256	-,059	-,416	,680
Self-evaluation content	,155	,037	,624	4,194	,000
Self-evaluation frequency	-,058	,051	-,169	-1,127	,267

4 CONCLUSIONS

As established by our research, self-evaluation encouraged students to improve their individual knowledge and to become more responsible for their study outcomes, which was important in the context of the learning paradigm. This issue was also significant for the maritime studies.

In the initial stage of the research, the importance of the frequency, the content, and the form of student self-evaluation was established. For better effectiveness of the maritime studies, different forms of student self-evaluation can be applied (written, oral, or combined). Self-evaluation can be used with different regularity (from 0 to 11 times) and have different content (self-evaluation of the progress, difficulties, learning gaps, or planning of the learning prospects). In accordance with the research outcomes, self-evaluation could be used more frequently and regularly; a reasonable number of self-evaluation events could be investigated further.

Although all three self-evaluation elements were significant in the context of the effectiveness of the subject mastering, self-evaluation of the content had the greatest influence on the latter. The effectiveness of studies increased when students had to self-evaluate their own individual progress in the written form and to indicate the difficulties they encountered in the studies of the subject.

REFERENCES

Andrade, H. 2000. Using rubrics to promote thinking and learning. Educational Leadership. No. 57(5): 13–18.

Andrade, H. & Ying, D. 2007. Student responses to criteria-referenced self-assessment. Assessment & Evaluation in Higher Education 32(2): 159-181.

Bartusevičienė, I., Rupšienė, L. (2010). Studentų pasiekimų vertinimo periodiškumas kaip studijų rezultatyvumo veiksnys: socialinės pedagogikos studijų programų studentų nuomonė. Tiltai 2 (51): 99-112.

Bartuseviciene, I., Rupšienė, L. 2011. Assessment Components Influencing Effectiveness of Studies: Marine Engineering Students' Opinion. In: A.Weintrit & T.Neumann (Eds.). Human Resources and Crew Resource management. Marine navigation and Safety of Sea Transportation. London: Taylor& Francis Group: 71-77.

Biggs, J.B. & Tang, C. 2007. Teaching for Quality Learning at University. Berkshire: Mcgraw Hill, Society for Research into Higher Education & Open University Press.

Boud, D. & Falchikov, N. 2007. Developing assessment for informing judgement. In D. Boud, N. Falchikov (eds.). Rethinking Assessment in Higher Education. London and New York: Routledge Taylor and Francis Group.

Brew, A. 1999. Towards autonomous assessment: using self-assessment and peer-assessment. In S. Brown, A. Glasner (eds.). Assessment Matters in Higher Education. Buckingham: Society for Research into Higher Education/ Open University Press.

Cambra-Fierro, J. & Cambra-Berdun, J. 2007. Students' self-evaluation and reflection, part 2: An empirical study. Education & Training 49(2): 103–111.

Edwards, R. 1989. An experiment in student self-assessment. British Journal of Educational Technology 20(1): 5–10.

Fitz-Gibbon, C. & Kochan, S. 2000. School effectiveness and education indicators. In T. & D. Reynolds (eds.). The international handbook of school effectiveness research. London: Falmer Press: 257–282.

Garcia, L. M., Roblin, N. P. 2008. Innovation, research and professional development in higher education: Learning from our own experience. Teaching and Teacher Education: An International Journal of Research and Studies 24(1): 104–116.

Hanrahan, S. J. & Isaacs, G. 2001. Assessing self- and peer-assessment: the students' views. Higher Educational Research and Development 20(1): 53–70.

Hart, P. D. 2008. How should colleges assess and improve student learning? Employers' views on the accountability challenge. Washington, DC: Research Associates, Inc. Association of American Colleges and Universities. Retrieved from http://www.aacu.org/publications/index.cfm.

Kalvaitienė, G., Bartuseviciene, I., Senčila, V. 2011. Improving MET Quality: Relationship Between Motives of Choosing Maritime Professions and Students; Approaches to Learning. TransNav - International Journal on Marine Navigation and Safety of Sea Transportation 5(4): 535-540.

Lefrancois, G. 1997. Psychology for Teaching. Belmont: Wadsworth Publishing Company.

Ljungman, A. G. & Silen, C. 2008. Examination involving students as peer examiners. Assessment & Evaluation in Higher Education 33(3): 289–300.

Malone, V. & Pederson, P. V. 2008. Designing assignments in the social studies to meet curriculum standards and prepare students for adult roles. Clearing House: A Journal of Educational Strategies, Issues and Ideas 81(6): 257–262.

McDonald, B. & Boud, D. 2003. The impact of self-assessment on achievement: the effects of self-assessment training on performance in external examinations. Assessment in Education 10(2): 209–220.

Orsmond, P., Merry, S. & Reiling, K. 2002. The use of formative feedback when using student derived marking criteria in peer and self-assessment. Assessment & Evaluation in Higher Education 27(4): 309–323.

Paris, S. G. & Paris, A. H. 2001. Classroom applications of research on self-regulated learning. Educational Psychologist 36(2): 89–101.

Petty, N. W. & Green, T. 2007. Measuring Educational Opportunity as Perceived by Students: A process indicator. School Effectiveness and School Improvement 18 (1): 67–91.

Ross, J. A. 2006. The reliability, validity, and utility of self-assessment. Practical Assessment, Research and Evaluation 11(10). Retrieved from Internet: http://pareonline.net/getvn.asp?v=11&n=10.

Ross, J. A. & Starling, 2008. M. Self-assessment in a technology supported environment: The case of grade 9 geography. Assessment in Education: Principles, Policy & Practice 15(2): 183-199.

Rupšienė, L., Bartusevičienė, I. (2011). Analysis of the concept of effectiveness of studies. Andragogika Mokslo darbai 1: 59-71.

Stallings, V. & Tascione, C. 1996. Student self-assessment and self-evaluation. Mathematics Teacher 89(7): 548–555.

Stellwagen, J. B. 1997. Phase two: using student learning profile to develop cognitive self-assessment skills. American Secondary Education 26: 1–8.

Maritime Education and Training (MET)
STCW, Maritime Education and Training (MET), Human Resources and Crew Manning, Maritime Policy,
Logistics and Economic Matters – Marine Navigation and Safety of Sea Transportation – Weintrit & Neumann (Eds)

Marine E-Learning Evaluation: A Neuroscience Approach

D. Papachristos, N. Nikitakos & M. Lambrou
University of Aegean, Chios, Greece

ABSTRACT: The evaluation with the use of neuroscience methods and tools of a student's satisfaction – happiness from using the e-learning system (e-learning platforms, e-games, simulators) poses an important research subject matter. In the present paper, it is presented a research on course conducted in the Marine Training Centre of Piraeus. In particular, this research with the use of a neuroscience tools-gaze tracker and voice recording (lexical-sentiment analysis), investigates the amount of satisfaction of the students using Engine room simulator (ERS 5L90MCL11, Kongsberg 2003 AS) by monitoring the users' eye movement and speech in combination with the use of qualitative and quantitative methods. The ultimate goal of this research is to find connection between satisfaction and usability by using non conventional methods (neuroscience tools) in e-learning marine systems.

1 INTRODUCTION

The use of non conventional methods and tools (biometric tools) is a useful contribution in the amelioration of Marine Education (ME). ME follows certain education standards (STCW'95) for each specialty (Captain, Engineer) and for each level (A', B', C'). Its scope is the acquisition of basic scientific knowledge, dexterities on execution (navigation, route plotting, engine operation etc.) as well as protecting the ship and crew (safety issues and environment protection issues). Specifically, the STCW'95 standard defines three competency levels: management, operational and support while at the same time it defines related dexterities. Every dexterity level suggests the totality of the learning goals and the goal definition is the basic characteristic of training. Complex competences are made up by simpler ones. This hierarchical progression in the dexterity level sets an austere framework for the educator course designer in each marine faculty. The introduction of simulators and other modern training tools constitutes an important research question related to the degree it can fulfil all the expectations set forth the STCW'95 (IMO, 2003, Papachristos et al., 2012a, Papachristos & Nikitakos, 2011, Tsoukalas et al., 2008).

We propose a research framework for educational and usability evaluation of marine e-learning systems that combines a neuroscience approach (biometric tools for gaze tracking, speech recording for measuring emotional user responses-lexical analysis) with usability assessment. Certainly, the proposed approach may require further adaptations in order to accommodate the evaluation of particular interactive e-learning systems. The main elements of the proposed approach include (Papachristos et al., 2012a, 2012b):

- Registration and interpretation of user emotional states
- Gaze tracking and interpretation
- Speech recording and lexical analysis (sentiment processing)
- Usability assessment questionnaires
- Wrap-up interviews.

This procedure is a primary effort to research the educational and usability evaluation with emotion analysis (satisfaction) of the users-students in marine e-learning environments.

2 THEORETICAL BACKGROUND

In the investigative field of psychology, the use of the English word *affect* is very popular, use to usually cover a plethora of concepts such as emotions, moods and preferences. The term emotion tends to be used for the characterization of rather short but intense experiences, while moods and preferences refer to lower intensity but greater duration experiences. In general, we could note that psychology considers the emotional mechanism as a determinist mechanism that pre-requires a stimulus – cause incited in the brain by use of the neural and endocrine system (hormonal), the response – emotion (Malatesta, 2009, Papachristos et al., 2012a, 2012b).

Modern scientific community suggests different views concerning comprehension of the emotional mechanism. There is the view that emotion is defined by the natural reactions caused in the body (sweating, pulse increase, etc.), while other researchers believe that it is purely a mind process, while there are also hybrid views that define, each one in a varying degree, the participation and the manner where the human functions are involved in the emotional experience (Vosniadou, 2001).

Many psychologists have claimed that the only way to interpret the totality of emotions is to suggest that there is a common evolutional base in the development of facial emotional expressions. But the biological approach cannot explain all the facets of a human's emotional behavior, (Vosniadou, 2001).

During the last 25 years, psychology focuses again in the sequence of events involved in the creation" of an emotion. Zajonc considers that experiencing an emotion happens often before we have the time to assess it, while contrarily, Lazarous considers that the thought precedes the emotional experience, assessing that instantaneous cognitive assessments of situations can happen at the same time alongside the emotional experience (Lazarous, 1982, Zajonc, 1984). The speed with which we assess a situation is influenced by our previous experiences. The age scale also seems to influence the creation of emotions. It must be noted that there are also emotions that do not require cognitive processes (thought). For example, loud noises or seeing a lion. Such emotional reactions can be important for the survival of the species and are related to certain stereotypical facial expressions which have global meaning (Vosniadou, 2001).

Also, another factor that can be investigated in relation to the emotional experience is the language process. The psychological research in the language production, comprehension and development is developed mainly after 1960 as a result of linguist's N. Chomsky research on generative grammar (Pinker, 2005). The psycholinguistic research showed that language comprehension and production is not influenced only from factors not related to their linguistic complexity but also from the speaker's/listener's existing knowledge for the world around him/her, as well as by the information included in the extra linguistic environment (Vosniadou, 2001).

Investigating the emotional gravity of words spoken by a speaker and defined its emotional state (current or past) constitutes a state of the art issue. Most of the emotional state categorization suggested concern the English language. To overcome this problem, studies have been conducted that approach the matter cross-culturally and study the assignment of the categories to various languages. This assignment has conceptual traps since the manner in which an emotional state is apprehensible; an emotional state is influenced by cultural factors as well. In a rather recent cross-cultural study done by Fontaine et al., (2007), 144 emotional experiences' characteristics were examined, which were then categorized according to the following emotional "components": (a) event assessment (arousal), (b) psycho physiological changes, (c) motor expressions, (d) action tendencies, (e) subjective feelings, and (f) emotion regulation.

International bibliography contains various approaches – techniques (sorting algorithms) concerning linguistic emotional analyses, which are followed and are based mainly in the existence of word lists or dictionaries with labels of emotional gravity along with applications in marketing, cinema, internet, political discourse etc (Lambov et al., 2011, Fotopoulou et al., 2009). There are studies also concerning sorting English verbs and French verbs that state emotions based on conceptual and structural-syntactical characteristics. For the Greek language there is a study on verbs of Greek that state emotions based on the theoretical framework "Lexicon-Grammar" that is quite old and doesn't contain data from real language use; there are also some studies concerning Greek adjectives and verbs that state emotions and comparison with other languages (French – Turkish) under the viewpoint: Structural-syntactical + conceptual characteristics. More recent studies in Greek conducted systematically the noun structures based on the theoretical framework of "Lexicon-Grammar" and the establishment of conceptual & syntactical criteria for the distinction and sorting of nouns based on conceptual-syntactical characteristics of the structures in which they appear (Papachristos et al., 2012a).

The observation of eye movement, as well as the pupil movement, is an established method in many years now and the technological developments in both material equipment and software, made it more viable as a practicality measurement approach. The eyes' movements are supposed to depict the level of the cognitive process a screen demands and consequently the level of facility or difficulty of its process. Usually, the optical measurement concentrates on the following: the eyes' focus points, the eyes' movement patterns and the pupil's alterations. The measurement targets are the computer screen areas definition, easy or difficult to understand. In particular the eyes movement measurements focus on attention spots, where the eyes remain steady for a while, and on quick movement areas, where the eye moves quickly from one point of interest to another. Moreover the research interest is focused in the interaction of gaze tracking during the presentation of information and content (internet) in a natural environment (Dix et al., 2004, Kotzabasis, 2011).

3 RESEARCH METHODOLOGY

The Research Methodology must fulfill all three requirements of the cognitive neuroscience: (a) experiential verification, (b) operational definition, and (c) repetition.

The main purpose of this research activity is the analysis of emotional state and the investigation of the standards that connect the user's Satisfaction-Happiness by use of the eye-head movement & oral text (as the basis for the situation) in the basic dipole: *happiness (satisfaction) – sad (non satisfaction)*.

We use a research protocol PR-AS (based on previous research with new research tools like voice recording, SUS tool etc.) (Papachristos & Nikitakos, 2011, Papachristos et al., 2012a, 2012b). It is defined in detecting, recognizing and interpreting the emotional information in conjunction with other information created during the execution of a scenario in an electronic learning marine system (simulators or training software). The emotional information comes from the user's emotional state before, during, and after the scenario/exercise. Its structure concerns the following sections (Papachristos et al., 2012a):
– the mood/emotion before the scenario/exercise (oral text)
– Behavioral action (head movement, gaze) during the scenario and
– the emotional post-experience – satisfaction (oral text).

Measuring the emotional information will be realized using the following processes:
– *Natural parameters' measurement*: Movement parameters (head movement, gaze movement) and oral text as text and
– *Registering* user opinion/viewpoint/view.

The suggested protocol (*Protocol Research of Affect Situation, PR-AS*) is comprised by the following sections (Fig. 1) (Papachristos et al., 2012a):

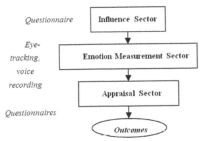

Figure 1. Structure of PR-AS

– *Influence Sector, IS:* it based on Action Tendency Theory (concern view) and on Practical Reasoning Theory. This theoretical processing is characterized as a Framework for User's Innate Stimuli. The influence's department consists of the following measurements that take place before the scenario execution by way of questionnaires: (a) profile (learning-medical), (b) personality, (c) expectations-interesting and (d) personal background (education, professional experience, computer using) (Bratman, 1987, Fridja, 1986, Potoker & Corwin, 2009) .
– *Emotion Measurement Sector, EMS:* Measurements concerns the happiness-sad (emotion-mood) in combination with the degree of activation-assessment by the user within the framework of this dipole, i.e. the measurement of dynamics in relation to the stimuli (sound, animation, schemas, etc.) received in total by the software-scenario (virtual relationship) considering that the user is always on a core emotional state (core affect) and the specific satisfaction for the scenario and software (evaluation process of the educational use for the software and scenario/exercise to the degree of satisfaction of the trainee-user) adopted by the Oatley approach that the (personal) goals have been achieved there is a sense of joy, while failures if followed by sadness and despair and is connected with the emotion of satisfaction. At the same time the natural parameters comprising the protocol's core are registered. These are the video and voice recording (emotional reasoning). This is based on the use of tools for recording head movement (distance from the monitor, left-right head movement, left-right head rolling), gaze tracking (x,y coordinates) and voice recording of spoken words (as reasoning of meta-emotional experience – lexicalization of emotional gravity). Additionally, the researcher records observations related to the physiological and non-physiological attitude of the user (mistakes, time of execution, execution success, the user's psychological state) (Oatley, 1992, Russel, 2003).
– *Appraisal Sector, AS:* In this section, the Satisfaction recording takes place but also

commenting related to the day when measurement is taking place & in total up to that moment, as far as the software tool is concerned, after the experimental conduct of the scenario/exercise (usability), personal self-evaluation, scenario evaluation (benefits) in combination with the weighed usability assessment tool (DEC SUS Tool) (Brooke, 1996, Tsai et al., 2012).

The detection of emotional information will be realized using the Technical & Theoretical Tools (Fig.2):

– *Tool-1 (T1):* In the protocol the optical data registration will be conducted by the "Face Analysis" software in connection with a Web camera set on the computer in which there is the subject of the research (engine simulator). That particular software records a large number of variables but we focus on the following parameters that refer to the user's eyes and head movement: (a) eyes movement (Eye gaze vector), (b) user's head position in regard to the eyes up/down – right/left movement (Head Pose Vector: pitch, yaw), (c) eye distance from the computer screen (Dist_monitor) and (d) rolling of the head (eye angle from a horizontal level) (Asteriadis et al., 2009).

– *Tool-2 (T2):* Use of a microphone for voice recording of spoken words (speech-text). This will be used for the registration of 3 temporal marks: (a) First Point (T_A) - the temporal mark before recording for measuring mood. This executes the voice recording (1 file) of the user where the user explains how he/she feels and why, (b) Second Point (T_B) - a temporal mark after the recording for the measurement of mood-emotion after the recording. This executes the voice recording (1 file) of the user where the user explains how he/she feels and why and (c) Third Point (T_C) - a temporal mark after the recording where the satisfaction choices are justified (software, scenario). This executes the voice recording (1 file) of the user where the user explains how he/she feels and why. The voice recordings consisting of 3 .WAV type files will be analyzed further during the processing section in three dimensions: *Lexicological* (emotional analysis), *Style analysis of linguistic characteristics*, and *Qualitative analysis* of the spoken word so that the user emotional state/satisfaction can be justified (He and Zhou, 2011).

– *Tool-3(T3):* Questionnaires using for opinion/attitudes/expectation/self-evaluation. It concerns 3 questionnaires: (a) T3-A, influence data, (b) T3-B, mood, scenario label, recording data and (c) T3-C, User assessment in combination with the usability tool SUS (Brooke, 1996, Ekman, 1999).

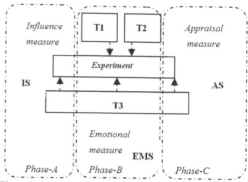

Figure 2. Structure of PR-AS

4 PARTICIPANTS

The initial randomized sampling was carried out between May and June 2012, in the Marine Engine System Simulator (MESS) Laboratory of the Marine Training Centre of Piraeus. 13 Merchant Marine officers were used as the experiment subjects; they underwent a specific procedure in the engine room simulator ERS 5L90MCL11 with video recording of ~23 minutes per student. They then completed the questionnaires and were interviewed follow the research methodology framework.

5 DATA ANALYSIS

The data of experiment come from three sources:
– questionnaires,
– optical data (eye tracking) and
– interviews (voice recording).

13 professional Merchant Marine officers, all male, were used as our experiment subjects. They underwent a specific experimental procedure (Tab.1).

Table 1. Structure of Sample

Merchant Marine Officer Order	A' (%)	B' (%)
Officers	38.5 (5)*	61.5 (8)*
Experience		
Sum (total of years)	37	46
Mean (years)	7.4	5.75
Max (years)	9	15
Min (years)	6	4

*(frequency)

The sample's age profile as shown in Table 2, prevail the younger (12-35 age).

Table 2. Sample's age profile

Age's scale	24-35 (%)	36-45 (%)	>45 (%)
Merchant Marine Officers	53.8 (7)*	30.8 (4)*	15.3(2)*

*(frequency)

The sample's medical and personality profile (5 Factor model) as shown in Table 3 and Table 4. The personality profile presents homogeneity (high – medium) and the medical profile has a proportional of the sample having diseases of eye (myopia, astigmatism, etc.).

Table 3. Medical profile

Medical Profile	Eye diseases (%)	Eye operation (%)
Merchant Marine Officers	46.1 (6)*	7.6 (1)*

*(frequency)

Table 4. Personality profile

Merchant Marine Officers (13 male)

	vey high (%)	high (%)	medium (%)	low (%)	very low (%)
Extraversion	-	23	61.5	-	7.7
Agreeableness	23	46.1	30.7	-	-
Conscientiousness	15.4	76.9	-	7.7	-
Neuroticism	-	7.7	38.4	7.7	23
Openness	15.4	38.4	30.7	15.4	-

The next table shows the educational and simulation background (Tab.5).

Table 5. Educational and Simulation background

Merchant Marine Officers (13 male)

	Positive (%)	Negative (%)
Education in Computers	54	46
Simulation experience		
Education	61.5	38.5
Home	15.4	74.6
Job	54	46

The next table shows the results of model motivation (based Vroom model) (Tab. 6).

Table 6. Motivation model Results

Merchant Marine Officers (13 male)

	Positive (%)	Negative (%)
Performance-Outcome Expectance		
Job search	69	21
Payment	38.5	61.5
Professional development	84.7	15.3
Valence		
Professional value	84.7	15.3
Social value	7.6	92.4
Effort-Performance Expectance		
Professional performance	77	23
Job Security	100	
Interesting		
New technologies	100	
Educational benefits	100	
Personal needs	7.6	92.4

The next table shows the results of training program evaluation (marine engine system simulator training) and Simulator as software tool (Tab.7).

Table 7. Training program evaluation

Merchant Marine Officers (13 male)

	vey high (%)	high (%)	medium (%)	low (%)	very low (%)
Training Program					
Educational goal	23	54	23	-	-
Time schedule	38.5	30.7	23.1	7.7	-
Total assessment	15.4	38.6	23	23	-
Simulator					
Navigation	38.5	46.1	15.4	-	-
Interface	23	61.6	15.4	-	-
Multimedia	23	53.9	15.4	7.7	-

The next table shows the user satisfaction gradation in 5th scale concerning the scenario and Marine Simulator (Tab.8).

Table 8. Simulator and Scenario Satisfaction

Merchant Marine Officer (13 male)

Satisfaction scale	vey high (%)	high (%)	medium (%)	low (%)	very low (%)
Scenario in Engine room	30.7	61.5	7.7	-	-
Simulator ERS	46.1	46.1	7.7	-	-

The following tables are contain the statistical measures of optical data (face analysis tool) for Gaze (vertical-y), Dist (distance from monitor) and Head roll per satisfaction (simulator & scenario):

Table 9. Gaze tracking parameter (satisfaction simulator)

Merchant Marine Officers (13 male)

	vey high (4 male)	high (8 male)	medium (1 male)
Satisfaction Simulator			
Mean	6.75	7.8	2.73
Max	225.06	332.6	137.7
Min	-252.2	-144.07	-84.1
STDEV	5.2	4.1	17.9

Table 10. Dist parameter (satisfaction simulator)

Merchant Marine Officers (13 male)

	vey high (4 male)	high (8 male)	medium (1 male)
Satisfaction Simulator			
Mean	1.11	1.06	1.03
Max	22.8	5.26	2.14
Min	0.05	0.13	0.32
STDEV	0.1	0.04	0.12

Table 11. Head Roll parameter (satisfaction simulator)

Merchant Marine Officers (13 male)

	vey high (4 male)	high (8 male)	medium (1 male)
Satisfaction Simulator			
Mean	0.58	1.45	-2.4
Max	89.1	88.9	55.2
Min	-86.1	-89.08	-28.0
STDEV	1.7	3.5	7.42

Table 12. Gaze tracking parameter (satisfaction scenario)

Merchant Marine Officers (13 male)	vey high (4 male)	high (8 male)	medium (1 male)
Satisfaction Scenario			
Mean	8.28	7.09	0.21
Max	225.06	332.6	-203.4
Min	-252.2	-131.4	-144.07
STDEV	6.6	2.95	20.8

Table 13. Dist parameter (satisfaction scenario)

Merchant Marine Officers (13 male)	vey high (4 male)	high (8 male)	medium (1 male)
Satisfaction Scenario			
Mean	1.12	1.07	0.99
Max	22.8	17.04	2.47
Min	0.09	0.05	0.17
STDEV	0.02	0.09	0.16

Table 14. Head Roll parameter (satisfaction scenario)

Merchant Marine Officers (13 male)	vey high (4 male)	high (8 male)	medium (1 male)
Satisfaction Scenario			
Mean	1.9	0.49	-2.16
Max	89.1	82.9	88.9
Min	-83.6	-86.1	-89.08
STDEV	3.4	2.44	6.68

The following table displays the measures of lexical data (sentiment & opinion analysis):

Table 15. Lexical Analysis

Merchant Marine Officers (13 answers-text)	vey high	high
Satisfaction Simulator		
Using Modifier[1]	83.3%	66.6%
Using Comparison degree[2]	-	16.6%
P_{top} [3]	1-16.6%	1-16.6%
	2-0%	2-16.6%
	0-83.4%	0-66.8%
Total words (all texts), TotN$_w$	250	129
Mean (all texts)	41.6	21.5
IndexWord$_{Satisf}$ [4]	0.19	0.12
IndexWord$_{NonSatisf}$ [5]	0.02	0.1
Satisfaction Scenario		
Using Modifier[1]	75%	75%
Using Comparison degree[2]	-	25%
P_{top} [3]	1-25%	1-37.5%
	2-25%	2-12.5%
	0-50%	0-50%
Total words (all texts)	181	236
Mean (all texts)	45.25	29.5
IndexWord$_{Satisf}$ [4]	0.16	0.17
IndexWord$_{NonSatisf}$ [5]	-	0.02

[1] Lexical phrase or word with sentiment volume
[2] Positive, Comparative, Superlative
[3] Topology of sentiment phrases in text:1 in fist ½ of text, 2 in second ½ of text, 0 homogeneity in all text
[4] IndexWord$_{Satisf}$ = $\sum (W_{S+} / N_w)_{text-i}$ / TotN$_w$
[5] IndexWord$_{nonSatisf}$ = $\sum (W_{S-} / N_w)_{text-i}$ / TotN$_w$
for W_{S+}, W_{S-} : $\sum$ Word with sentiment or opinion load per text (positive polarity+ or negative polarity-)

These results are based on the Greek Lexicon of Emotions (Vostantzoglou, 1998 2nd edition revised).

Finally, the following table contains the results from the System Usability Scale (DEC SUS) usability assessment tool:

Table 16. SUS score results[1]

Merchant Marine Officers (13 male)	full sample	very high	high
Satisfaction Simulator			
Mean	73.2	79.1	69.1
Max	92.5	92.5	82.5
Min	62.5	62.5	62.5
STDEV	10.3	10.8	7.3
Mode	62.5	-	62.5

[1] 100-80 high score, 80-60 satisfactory rating, <60 low usability

6 CONCLUSIONS

The experimental data process results establish that:

– Visual attention (VA) from the "Face Analysis tool" shows attention increases as satisfaction for the scenario increases. This is observed in the dist parameter (distance from monitor, >1 close to the screen), The Head Roll parameter (rolling of the head – eye angle from horizontal level, <10 attention depending on the scenario, >10 high mobility), and the Gaze tracking parameter (Gaze vertical parameter >1 view the screen).

– The Engine room Simulator training increases the sense of job security and performance, has professional value and help to professional development (motivation model).

– In lexical-sentiment analysis shows the total words (mean) increases as Satisfaction for Simulator and Scenario increases and the IndexWord$_{Satisf}$ as Satisfaction for Simulator too (from high → very high satisfaction).

– In the SUS score has satisfactory rating and growing the score from high → very high satisfaction.

– High usability (easy to use, easy to learn).

We found connection between lexical data (from sentiment analysis of answers' users) and eye tracking parameters (gaze, head roll, distance from monitor) with satisfaction scale. It shows increases the lexical parameters (total words per users' interview, Index Word$_{Satisf}$) and Distance from Monitor (mean) as satisfaction for simulator or scenario increases, which means that they both fully watching the scenario and have difficulty in using simulator and a non-balance in the inclination of the head (Head roll ~0) in very high satisfaction for simulator, which probably suggests a deeper inspection of the Engine Simulator interface (that is probably ought to the lack of more practice time). The gaze parameter (mean) increases and lexical data (total words mean per users' interview) as satisfaction for scenario increases witch means the students watch inside the screen and not outside of it (there is an interest on the scenario execution).

Additionally, we observe connection between lexical data form sentiment analysis, eye tracking parameters (distance from monitor, head roll) and SUS score (usability assessment). The results confirm the connection between satisfaction and usability by using non conventional methods (neuroscience approach: eye and head tracking, voice recording) (SUS score, lexical data, eye parameters increases as satisfaction for simulator increases).

The connection between all above elements resulted from the processing of the optical registration data and the users' interview (voice recording) & questionnaires.

The approach is general in the sense that it can be applied in various types of e-learning marine systems. It is also pluralistic in the sense that it provides the evaluator with complementary sources of data that can reveal important aspects of the user experience during ship control. Certainly, the proposed approach may require further adaptations to accommodate evaluation of particular interactive systems.

REFERENCES

Asteriadis, S. Tzouveli, P. Karpouzis, K. Kollias, S. 2009. Estimation of behavioral user state based on eye gaze and head pose—application in an e-learning environment, *Multimedia Tools and Applications*, Springer, Volume 41, Number 3 / February, pp. 469-493.

Bratman, M. E. 1987. *Intention, Practical Reasoning*, Cambridge, MA, Harvard University Press.

Brooke, J. 1996. SUS: A "quick and dirty" usability scale. In: Jordan, P. W., Thomas, B., Weerdmeester, B. A., McClelland (eds.) Usability Evaluation in Industry, Taylor & Francis, London, UK pp. 189-194.

Dix, A. Finlay, J. Abowd, G. D. Beale, R. 2004. *Human-Computer Interaction*, UK:Pearson Education Limited.

P. Ekman, 1999. *The Handbook of cognition and emotion*, J. Wiley & Sons Ltd..

Fontaine, J. R. Scherer, K. R. Roesch, E. B. Ellsworth, P. C. 2007. The world of emotions is not two-dimensional , *Psychological Science*, 18(2), pp. 1050-1057.

Fotopoulou, A. Mini, M. Pantazara, M. Moustaki, A. 2009. "La combinatoire lexicale des noms de sentiments en grec moderne", in *Le lexique des emotions*, I. Navacova and A. Tutin, Eds. Grenoble: ELLUG.

Frijda, N. H. 1986. *The emotions*, Ed. De la Maison des Sciences de l'Homme.

He, Y. Zhou, D. 2011. Self-trainning from labeled features for sentiment analysis, *Information Processing and Management*, 47, 2011, pp. 606-616.

IMO-International, Maritime Organization, 2003. Issues for training seafarers resulting from the implementation on board technology, *STW 34/INF.6*.

Kotzabasis, P. 2011. *Human-Computer Interaction: Principles, methods and examples*, Athens, Kleidarithmos (in Greek).

Lambov, D. Pais, S. Dias, G. 2011. Merged Agreement Algorithms for Domain Independent Sentiment Analysis, Pacific Association, For Computational Linguistics (PACLING 2011), *Procedia - Socila and Behavioural Sciences*, 27, pp. 248-257.

Lazarous, R. S. 1982. Thoughts on the Relation between Emotion and Cognition, *American Psychologist*, 24, pp. 210-222.

Malatesta, L. 2009. " Human – Computer Interaction based in analysis and synthesis optical data", *Phd Thesis*, Athens (in Greek), NTUA.

Oatley, K. 1992. *Best Laid Schemes: The Psychology if Emotions*, Cambridge University Press.

Pinker, S. Jackendorff, R. 2005. The faculty of language: what's special about it?, *Cognition*, 95, pp. 201-236.

Papachristos, D. Nikitakos, N. 2011. Evaluation of Educational Software for Marine Training with the Aid of Neuroscience Methods and Tools, *International Journal of Marine Navigation and Safety of Sea Transportation*,Vol.5, No4, December 2011, pp. 541-545.

Papachristos, D. Alafodimos, K. Nikitakos, N. 2012a. Emotion Evaluation of Simulation Systems in Educational Practice, *Proceedings of the International Conference on E-Learning in the Workplace (ICELW12)*, 13-15 June, NY: Kaleidoscope Learning, www.icelw.org.

Papachristos, P. Koutsabasis, K. Nikitakos, N. 2012b. Usability Evaluation at the Ship's Bridge: A Multi-Method Approach, Proceedings of The 4th International Symposium on "Ship Operation, Management and Economics - SOME12" SNAME (Greek Section), Dec 8-9, Athens, Greece.

Potoker, E. S. Corwin, J. A. 2009. Predicting Emotional Intelligence in Maritime Management: Imperative, Yet Elusive, *International Journal on Marine Navigation and Safety of Sea Transportation*, Vol.3 No 2, June 2009, pp. 225-229.

Russel, J. A. 2003. Core affect and the psychological construction of emotion, *Psychological Review*, 110(1), pp. 145-173.

Tsai, M. J. Hou, H. T. Lai, M. L. Liu, W. Y. 2012. Visual attention for solving multiple choice science problem: An eye tracking analysis, *Computers & Education*, 58, pp. 375-385.

Tsoukalas, V. Papachristos, D. Mattheu, E. Tsoumas, N. 2008. Marine Engineers' Training: Educational Assessment of Engine Room Simulators, *WMU Journal of Maritime Affairs*, Vol.7, No.2, pp.429-448, ISSN 1651-436X, Current Awareness Bulletin, Vol. XX-No.10, Dec. 2008, IMO Maritime Knowledge Centre, pp.7.

Vosniadou, St. 2001. *Introduction in Psychology*, Vol. I, Athens, Gutenberg (in Greek), 2001.

Zajonc, R. B. 1984. On the Primacy of Affect, *American Psychologist*, 39, pp. 117-123.

Maritime Education and Training (MET)
STCW, Maritime Education and Training (MET), Human Resources and Crew Manning, Maritime Policy,
Logistics and Economic Matters – Marine Navigation and Safety of Sea Transportation – Weintrit & Neumann (Eds)

Administrative Model for Quality Education and Training in Maritime Institution

R.A. Alimen
John B. Lacson Foundation Maritime University-Molo, Iloilo City, Philippines

ABSTRACT: The study was conducted to determine the administrative model for quality education and training in maritime education in the Philippines. The researcher employed quantitative-qualitative research design involving data-gathering instrument, interview, and narrative description that influence in teaching maritime students towards quality education and training. Respondents of the study were categorized according to different categories. Descriptive statistical tools were used in this study mean, frequency count, and percentages while t-test, ANOVA, and Pearson's r were used as inferential statistics. Quantitative data were analyzed using to achieve the objectives of the present study. The data-gathering instrument was subjected to reliability and validity procedures evaluated by experts in research, statistics, instrumentation, and maritime education and training. Qualitative data were captured by using interviews among the participants of this study. Comments, suggestions, and remarks of the respondents were also cited in this study. Results revealed the type of administrative model needed for quality education and training in the Philippines as an entire group and as to different categories among maritime students and seafarers. Qualitative information and comments were captured in this study and its influence in teaching and training in maritime university in the Philippines.

1 BACKGROUND AND THEORETICAL FRAMEWORK OF THE STUDY

Filipino Seafarers play a major role in the country's economy. Mitsunobe (1999) emphasizes that more than 193,000 Filipino seafarers today are working onboard foreign ships. In 1998 alone, the Filipino seafarers remitted $1.6 billion out of the estimated $6 billion to the Philippine economy.

The report of Philippine Overseas Employment Administration (POEA) states that from 1984 to 2001, there was a yearly increase in the deployment of Filipino seafarers (R.A. No. 8042, (1995). It was also discussed in the study of Jaleco (2004) that the Philippines is recognized as the manning capital of the world. It supplies almost every vessel that sails the seven seas with Filipino marines and marine engineers on board. This is the reason why the Philippines is considered the biggest supplier of seafarers in the global market for the past several years.

Competition with other nationalities like Indonesia, India, and Eastern European countries is also increasing (Querol, 2002). Francisco has the same observation on the prospects and contributions of the Philippines in the world's manning industry when he said that: "The bright prospects and the great contributions that the manning industry gives to the economic growth of the nation prompted other developing countries like Crotia, China, Vietnam, India, Myanmar, and even now Korea, to develop and strengthen their own seamen resource base (Francisco, 2000)." Thus, the concern of this study was to design an administrative model to assure for quality education and training in a maritime institution.

The conceptual framework and its relationships are illustrated in Figure 1 below.

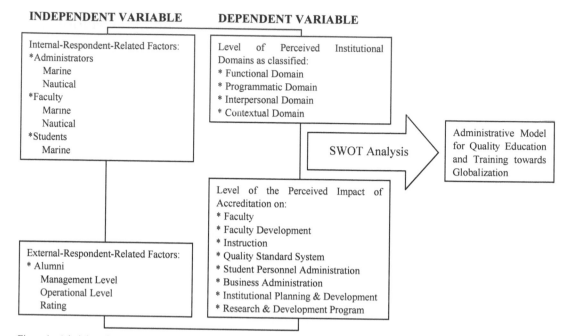

Figure 1. Administrative model for quality education and training towards globalization by employing the SWOT analysis technique on the level of perceived institutional domains and perceived level of impact of accreditation and as influenced by respondent-related factors.

2 OBJECTIVES OF THE STUDY

Specifically, the study presents the following objectives:

1 To discuss the perceived institutional domains such as functional, programmatic, interpersonal, and contextual of a maritime institution in the Philippines
2 To identify the perceived impact of accreditation of a maritime institution in the areas of faculty, faculty development, instruction, quality standards system, student – personnel administration, business administration, institutional planning and development, and research and development program.
3 To determine the difference in the level of the perceived institutional domains of a maritime institution when the respondents are classified as administrators, faculty, students, and alumni.
4 To determine the difference in the level of perceived impact of accreditation of maritime institution when the respondents are classified as administrators, faculty, students, and alumni.
5 To discuss the respondent-related factors as predictors of institutional domains and accreditation for quality education and training.
6 To present the relationship between the level of institutional domains and level of impact of accreditation as perceived by the respondents.

7 To unveil the strengths, weaknesses, opportunities, and threats of the maritime institution in terms of institutional domains and areas of accreditation?
8 To explain the administrative model as needed for quality assurance in a maritime institution, by utilizing the SWOT framework.

3 RESEARCH DESIGN

The researcher employed qualitative-quantitative research design for data collection. Two Likert-type data gathering instruments were used to come up with the administrative model employing SWOT framework. The "Rating Scale for Institutional Domains," which was used to find out the level of perceived domains, and the "Rating Scale on the Impact of Accreditation" was also utilized to ascertain the level of impact of accreditation.

4 DATA-GATHERING INSTRUMENT AND STATISTICAL TOOLS

The research employed quantitative mode of data collection. Two Likert-type data gathering instruments were used to come up with the administrative model employing SWOT framework. The "Rating Scale for Institutional Domains," which was used to find out the level of perceived domains, and the "Rating Scale on the Impact of Accreditation" was also utilized to ascertain the level of impact of accreditation. The descriptive analyses through the use of frequency (f), percentage (%), averages or means (Ms), and standard deviation (SDs) were used. For inferential analyses, t-test, ANOVA, Pearson's r, and Multiple Regression Analysis methods were employed in this investigation. The significant level was set at .05 alpha for two-tailed tests.

5 RESULTS OF THE STUDY

The results revealed that:
    The following are the findings of the present study:
1 The level of perceived programmatic domain, interpersonal domain, and level of perceived contextual domain was "very effective." However, the level of perceived functional domain in the maritime institution was "effective."
2 The level of the perceived impact of accreditation of the maritime institution in the areas of faculty, faculty development, instruction, quality standard system, student personnel administration, institutional planning and development, and the area of research development program was "very influential." However, the level of perceived impact of accreditation on the area of business administration (financial governance) was "influential."
3 There was a significant difference in the level of institutional domains when respondents were classified according to administrators, students, and alumni. No significant difference existed when respondents were grouped according to faculty.
4 There was significant difference in the perceived impact of accreditation in the institution when the respondents were classified according to students, and alumni. No significant difference existed when respondents were grouped according to administrator and faculty.
5 The students and alumni were significant predictors of institutional domains.
6 The students and alumni were significant predictors of the impact of accreditation.
7 There was a positive significant correlation that existed between institutional domains and impact of accreditation.
8 The strengths of the functional maritime university in terms of the institutional domains were identified as the following: "good in planning especially in addressing STCW standards", "school culture for excellence and competences, professionalism of the instructors and staff," "innovative teaching and classroom management, strategic vision of school, laboratory activities and simulation, linkages of the school, philosophy of school like discipline, honest, hard work." However, the weaknesses identified are the following: "weak collaboration and coordination of task with regard to different units," "inconsistency of school policies" and "late dissemination of information and decision."
The opportunities of the functional university in terms of institutional domains were the following: "the placement program of the institution and linkages," "international recognition of the institution," "qualification and expertise of the faculty" and "competency program and assessment."
The threats were identified by the external respondents as the following: "standards required by shipping companies," "poor attitudes of graduates toward work," "selected negative Filipino culture," "nature of the maritime curriculum," and "support of government to maritime education and training of Eastern European countries and China."
On impact of accreditation, the strengths identified in the study were the following: "shipboard training activities of the students," "quality standard system of the school as recognized by international certifying body," "leadership and officership," "training in school," and "professional upgrading of faculty and staff."
The weaknesses identified were the following: "poor participation in the planning of budget of faculty, staff, and students" and "problem in prioritizing of projects and programs with respect to annual budget."
The opportunities reflected in the study were the following: "advanced shipboard training and computerization," "more opportunities for students to develop their leadership potentials," "professional upgrading of faculty and laboratory assistants," and "postgraduate studies of the faculty and staff in their fields of specialization."
The threats identified were the following: "problem of participation in the planning process of different entities in the academic community -- faculty, staff, administration, and students," "requirements for the upgrading of the salaries of teaching and non-teaching personnel," "problem of allocation of resources" and "prioritization of various objectives and policies."
9 The administrative model utilized Richard A. Swanson's "Systems Model for Performance Improvement" modified by the researcher to fit the new model considering the different components: SWOT, Institutional Domains, and Impact of Accreditation.

6 ADMINISTRATIVE MODEL FOR QUALITY EDUCATION AND TRAINING

The administrative model needed to assure quality in education and training at maritime institution utilizing the SWOT framework was done by presenting the two major parts: (1) synthesis of the different components leading to the design of the administrative model, and (2) the presentation of the administrative model patterned after Richard Swanson's System Model for Performance Improvement (Swanson, 1999). The diagram below shows the interrelationships of the components considered in the designing of the administrative model to be used by administrators in maritime institutions in the Philippines.

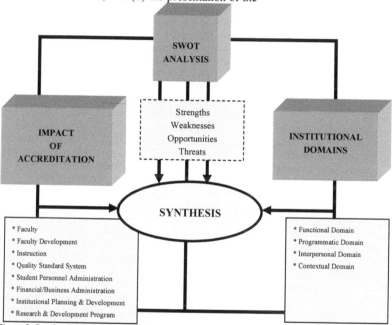

Figure 2. Synthesis of the Different Components

The researcher considers all the components involved in this inquiry in the formulation of the administrative model to assure quality in education and training in the Philippines. Below is the administrative model designed by the researcher modified from Swanson's framework.

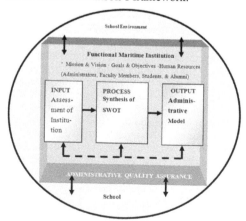

Figure 3. An Administrative Model for Quality

REFERENCES

Francisco, Josephine J. An Article. JBLF Maritime Education Review. 9(1), S.Y. 1999-2000.
Jaleco, Victor B. Teaching-Learning Situation in Maritime Schools. Unpublished Dissertation, University of San Agustin. Graduate School, Iloilo City. 2004.
Mitsunobe, 2000. in Shipmate, Shipmate-The Filipino Seamen's Digest. Vol. 2 No.12. January 1999. e-mail to shipmate@impactnet.com. or shipmatephil@hotmail.com
Querol, Alexander E., AVP-Magsaysay Maritime Corporation. Commencement Address delivered to the Graduating Class of 2002 of JBLCF-Bacolod. JBLF Maritime Education Review. 12 (1), SY 2001-2002.
SWOT 1. htm. Quick MBA.com.

Maritime Education and Training (MET)
STCW, Maritime Education and Training (MET), Human Resources and Crew Manning, Maritime Policy,
Logistics and Economic Matters – Marine Navigation and Safety of Sea Transportation – Weintrit & Neumann (Eds)

A Concept of Examinations for Seafarers in Poland

J. Hajduk & P. Rajewski
Maritime University of Szczecin, Poland

ABSTRACT: So far, seafarers wishing to take an exam in Poland have had to stand before an examination board of one of the maritime offices, in compliance with provisions of the Law on Maritime Safety and applicable directives on seafarer qualification requirements issued by Infrastructure Minister. There is an important need for reviewing and standardizing these exams now that the law on maritime safety has been revised and a National Maritime Examination Board has been established. These authors present their concept of administering exams for seafarers that goes in line with worldwide trends in this respect, is objective, meets STCW Convention standards and offers a possibility of benchmarking the quality of education at various training centres.

1 INTRODUCTION

In the context of education, an examination (Latin *examen*) till recently was perceived as a form of checking one's knowledge. At present, when it comes to verifying one's competences that involve practical skills the term *assessment* seems more proper, as it refers to both knowledge and skills. Accordingly, the scope of examining has been extended to include practical tasks, so that the term examination evaluates skills an applicant has at a required level of competence. Competence is understood as theoretical knowledge and practical skill distinguishing a person by his/her ease of efficient, effective and quality-satisfying performance of tasks. Additionally, the above definition of competence is broadened with expected attitudes and personal qualities of the applicant. Actions of a person competent in a given field should meet criteria adopted in a given community / organization (Guide 2011). According to the National Qualifications Framework adopted in Poland, the process of assessment should confirm that the assumed learning outcomes have been achieved (ERK 2009).

"Till 2012 each officially issued certificate endorsing a qualification should include a note indicating its place within the national, thereby in European Qualifications Framework" (Chmielecka 2010). The existing seafarer examining practice in Poland is based on the tradition of evaluating knowledge, without attention paid to the practical component. Besides, the preferred oral form of exams has been burdened more or less with examiner's subjectivity. There may have been cases where the final grade (assessment) was stained with biased selection of questions, or even questions improperly formulated by examiners. Non-substantial factors, for instance examinee's appearance, might have had an impact on the evaluation by the examiner. If we consider recommendations resulting from the STCW Convention, including the requirement of practical training for acquiring skills, then one conclusion is that the methods of examining seafarers so far has become conservative and does not suit modern methods of verifying competences, or to put the message clear: it does not satisfy international standards in this respect.

The concept of seafarer examination herein presented takes into consideration long experience of both authors (Hajduk 1997a, b, c, Wawruch 1997), as well as national and international trends and STCW Convention requirements for skills verification (Guide 2011, STCW 2011). The concept combined with the established National Maritime Examination Board can be implemented in the whole country without a division into regions. In the long run, the modified approach will permit to assess and verify the level of training provided at various centres running qualification upgrading courses for all levels of competence. The authors' concept clearly separates exams evaluating theoretical knowledge and practical skills, as there are objective assessment criteria for either type of exam (Chmielecka 2010, ERK 2009, Guide 2011, Hajduk 1997a, b, c, Projekt 2012c, Rajewski 2012, STCW 2011, Ustawa 2011, Wawruch 1997).

2 ASSUMPTIONS

While developing the form and structure of the exams, the authors assumed that a new form of examining seafarers has to ensure that:
– the examining system is uniform all over the country;
– exam assessment is objective by introducing system-induced supervision, eliminating subjective personnel supervision;
– exam principles and procedures are transparent;
– the form of exam is clear and examinee-friendly.

The examining methods listed below are components of an overall examination that consists of modules needed for checking both knowledge and skills. A module contains all or only selected forms of assessment specified in enclosed examination sheets. Each module includes functions provided by the STCW Convention and course unit (subject) names, as established in draft regulations of the Ministry of Transport, Construction and Maritime Economy (Projekt 2012a, b). For exams not included in new draft regulations the provisions of the existing regulations have been used (Rozporządzenie 2005a, b, c). Depending on a qualification, an exam structure consists of one to three modules. Each module comprises required forms of exam for an evaluated competence.

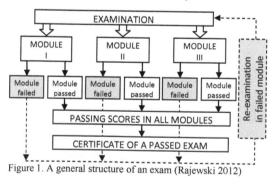

Figure 1. A general structure of an exam (Rajewski 2012)

The following methods of assessment may be applied in the proposed examining concept:
1 Theoretical exam, divided into:
  – a multiple choice test;
  – a written exam.
2 Practical exam, divided into:
  – exam conducted on real objects;
  – exam on a simulator.
3 Partial exams.

A general structure of examinations is shown in Figure 1.

A module is a basic unit in the exam structure. Each module makes up a separate whole. Depending on exam requirements covering various course unit topics, a module contains various examining forms. The module has functions and associated course units to be examined. Each course unit, component of a function and a module, requires a specific method of assessing relevant knowledge and skills by assigned to it appropriate forms of exams conducted within a module. An example structure of a module is shown in Figure 2.

Module name, e.g. **MODULE I**

Description of course units incl. in module I

Course unit name, e.g. Course A1
1. Course unit name
2. Number of test items in the test bank: ....
3. Number of test items in a module test: ....
4. Share of course unit tasks in the written exam: ....

Course unit name, e.g. Course An
1. Course unit name
2. Number of test items in the test bank: ....
3. Number of test items in a module test: ....
4. Share of course unit tasks in the written exam: ....

Description of exams incl. in module I

Module I, contents of the test component
1.1. Number of test items in a module test in course unit A1
1.2. ....
1.n. Number of test items in a module test in course unit An
2. Total number of test items
3. Numbering of test items drawn from a test base

Module I, contents of the written exam component
1. Number of written / oral tasks covering topics of course units A1–An in an exam task base
2. Number of written / oral tasks per one exam in module I
3. Numbering of tasks drawn from an exam task base
4. Exam duration

Module I, contents of a practical exam on a real object
1. Number of tasks covering topics of course units A1–An in an exam task base
2. Number of tasks per one exam, module I
3. Numbers of tasks drawn from a base
4. Exam duration
5. Equipment requirements for conducting an exam
6. Examiner's qualifications
7. Applicant's assessment criteria

Module I, contents of a practical exam on a simulator / ship
1. Number of scenarios covering topics of course units A1–An in an exam scenarios bank
2. Number of scenarios per one exam, module I
3. Numbers of scenarios drawn from a base
4. Exam duration
5. Equipment requirements for conducting an exam
6. Examiner's qualifications
7. Applicant's assessment criteria

Figure 2. An example module structure containing all exam components (Rajewski 2012)

Each test or exam form to be implemented under a module has a specific number of test items, written tasks and scenarios (the latter for a practical exam on a real object and/or a simulator) that are stored in a relevant exam base. The weight of a course unit is accounted for by the exam form and number of test items, tasks and scenarios given on exam sheets. The English language skills are assessed by two methods: tasks (written exam) and required commands and instructions to be given (spoken exam).

3 THE STRUCTURE OF THE EXAMINATION PROCESS

Based on the module structure, three types of exam arrangement have been distinguished, including one, two or three modules. Figure 3 shows how an exam with two modules is arranged.

The structure of an exam may comprise varied modules, depending on the scope of assessed skills and knowledge. Four basic forms of exam are taken into account:
- a test and a written exam, making up the theoretical part of the exam;
- an exam with the use of real objects and on a simulator or ship, assessing practical skills and abilities to use knowledge in practice.

Theoretical exam

Test

Two methods of testing are envisaged:
1. electronically recorded by a computer, organized in a room equipped with individual computer stations, one station for one applicant (examinee);
2. recorded on paper exam sheets, organized in a room with traditional desks for applicants, a computer with an access to exam task base and a fast printer.

Written exam

Similarly, two methods of written exams can be implemented:
1. electronically recorded by a computer, organized in a room equipped with individual computer stations, one station for one applicant (examinee);
2. recorded on paper exam sheets, organized in a room with traditional desks for applicants, a computer with an access to exam task base and a fast printer.

There may be a transition period adopted, when test items and/or exam tasks will be retrieved from an exam base and printed outside the specific exam room under a supervision of an examination board.

Practical exam

Exam on a real object

The exam is conducted with the use of a real object specified in tasks covering a certain scope of topics, recorded in exam sheets (e.g. AIS receiver, fuel purifier).

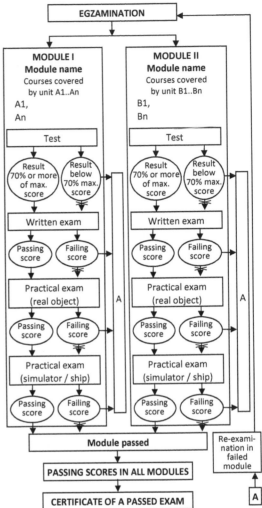

Figure 3. A diagram of an exam consisting of two complete modules (Rajewski 2012)

Exam on a simulator / ship

The exam is conducted on a ship or a simulator. If the latter is used, it has to satisfy standards of an operational simulator. If a practical exam takes place on a ship, it has to carry equipment the handling of which is to be examined.

Partial exams

A partial exam is understood as a part of qualification exam for an operational level, resulting from the division imposed by the study program based on course units (subjects), in compliance with the Regulation on the Framework of Training Programs and Exam Requirements (specific to each of the departments).

Partial exams (Projekt 2012a, b, d) are conducted during the training as it is defined in Art. 64 par. 1.3 of the Act on maritime safety of 18 August 2011 (Ustawa 2011). The time, place and range of exams is established in the study program. Partial exams for the operational level refer to persons who attend training at a maritime education institution as specified in Art. 74 par. 2.1 (Ustawa 2011). The prerequisite for passing a qualification exam conducted by a maritime education institution as specified in Art. 74 par. 2.1 is completion of studies as defined by the Act on Higher Education.

4 ORGANIZATION OF THE EXAMINATION PROCESS

The most comprehensive examinations cannot last longer than three days, with dates assigned to an applicant by an examination board secretary. An applicant takes the exam in subsequent modules, as they are established in exam sheets (Figs 3 and 4).

An applicant has to pass a test or exam included in a module to be allowed to take the next exam in the same module. However, failing a module does not prevent an applicant from taking exams or tests in other modules.

The first part is a theoretical exam composed of a test and a written exam that should last one or two examination days. If the theoretical exam (test and written exam) is conducted in a room with individual exam computer stations, a set of test items or written tasks for an applicant are drawn at random by the computer program started by an examination board member supervising an exam session. If a theoretical exam is taken in a room equipped with traditional classroom type exam desks, a computer with a quick connection to an exam base and a quick printer, applicants are handed in paper exam sheets. Sets of test items / written tasks are also drawn by a computer program started by an examiner supervising that exam. All exam papers are coded.

Applicants take tests on the first exam day. The test results make up the first indicator of applicants' theoretical knowledge and skills. A test in one module should not take longer than 90 minutes. A single module test can have a maximum of 90 test items. It is assumed these tests will have a form of multiple choice tests. Applicants are asked to indicate one correct answer out of minimum three options. The tests are solved individually by each applicant. There should be thirty-minute breaks between particular exam sessions. In each module a test is regarded as passed if an applicant gives a preset minimum number of correct answers, which permits him/her to take the subsequent part of the module. If an applicant fails a test, s/he may not take further exams of the module. Applicants should be informed on the test. The sessions, each not longer than 90 minutes, are separated by 30-minute breaks. A single written exam contains one theoretical problem that has an allocated time for solving. A written exam can last up to 90 minutes. Applicants

attempt to solve the problems on their own. They are informed on the written exam results not later than one day after completion of this part. The passing of a written exam allows an applicant to continue the overall exam by taking the other exams assigned to a given module.

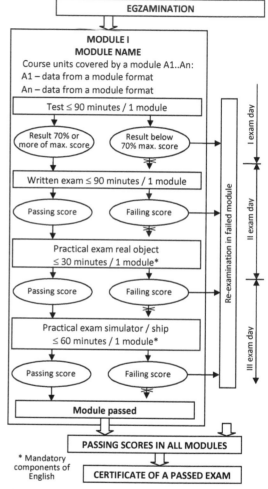

Figure 4. A sequence schedule of a three-module exam (Rajewski 2012)

The last exam day (for an applicant it can be third, seventh or another day counting from the beginning of an examination session) is the day of practical exams, if such are established. In a practical exam, an applicant is to perform a task or tasks following a scenario either on a real object and/or a simulator that satisfies operational simulator requirements. Tasks are drawn at random by an examiner from a relevant exam task base.

A practical exam on a real object consists of one practical task that can be completed in the allocated time. An applicant performs tasks on his or her own.

The exam duration for one applicant does not exceed 30 minutes. The result is announced to the applicant by an examiner directly after that part of the exam is finished.

A practical exam on a simulator or ship comprises one scenario that can be implemented in an allocated time. Scenarios are drawn at random by an examiner from a relevant exam scenario base. 60 minutes is a period allocated to this type of exam. Tasks included in a scenario are performed by an applicant on his or her own. The result is announced to the applicant by an examiner directly after that part of the exam is finished.

Exams covering a narrower scope of problems may have course units grouped in one or two modules, and their forms of examining may also be limited, for instance only a test and a written exam, or a test and a practical exam, all in accordance with exam sheets. Then such exams will last one or two exam days.

Following the European and National Qualifications Frameworks, the authors have assumed that passing a module means an applicant's skills and knowledge covered by the module are recognized, so that module will not be included during a re-examination. If an applicant fails a re-examination, the subsequent exam will include all the modules covering the required scope of knowledge and skills.

5 EXAMPLES OF EXAM SHEETS

The transparency of examinations strictly depends on the transparency of the various exam forms and the scope of knowledge and skills assessed, available to both examiners and examinees. Exam sheets have been drawn up in the same format for each exam comprising a required level of competence. An example exam sheet for the operational level exam in the deck department is presented in Tables 1 – 2.

Table 1. An exam sheet for the deck department, operational level (Rajewski 2012)

5.1. International shipping. Deck department										
5.1.3. Operational level										
			Theoretical exam				Practical exam			
			Multiple-choice test		Written exam		Real object		Simulator / ship	
Module	Function	Course unit	Test items	Time [min]	Number of tasks	Time [min]	Number of tasks	Time [min]	Number of practical scenarios on a simulator	Time [min]
I Type A Fig. 2	Navigation	Navigation	30	90	1	90	1	30	1	60
		Meteorology and oceanography	5							
		Navigational equipment	10							
		Ship handling	5							
		Maritime rescue	5							
		Marine communications	5							
		Navigation safety	20							
		English	10							
II Type B Fig. 3	Cargo handling and stowage Ship and personnel care	Ship construction and stability	20	70	1	60	1	30	none	
		Maritime transport	15							
		Ship management	5							
		Ship safety	10							
		Maritime law	10							
		Marine environment protection	5							
		Computer science	5							

Table 2. Exam topics for the deck department, operational level (Rajewski 2012)

	Topics covered in a written exam	Topics covered in a real object exam	Topics covered in an exam held on a simulator and/or ship:
Module 1	Comprehensive tasks done on a navigational chart	Handling navigational equipment	Keeping a navigational watch
	Tidal calculations	Observation and meteorological data interpretation	
	Calculations for rhumb line (loxodromic) navigation – comprehensive tasks	Voyage planning, navigation and monitoring (ECDIS)	
Module 2	Calculating ship's intact stability	Use of a loading instrument	None

6 SUMMARY

As the revised Act on maritime safety is now in force, a number of regulations relating to it should be replaced by new instruments adjusting the system of education and training, examining and certification in Poland to mandatory requirements. At present a few draft regulations have been published and remain open to public consultations. Only one regulation is excluded, namely the one concerned with candidates for the chairperson, vice-chairpersons and examiners, members of National Maritime Examination Board; the regulation in fact sanctions the activity of the Board.

Expectations of the shipping community in Poland, of marine training centres in particular, are focused on attaining full harmony between all revised regulations so that the training, examining and certification of seafarers will make up a consistent system, free from any disturbances in the continuity of professional qualification upgrading process. One problem is the required deadlines for amending certain regulations and a lack of logic chronology in drawing up particular draft documents. Instead of a sequence of actions, we are observing simultaneously presented concepts and actual draft regulations, which often make the maritime community concerned if the proposed system will be coherent.

The herein presented concept is compatible with known draft regulations relating to training, examining and certification of seafarers in Poland. It is also in conformity with relevant provisions of the STCW Convention, National Qualifications Framework implemented in Polish education and higher education system and the experience of other countries that have already implemented centrally administered examinations. The concept is methodologically consistent and excludes examiners' subjectivity. In the future it will allow to verify the quality of training at various training centres. In authors' opinion the implementation of the described concept is presently indispensible. It will be possible to modify various components of exams, topics of written tasks, topics demonstrated on real objects and scenarios executed on simulators or a ship can be made, following a detailed analysis of the system in action, at least two years after its implementation.

REFERENCES

Chmielecka, E. 2010. Europejskie ramy kwalifikacji, Część I. Forum Akademickie.

ERK 2009. Europejskie ramy kwalifikacji dla uczenia się przez całe życie (ERK). Urząd Oficjalnych Publikacji Wspólnot Europejskich, Luksemburg.

Guide 2011. Guide for Mariners — Merchant Marine Examinations (Deck & Engineering Guidance). United States Coast Guard. National Maritime Center. Martinsburg, WV 25404. July.

Hajduk, J. (ed.) 1997a. Wymagania egzaminacyjne na poziom operacyjny. WSM Szczecin & WSM Gdynia.

Hajduk, J. (ed.) 1997b. Wymagania egzaminacyjne na poziom operacyjny w żegludze przybrzeżnej. WSM Szczecin & WSM Gdynia.

Hajduk, J. (ed.) 1997c. Wymagania egzaminacyjne na poziom pomocniczy. WSM Szczecin & WSM Gdynia.

Kubat, M. 2013. Kompetencje zawodowe. www.wup.lodz.pl/files/ciz/ciz_Kompetencje_zawodowe.pdf; Access: 17.01.2013.

Projekt 2012a. Projekt Rozporządzenia Ministra Transportu, Budownictwa i Gospodarki Morskiej z dnia 27.02.2012 r. w sprawie wyszkolenia i kwalifikacji zawodowych marynarzy.

Projekt 2012b. Projekt Rozporządzenia Ministra Transportu, Budownictwa i Gospodarki Morskiej z dnia 26.06.2012 r. w sprawie ramowych programów szkoleń i wymagań egzaminacyjnych dla marynarzy działu pokładowego.

Projekt 2012c. Projekt Rozporządzenia Ministra Transportu, Budownictwa i Gospodarki Morskiej z dnia 22.10.2012 r. w sprawie trybu i sposobu przeprowadzania egzaminów kwalifikacyjnych oraz warunków wynagradzania członków zespołu egzaminacyjnego.

Projekt 2012d. Projekt Rozporządzenia Ministra Transportu, Budownictwa i Gospodarki Morskiej z dnia 19.11.2012 r. w sprawie wyszkolenia i kwalifikacji zawodowych marynarzy.

Rajewski, P. (ed.) 2012. Opracowanie jednolitego ramowego systemu wymagań egzaminacyjnych do projektów rozporządzeń w sprawie ramowych programów szkoleń. Praca wykonana na zlecenie MTBiGM. Akademia Morska w Szczecinie.

Rozporządzenie 2005a. Rozporządzenie Ministra Infrastruktury z dnia 4 lutego 2005 r. w sprawie wyszkolenia i kwalifikacji zawodowych marynarzy.

Rozporządzenie 2005b. Rozporządzenie Ministra Infrastruktury z dnia 13 lipca 2005 r. w sprawie szkoleń i wymagań egzaminacyjnych w zakresie kwalifikacji zawodowych marynarzy.

Rozporządzenie 2010. Rozporządzenie Ministra Infrastruktury z dnia 3 grudnia 2010 r. zmieniające rozporządzenie w sprawie wyszkolenia i kwalifikacji zawodowych marynarzy.

Rozporządzenie 2012. Rozporządzenie Ministra Transportu, Budownictwa i Gospodarki Morskiej z dnia 2 marca 2012 r. w sprawie kandydatów na przewodniczącego, zastępców i egzaminatorów Centralnej Morskiej Komisji Egzaminacyjnej.

STCW 2011. STCW Convention and STCW Code including 2010 Manila Amendments. International Maritime Organization. London 2011.

Ustawa 2011. Ustawa z dnia 18 sierpnia 2011 r. o bezpieczeństwie morskim.

Wawruch, R. (ed.) 1997. Wymagania egzaminacyjne na poziom zarządzania. WSM Szczecin & WSM Gdynia.

Maritime Education and Training (MET)
STCW, Maritime Education and Training (MET), Human Resources and Crew Manning, Maritime Policy,
Logistics and Economic Matters – Marine Navigation and Safety of Sea Transportation – Weintrit & Neumann (Eds)

Implementation of the 2010 Manila Amendments to the STCW Convention and Code in Ukrainian MET System

M.V. Miyusov, V.M. Zakharchenko & D.S. Zhukov
Odessa National Maritime Academy (ONMA), Ukraine

ABSTRACT: In June 2010 significant changes to the STCW (Standards of Training, Certification and Watchkeeping) Convention and Code were agreed at IMO Diplomatic Conference in Manila in order to bring the Convention and Code up-to-date with new developments. These changes, known as "The Manila amendments to the STCW Convention and Code" were due to enter into force on 1st January 2012.
The above mentioned Manila amendments are the first major revisions since the Convention and Code were updated in 1995. These amendments were necessary to keep training standards in line with new technological and operational requirements that require new shipboard competencies.
According to the Regulation I/15 of the Convention – "Transitional Provisions" give us five years to adopt national and international MET system in accordance with new STCW requirements.

1 INTRODUCTION

Process of comprehensive review of the Convention and the STCW Code was actually initiated in 2006 at the 37-th session of Subcommittee on Standards of Training and Watchkeeping (STW Subcommittee) of International Maritime Organization (IMO). Suggestions of parties and non-governmental organizations concerning amendments to the STCW Convention and STCW Code were considered at the 38th, 39th, 40th and 41st sessions of STW Subcommittee, two ad hoc intersessional meetings of the STW working groups in 2008 and 2009. From the very beginning of process of review basic principles were defined as: retain the structure, goals and number of chapters of the Convention of STCW; not to down scale existing standards; address inconsistencies, interpretations, MSC instructions, clarifications already issued, outdated requirements and technological advances; address requirements for effective communication; provide for flexibility in terms of compliance and for required levels of training and certification and watchkeeping arrangements due to innovation in technology; address the special character and circumstances of short sea shipping and the offshore industry; and address security-related issues.

Long activity of International Maritime Organization was crowned by a significant event – Diplomatic conference of the countries being the parties of the STCW Convention which took place in June, 2010 in the capital of Philippines Manila and where amendments to the STCW Convention and the STCW Code called Manila amendments were approved.

Essential reforming of standards of training of seafarers became as a result of comprehensive review of the Convention and the STCW Code taking into account comprehensive character of Manila amendments, we can speak about almost new edition of the STCW Convention and STCW Code.

One of the key aspects of implementation of new provisions of the STCW Convention and the STCW Code is the Maritime Higher Education System which is based on Maritime higher educational institutions. With reference of that we focus our attention on amendments which have a direct relation to development of professional standards and educational standards in Maritime higher educational institutions, and factors which will have in the nearest future direct impact on implementation of Manila amendments in system of training of seafarers in Ukraine.

2 NEW REQUIREMENTS TO THE STANDARDS OF TRAINING

The provisions that define the main contents of training (study) programs in Maritime higher educational institutions are established in chapters II and III of the STCW Code. In this context amendments to these chapters of the STCW Convention and STCW Code can be summarised to the following:
- inclusion of new regulations (II/5, III/5, III/6 and III/7) in the STCW Convention and appropriate sections in the STCW Code;
- inclusion of new competences into existing specifications of minimum standards of competence in the tables A-II/1, A-II/2, A-III/1 and A-III/2 of the STCW Code;
- inclusion of new components in specifications of knowledge, understandings and proficiency within existing competences;
- change of approaches to practical training of engine department personnel at the operational level.

Taking into account that some of new regulations define standards of training of seafarers at the Support level (II/5, III/5 and III/7), and with a view that a subject of this article is the analysis of required changes in standards of training in maritime higher educational institutions, further the authors focus main attention on the innovations relating to standards of competence of ships' officers.

2.1 *Amendments to the standards of training for deck department*

First of all, lets pay attention to the standards established in chapter II "Master and deck department".

As for mandatory minimum requirements for certification the Operational level deck officers on ships of 500 gross tonnage or more, established in the table A-II/1 of the STCW Code, the inclusion in the specification of the minimum standards of competence of new competences connected with:
- use of ECDIS to maintain the safety of navigation;
- application of leadership and teamworking skills;
- contribution to the safety of personnel and ship.

It should be noted also the inclusion in the specification of the minimum standards of competence of additional obligatory knowledge and proficiency which support the competence "Maintain a safe navigational watch" (to use of information from navigation equipment; knowledge of blind pilotage techniques; the use of reporting in accordance with the General principles for Ship Reporting Systems and procedures of Vessel Traffic Systems; Knowledge of bridge resource management principles. The section B-II/1 of the STCW Code was added with recommendations about training of the chief mates in Celestial navigation.

Regarding amendments to standards of competence of masters and chief mates on ships with a gross tonnage of 500 or more (Section II/2) we will pay attention to the following changes in the specification of the minimum standards of competence:
- inclusion competence on maintain the safety of navigation through the use of ECDIS and associated navigation systems to assist command decision-making;
- replace the competence "Organize and Manage of Crew" to competence "Use of leadership and managerial skills" with considerably expanded content of necessary training.

In specifications of the minimum standards of competence as management level and operational level (tables A-II/1 and A-II/2 of the STCW Code), the competences relating to ability of working in team, to leadership and management skills are included. Similar competences are included as well in specifications of the minimum standards of competence of engine department at operational and management levels.

Also it is necessary to pay attention to emergence within competence "Maintain a safe navigational watch" of the part of knowledge, understanding and skills in "Bridge resource management". Noted amendments testify to considerable attention which the International Maritime Organization pays to a role of a human factor to the safe navigation.

2.2 *Amendments to the standards of training for engineers*

The requirements to standards of training of the engine department personnel are established in the Chapter III "Engine department". It's note that this chapter underwent essential changes. In general, amendments to the standards of training and certifying of engine department personnel can be summarised to the following:
- inclusion to the STCW Convention of three new regulations and sections of the STCW Code corresponding to them;
- restructuring of tables of specifications of minimum standard of competence of officers in charge of an engineering watch, second engineer and chief engineer officers (tables A-III/1 and A-III/2 of the STCW Code);
- inclusion of new competences in the specifications of the minimum standards of competence established in the tables A-III/1 and A-III/2 of the STCW Code;
- change of approaches to practical training of engine department personnel.

However, first of all, we will note essential changes in the text of the regulation III/1 of the Convention of STCW establishing requirements to training of officers in charge of an engineering watch, namely:

- withdrawal requirements for the minimum 30-month duration of the approved education and training in the new edition;
- change of requirements to practical training of candidates for obtaining the Officer in charge of an engineering watch Certificate of Competence (new edition requires 12-month practical training as a part of the approved training program with registration in the approved Training Records Book, or otherwise - 36 months period of an approved seagoing service).

Let's notify also that the tables A-III/1 and A-III/2 of the STCW Code were essentially restructured. But the substantial of competences which being presented at the previous edition of the STCW Code was kept or "modernized" or more logical compliance of competence and functions provided for engine team.

From amendments to standards of training for officer in charge of an engineering watch we will mark the entering in the specification of the minimum standard of competence (the table A-III/1 of the STCW Code) the competences related to:

- use internal communication systems;
- maintenance and repair of electrical and electronic equipment;
- application of leadership and teamworking skills;
- contribute to the safety of personnel and ship.

Two last competences are similar to the new competences from the table A-II/1 of the STCW Code (the operational level Deck officer).

Regarding standards of training for chief engineer officers and second engineer officers on ships powered by main propulsion machinery of 3000 kW propulsion power or more (the table A-III/2 of the STCW Code), we will note the new competences related to:

- manage operation of electrical and electronic control equipment (essential change of substantial part in comparison with the previous edition);
- manage troubleshooting restoration of electrical and electronic equipment to operating condition;
- use leadership and managerial skills (by analogy to standards of competence for masters and chief mates).

Especially note the "strengthening" for management and operational levels engineer officers specifications of the minimum standards of competence related to operation, maintenance and repair of electrical and electronic equipment and electronic control systems.

2.3 *Standards regarding electro-technical officers*

One of the most "revolutionary" innovations of Manila amendments is entered in the STCW Convention and Code standards of training and certification for the electro-technical personnel: electro-technical ratings and electro-technical officers established by the regulations III/6 and III/7 of the STCW Convention and the sections A-III/6, B-III/6 and A-III/7 of the STCW Code respectively.

The particular interest is represented in standards of training for the electro-technical officers established by the regulation III/6 of the STCW Convention and the sections A-III/6 and B-III/6 of the STCW Code. Ukraine represented by the Odessa National Maritime Academy was the co-sponsor (together with Germany, Malaysia, Poland, the United Kingdom and the United States of America) of the standards of competence for electro-technical officers containing in the table A-III/6 of the STCW Code and approved by Manila STCW Conference.

New edition of the STCW Code established three functions for electro-technical officers at the operational level:

- Electrical, electronic and control engineering;
- Maintenance and repair;
- Controlling the operation of the ship and care for persons on board.

The most of the competences established for electro-technical officers are traditional for Ukraine. At the same time, we would like pay attention to new mandatory competences of electro-technical officers. These competences are related to:

- operation and maintenance power systems in excess of 1000 volts;
- operation computers and computer networks on ships;
- maintenance and repair of bridge navigation equipment and ship communication systems;
- application of leadership and team working skills.

Also it is necessary to note the recommendation of the section B-III/6 of the STCW Code to develop training programs for electro-technical officers with taking into account the IMO Resolution A.702(17) concerning Radio maintenance guidelines for the GMDSS.

2.4 *Transitional period*

As mentioned above the STCW Convention and the STCW Code Manila amendments essentially reform training and certification standards as for deck department and for engine department also. Its obviously fact that for full implementation of standards of training established by new edition of the STCW Convention and STCW Code we need revision of the corresponding description of professional qualifications (professional standards) of marine transport workers, development and implementation new study (training) programs in maritime educational institutions for all appropriate specializations and educational degrees.

The most important issue - terms of the transitional period. The regulation 1/15 "Transitional Provisions" of new edition of the STCW Convention established that until 1 January 2017 parties may continue to issue, recognize and endorse certificates in accordance with the provisions of the Convention which applied immediately prior to 1 January 2012 in respect of those seafarers who commenced approved seagoing service, an approved education and training programme or an approved training course before 1 July 2013.

Thus, the necessity of transition of all maritime educational institutions to the programs of training updated according to Manila amendments not later than July 1, 2013 is unambiguously established.

3 IMPLEMENTATION IN UKRAINIAN MET SYSTEM

The Manila amendments process of implementation in Ukrainian national MET system practically synchronizes with processes of reforming of the higher education system of Ukraine and its adaptation to the European Higher Education Area.

So, one of the declared top-level making reforms of the Ukrainian system of the higher education is modification of the structure of educational degrees, and in particular – abolition of the "Specialist" degree that is traditional degree for Ukrainian higher education system. It, in turn, can cause the necessity of revision of approaches to defining the competences and levels of responsibility in accordance to requirements of the STCW Code for Maritime higher education institutions graduates of remained educational qualifications (degrees). And it is natural, that these approaches will be reflected on all new professional and educational standards.

In this regard the one of most important impact factors is approving of the National Qualification Framework in Ukraine. National Qualification Framework established systematic general description of the competences corresponding to qualification levels and should become a basis for defining the competences and levels of responsibility for different levels of qualifications of seafarers and for different educational qualifications (degrees).

All issues of implementation of the Manila Amendments in Ukraine are under keeping of Coordination council on training and certification of seafarers. The representatives of Ministry of infrastructure, Ministry of education and science, youth and sport of Ukraine and Ministry of Health of Ukraine are members of Coordination council.

The most important issue for Manila Amendments implementation in Ukraine was development the "Consolidated Action Plan of implementation of requirements of the STCW Convention and STCW Code with the 2010 Manila Amendments into National system of training and certification of seafarers".

This Action Plan was signed by Deputy Minister of Infrastructure of Ukraine, Deputy Minister of education and science, youth and sport of Ukraine, Deputy Minister of Health of Ukraine, Head of State Agency of Fishing and approved by Coordination council on training and certification of seafarers.

This plan covers all aspects of Manila Amendments implementation in Ukraine and defines actions and responsibilities for period of 2012 – 2013 years.

From the many actions of this Action Plan we would like to note the follows, which are successfully realized for present time:
− new educational program for Bachelor of navigation;
− new educational program for Bachelor of marine engineering;
− new educational programs for Bachelor of marine electrical engineering.

All new Bachelor programs mentioned above were developed by Methodological Commission for education in the field of maritime and river transport and approved by Ministry of education and science, youth and sport of Ukraine. These new Bachelor programs implemented all new requirements of the STCW Convention and STCW Code. New Bachelor programs are implemented in the Ukrainian Maritime higher education institutions from 2012-2013 study year.

New educational programs for "Junior Specialist" degree that corresponds to the Short cycle of higher education now is developed and their approval is expected nearest time. Implementation of these programs into study process is planned since year of 2013.

New edition of Training record books and eleven new refresh and update programs also are developed by Methodological Commission for education in the field of maritime and river transport and on the way to approval.

4 CONCLUSION

Implementation of Manila Amendments in Ukrainian MET system is the complex and multi-aspects task that demands taking into account of many factors.

All issues of implementation are supervised by the Methodological Commission for education in the field of maritime and river transport and Coordination council on training and certification of seafarers. Content of new educational programs and terms of their implementation conform to the new requirements of the STCW Convention and STCW Code.

REFERENCES

[1] Report to the Maritime Safety Committee / Subcommittee on standards of training and watchkeeping - 38th session – Doc. STW 38/17, (2007), p. 35.

[2] Adoption of amendments to the annex to the International Convention on Standard of Training, Certification and Watchkeeping for Seafarers (STCW), 1978 / Conference Resolution 1 // Conference of parties to the International Convention on Standard of Training, Certification and Watchkeeping for Seafarers (STCW), 1978 – Doc. STCW/CONF.2/DC/1, (2010).

[3] Adoption of amendments to the Seafarers' Training, Certification and Watchkeeping (STCW) Code / Conference Resolution 2 / Conference of parties to the International Convention on Standard of Training, Certification and Watchkeeping for Seafarers (STCW), 1978 – Doc. STCW/CONF.2/DC/2; STCW/CONF.2/DC/3, (2010).

[4] M. Miyusov, V. Zakharchenko. Education and training of electro-technical officers and STCW Convention and Code new standards implementation / Technical cooperation in Maritime Education and Training / Proceedings of the 11-th Annual General Assembly International Association of Maritime Universities. – Bursan Korea, Korea Maritime University 2010. – p. 159 – 164.

[5] M. Miyusov, V. Zakharchenko. Study Programmes for Electro-Technical Officers Development^ Two-Level Based Approach / Expanding Frontiers. Challenges and Opportunities in Maritime Education and Training / 13th Annual General assembly of International Association of maritime Universities // Fisheries and Maritime Institute of memorial University of Newfoundland – October 15-17, 2012 – p. 105 – 112.

[6] V. Zakharchenko. Implementation of the Manila amendments to the STCW Convention and STCW Code into standards of maritime higher education / All-Ukrainian Scientific and Methodological Journal "Maritime education", N 1-2' (Jan – Jul) 2011 - Odessa, 2011, p. 12 - 15.

Maritime Education and Training (MET)
STCW, Maritime Education and Training (MET), Human Resources and Crew Manning, Maritime Policy,
Logistics and Economic Matters – Marine Navigation and Safety of Sea Transportation – Weintrit & Neumann (Eds)

The Marine Transportation Program of a Maritime University in the Philippines: An Initial Evaluation

G.M. Eler

John B. Lacson Foundation Maritime University-Arevalo, Inc., Philippines

ABSTRACT: This study was an initial evaluation of the Marine Transportation Program of John B. Lacson Foundation Maritime University (JBLFMU)) as viewed by the graduates. It aimed at determining the inter-play of adequacy of curricular experiences in the degree preparation at JBLFMU with the level of competency on the tasks prescribed by Standard for Training & Watchkeeping(STCW'95) which consists of obligatory and recommendatory requirements which a member country of International Maritime Organization (IMO) must implement in her respective jurisdiction and for her seafarers to be competent and qualified to undertake the different levels of functions on board a sea going vessel in international and local routes ; job satisfaction, work-related benefits and problems met aboard the ship. The participants of the study were the 300 graduates who were grouped according to the year they graduated (BSMT).The data needed for the research were ob-tained through the use of the researcher- made questionnaires: The Level of Adequacy of Curricular Experi-ences in the degree preparation at JBLFMU Questionnaires, Level of Competency in the Training Tasks Prescribed by STCW '95, Job Satisfaction Aboard Ship, Work-related Benefits Survey Form and Problems Encountered Aboard Ship, all of which were constructed, validated, pilot tested for the present investigation. The data was analyzed quantitatively. All statistics were computed-processed through the Statistical Package for the Social Science (SPSS) Software. The study found out that the graduates considered their curricular experiences particularly in general education subjects, professional subjects and basic safety training in the degree preparation at JBLFMU as very adequate ;their level of competency of tasks prescribed by STCW'95 as high and a high level of satisfaction in their jobs. The graduates considered financial stability and sense of security for the family as their priorities in the work-related benefits and they rarely encountered personal, so-cial, professional and environmental problems. Significant differences were noted among the different batches of graduates in terms of their level of competency, job satisfaction and problems encountered aboard the vessel. The Adequacy of Curricular Experiences in the degree preparation at JBLFMU, Level of Competency of tasks prescribed by STCW, job satisfaction, work-related benefits, & problems encountered aboard ship are significantly and positively correlated. The Level of Adequacy of Curricular Experiences is a significant factor that affects Level of Competency, Job Satisfaction, work-related benefits and problems encountered aboard ship.

1 BACKGROUND OF THE STUDY

The dominance of the country in supplying the world's manpower to maritime fleet, to the extent that they comprised more than 30% of the world's total number of seafarers, made Manila the manning capital of the world. All these Filipino seafarers employed in the country and the world are graduates of the country's maritime education institutions. They are also instrumental for the social and economic upliftment of their families and social amelioration of the people in their respective communities. Because of the magnitude of contribution to the country, the seafarers are labeled as the "modern day living heroes" of the Philippines in the millenium (Arcelo,2002).

John B. Lacson Foundation Maritime University (Iloilo Maritime Academy) took the lead among other maritime institution in the country as it placed itself among the thirty (30) schools for higher learning given the full autonomy to operate by the Commission on Higher Education (CHED Memorandum Order No. 32, S. 2001)

This goes to show that John B.Lacson Foundation Maritime University is indeed at the forefront of leadership where maritime education is concerned. And for over six decades of providing maritime training, it was able to sustain the excellent supply of global and technical manpower through the delivery of quality maritime education.

Hence, it has expanded into providing graduate education for maritime studies through conventional on- campus classroom instruction and unconventional borderless learning-distance education.

Evaluation has always been an important education component in programs and curriculum planning. Studies have indicated that student perceptions, most specifically graduates can provide vital inputs for the educational evaluation process. Ohlsen (1964) States that a well-coordinated follow-up studies is needed to obtain from recent graduates a realistic picture of what lies ahead for the present students, to gain insights from and to help new students appraise their educational and vocational plans. Similarly, such follow up could eventually reappraise the school program, and generate ideas for improving the program. This study is anchored on the modified form of Daniel Stufflebeam's CIPP (Context, Input, Process, Product) model for curriculum evaluation. Each type of evaluation is tied to a different set of decisions that must be made in planning and operating a program. The specific application for curriculum evaluation of product component is to compare actual outcomes against a standard of what is acceptable to make judgments to continue, terminate, modify, or refocus activity. The product component on the model will be given emphasis here in order to provide baseline information for decision makers and major stakeholders. The products are the graduates of the program who actually gone on board the ship. These graduates are the respondents of this study.

2 STATEMENT OF THE PROBLEM

This study is an evaluation of the Marine Transportation Program of John B. Lacson Foundation Maritime University by the graduates. Specifically, this study sought answers to the following questions:

1 How adequate are the curricular experiences in General Education subjects, Professional subjects and Training acquired from the degree preparation as viewed by the graduates when taken as a whole and classified according to year of graduation?

2 What is the level of competence in the trainings tasks prescribed by the STCW '95 of the graduate when taken as a whole and when classified according to year of graduation?

3 What is the level of job satisfaction of the graduates when taken as a whole and classified according to year of graduation?

4 What are the work-related benefits of the graduates when taken as a whole and classified according to year of graduation?

5 What are the problems encountered aboard ship when taken as a whole and classified according to year of graduation?

6 Are there significant differences in the curricular experiences in General Education subjects, Professional subjects and trainings acquired from the degree preparation as viewed by the graduates when classified according to year of graduation?

7 Are there significant differences in the level of competence in the training tasks prescribed by the STCW '95 of the graduates when classified according to year of graduation?

8 Are there significant differences in the level of job satisfaction of the graduates when classified according to year of graduation?

9 Are there significant differences in the work-related benefits of the graduates when classified according to year of graduation?

10 Are there significant differences in the problems encountered aboard ship of the graduate when classified according to year of graduation?

11 Is there a relationship between adequacy of curricular experiences acquired from degree preparation and a) level of competence in the trainings tasks prescribed by STCW 95, b) job satisfaction, c) work-related benefits, and d) problems encountered aboard the ship?

12 Is adequacy of curricular experiences acquired from degree preparation a significant factor that affects level of competence, job satisfaction, work-related benefits, and problems encountered aboard the ship?

3 RESEARCH DESIGN AND METHODOLOGY

This study is an evaluation of Marine Transportation Program of the JBLFMU System as viewed by the graduates.

The main purpose of this study determined the interplay of adequacy of curricular experiences in the degree preparation at JBLFMU graduates particularly in the General Education Subjects (English Language & Marine Vocabulary, Math, Computer Operations ,Swimming, Rowing,Boat Commands, Sailing); Professional subjects (Seamanship,Navigation, Deckwatchkeeping,et.); Basic Safety Training (PSSR,Fire prevention & fire fighting, First aid,Personal Survival Techniques) towards level of Competency in the training tasks prescribed by STCW'95, job satisfaction, work-related benefits and problems encountered aboard ship. This will enable the researcher to find out how adequate is their preparation to meet the required competences aboard the vessel.

4 RESEARCH RESPONDENTS

The participants of the study were the 300 graduates of the Marine Transportation. The sample size of 300 was chosen by stratified random sampling with 60 respondents per year, for 5 years. The sample size of 302 from a target population of N of 1428 was recommended by Krecjeie and Morgan (1970), however, the investigator decided to use 300 samples.

5 THE RESEARCH INSTRUMENT

This investigation utilized a researcher-made data-gathering instrument to obtain the data needed for the research. The data-gathering instrument was divided into four parts.

Part 1a. Rating Scale on Adequacy of Curricular Experiences from the degree preparation

1b. Level of Competence in the training tasks prescribed by STCW'95.

Part 2, Rating Scale on Seafarer's Job Satisfaction

Part 3, Rating Scale on Marine Transportation Graduate's Work-Related Benefits

Part 4, Problems Encountered Aboard Ship Survey

These four (5) data-gathering instruments were accompanied by a brief information form to gather data on the participants' personal factors- age, seafaring experience, deck job classification, principal flag registration, type of vessel and year graduated (BSMT).

Data were gathered through the use of questionnaires. The data gathering instruments were submitted for validation to 5 jurors. The items in the instruments were all rated very acceptable by the jurors.

6 STATISTICAL TOOLS

The descriptive statistics used were frequency, means, standard deviations and ranks. The inferential statistics were, Analysis of variance (ANOVA), the Pearson's r, and Linear regression.

All inferential statistics were set to .05 alpha level. All statistics were computed-processed through the Statistical Package for the Social Science (SPSS) Software.

7 RESULTS OF THE STUDY

1 As an entire group, the graduates considered their curricular experiences in the degree preparation at JBLFMU as "Very Adequate." The graduates believed that they had been given "Very Adequate" curricular preparation in general education subjects, professional subjects and basic safety training.

2 The graduates believed their level of competency of task prescribed by STCW '95 is "High". The results of the level of competency on Navigation at the Operational Level, Cargo Handling and Stowage and Controlling the Operation of the Ship and Care for the Person on Board is also "High".

3 When taken as a whole and according to year of graduation, the graduates believed that they have a "High Satisfaction" in their jobs. The same result is true to all components of the job satisfaction as Achievement Value and Growth, Job Itself, Organization Design and Structure, Organizational Process, and Personal Relationship.

4 The first three in rank of the personal work-related benefits are: "The best facilities are enjoyed by my family". (Rank 1)," My family can buy clothesware they want" (rank 2) and "My family are covered by life insurance". The least personal-work related benefits for them are: "My siblings study in private institutions" (rank 8), "I provide business capital for my family" (rank 9) and the last is : "My family can travel in any place they want to go". On the other hand, the first three in rank for Social benefits are" My family enjoys the community respect" (Rank 1)," We have made new friends" (Rank 2) and " My family is known in the community". The least priority among the social benefits are: "I am invited to judge prestigious contest", "I am asked to sponsor a scholarship program in my alma mater" , "I am asked to sponsor a seminar, symposium, etc as an alumni".

5 Problems encountered by the graduates aboard the vessel were considered "rarely a problem." However, the problems encountered aboard the vessel were ranked by the graduates . Rank 1 is Personal problem, Professional problems as Ranked 2, Environmental problem as Ranked 3 and Social problems as rank 4. As regards to how often these problems are encountered aboard ship,the graduates found their Personal, Social, Professional, and Environmental problems encountered aboard ship as "Rarely a Problem." The 2002 and 2003 graduates perceived the Social Problems as "Never a Problem."

6 Significant differences were noted only on the following:
 – Among the different batches of graduates in terms of their Level of Competency in the Training Tasks Prescribed by STCW '95
 – Among the different batches of graduates in term of their Job Satisfaction.

- Among the different batches of graduates in term of their Problems Encountered Aboard Ship.
7 The level of Adequacy on Curricular Experiences in the degree preparation at JBLFMU and a.) Level of Competency in the Training Tasks Prescribed by STCW '95, b.)Job Satisfaction, c.) Work-Related Benefits, and d.) Problems Encountered Aboard Ship significantly and positively correlated.
8 The level of Adequacy on Curricular Experiences in the degree preparation at JBLFMU is a significant factor that affects Level of Competency in the Training Tasks Prescribed by STCW '95, Job Satisfaction, Work-Related Benefits, and Problems Encountered Aboard Ship.

8 CONCLUSIONS

From the findings, the following conclusions were drawn.

The curricular experiences in the General Education, subjects, professional subjects and Basic safety training acquired from the degree preparation at JBLFMU provided the graduates knowledge, concepts and skills that made them ready for the field of work.

The inclusion and the implementation of the provisions contained in the STCW'95 in the Marine Transportation curriculum have helped raised the level of competence of the graduates.

Different batches of graduates were satisfied of their jobs because their curricular experiences match or fit with what the industry is looking for. The graduates generally enjoyed working at sea and felt challenged by many aspects of the job, like the achievement value and growth (*measures how an individual perceives their current scope for advancement*), job itself(*measures the satisfaction experience with the type and scope of job tasks*), organization design and structure (*measures the satisfaction of several different structural aspects of organization*), organizational processes(*measures the satisfaction with internal processes within the organization)* and the personal relationship (*measures the satisfaction with interpersonal contact within organization*).

The facilities needed, economic, financial stability and sense of security for the family are the top priorities of the graduate as regards to personal-work related benefit. On the other hand , their top priorities for social work-related benefits are respect and social status.

The very adequate curricular experiences of the graduates prepared them for the nature of the work and the problems they may encounter aboard the vessel. Likewise, when the degree programs provided very adequate experiences to the graduates, it will assure a high level of competency in the training prescribed by STCW 95, gained benefits and less problems.

Finally, curricular experiences in three areas: general education subjects, professional subjects and basic safety training are very important among the graduates because they provide details about the specific skills and they are really used in the applied contexts. Without these, graduates cannot be in the job. These are entry points that lead graduates to be satisfied in their jobs. Indeed the adequacy of curricular experiences is a significant factor that affects Level of competency on tasks prescribed by STCW, job satisfaction, work-related benefits and problems encountered aboard ship.

9 RECOMMENDATIONS

1 The institution can include programs on strengthening the cadet's personhood and human worth. In addition, through the guidance programs, they can help the deck cadets to learn and develop coping, social and other related skills
2 To improve job satisfaction mainly related to the provision of increase training opportunities on computerized equipment and improvement in administrative and management procedures.
3 On problems: In this study rank 1 in the social problem is : I experience racial discrimination". It showed that the different cultural background of the crew manning the ship may present a problem in the safe and efficient operation of a ship. Training Institutions and other agencies will have to do their share to address this problem.
4 For Training Institutions:
- Development of the cultural awareness training program to cope with problems onboard. This training would develop and enhance Filipino appreciation of their own culture thereby strengthening or enhancing their personal values, with care useful mechanism in dealing with mixed-cultural problems. It will also develop confidence on themselves so they would not feel inferior and be able to face other nationalities on the equal basis footing.
Attitudes, like cultural relatively, and core values, such as freedom, quality, opportunity, respect and impartial justice should be developed and inculcated in the seafarers. These attitudes and values should be integrated in the training program.
Government Agencies:
Government agencies such as Philippine Overseas Employment Administration (POEA), should review and enhance their orientation policies in the deployment of

seafarers of the nationalities wherein Filipino will encounter on board.

Formulation of the Code of Conduct for mixed nationality crew so that the seafarers will be guided on the proper decorum on board ship.

5 The result of this study must be disseminated to all persons concerned in the maritime profession. Through this, they may realize their respective contribution for the advancement of the plight of seafarers.

6 The findings may provide information to shipping companies about the problems of seafarers so that company may understand what affects the performance of seafarers, thus, they may provide appropriate measures to help seafarers cope their problems aboardship.

7 The families and loved ones of the graduates will be guided by the present findings. They may tactfully motivate and continuously support their son, husband or relative seafarer to finish his employment contract.

8 Sustain a good program. There must be a regular curricular review to sustain the quality of graduates who can meet the challenging world of seafaring.

REFERENCES

Abadi, J.M. (2003), Satisfaction with Oklahoma State University among selected groups of international students, Dissertation Abstract International , 6V (8-A), 2821

Arcelo, Adriano. A., Direction in maritime education research program for global competitiveness.

Arcelo, Mary Lou, JBLCF Maritime Education Review, Vol. 6, Number 1 and 2, 1994-1995.

Best, J.W. and Khan J.V. (1998) "Research is Education" 8th ed. Egleword Cliffs, NS: Allyn and Bacon.

Bilbao, Purita,(1991), The teacher education program of West Visayas Sate University: an evaluation study

Drilon, Carlito (1995) The Dolphin

Evidente, Joaquin, (1991) JBLCF Maritime Education Review.

Flora, Efren, (2004), Filipino seafarers: their fears,problems and lifestyles, unpublished Masters thesis, JBLCF

Fritch, L.A. (1998). A follow-up of High School Graduates in defiance Country schools, University of Toledo.

Gasalao, Melchor (2002), Seafarer's work performance as influenced by their shipboard attitude an problems, Unpublished Masters thesis.

Helmreich Robert L.,Wilhelm,John A., et.al. (1981),"Motivation, organization and satisfaction aboard ship ", The University of Texas at Austin

Krejeie , R.V. and Morgan, D.W. (1970) "Determining the sample size for research activities, Educational and Psychological Measurement".

Kronberg,I (2003), Employee satisfaction on cruise ship, University of Surry, Great Britain,Masters Thesis

Magramo, Melchor (2003), "Quality Standards System: Status, Compliance with And Adherence to Among Maritime Schools in Western Visayas, an Unpublished Dissertation.

Nilssen, Bjorn Klerck (1997), ILO/WHO Consultation onGuidelines for conducting Pre-sea and Periodic Medical Fitness Examination of Seafarers. (Chief of the Maritime Industries Branch of the Sectoral Department of the International Labour Office).

Ohlsen, M. (1964), Guidance Service in the Modern School, New York; Haircourt, Brace and World Inc.

Maritime Education and Training (MET)
STCW, Maritime Education and Training (MET), Human Resources and Crew Manning, Maritime Policy,
Logistics and Economic Matters – Marine Navigation and Safety of Sea Transportation – Weintrit & Neumann (Eds)

New Laboratories in the Department of Marine Electronics in Gdynia Maritime University

D. Bisewski, K. Bargieł, J. Zarębski, K. Górecki & J. Dąbrowski
Gdynia Maritime University, Poland

ABSTRACT: The paper presents a new research and student laboratories in the Department of Marine Electronics in Gdynia Maritime University. The concept of these laboratories as well as basic features of laboratory equipment are described.

1 INTRODUCTION

Nowadays, the safety of maritime transportation is strongly dependent on electronic equipment used in navigation and communication systems (Bibik et al. 2008). The importance of ship officers having qualifications of electrical engineering and electronics, has been appreciated by the IMO by changing of the STCW convention (Wyszkowski & Mindykowski, 2012). The Faculty of Electrical Engineering in Gdynia Maritime University is one of the leading institutes in training of electrical and electronic engineers for maritime industry and economy. To ensure a high level of education of graduated students, didactic laboratories of the Faculty, have to be constantly modernized.

A construction of modern electronic circuits used in marine electronic systems requires contemporary technology and suitably equipped research laboratories. For the purposes of maritime economy the new technologies are especially required during education and practical training of future engineers.

The dynamic development of electronic technology requires the use of up-to-date equipment for the manufacture of electronic circuits. The main problem concerns on reduction the size of electronic circuits in order to decrease overall dimensions and power losses of electronic systems. Therefore, more and more electronic components are packed as a SMD (surface mount device) and dedicated for surface mount technology (SMT). SMD components are characterized by a smaller terminal outputs spacing due to a reduction of the size of device packaging (Kisiel, 2005).

On the other hand, an important aspect of manufacturing electronic systems are measurements of characteristics of electronic components, which are generally time-consuming and require the use of multiple measuring instruments. It is required to use instruments that allow to perform various measurements in short time with high accuracy.

The paper describes properties and basic parameters of the equipment used in the Laboratory of Technology and Constructions of Electronic Circuits as well as the Laboratory of Measurements Electronic Devices and Circuit located in the Department of Marine Electronics in Gdynia Maritime University.

2 LABORATORY OF TECHNOLOGY AND CONSTRUCTION OF ELECTRONIC CIRCUITS

The Laboratory of Technology and Construction of Electronic Circuits is primarily intended for the manufacturing and testing PCB (printed circuit board) along with milled or printed connections. The laboratory is used by the scientific workers in order to realize didactic process in the subject "Design and Construction of Electronic Equipment" as well as by graduate students realizing their own projects.

The design phase of PCB requires specialized processing equipment and computer software. In the designing process the Eagle software is used. This software can be downloaded from the producer website (www.cadsoftusa.com), whereas the description of the Eagle features are presented in the literature, e.g. (Dąbrowski & Posobkiewicz, 2010). The student version of the program enables designing of two-layer circuit board of the dimensions not exceeding 80x100 mm. During designing process the copper connections of the PCB, using Eagle, are created. In performing PCB process a milling method (for research projects) or a photochemical method (for student projects only) are used. For milling LPKF ProtoMat S62 Laser & Electronics AG (miling machine) (see Fig. 1), is used (www.lpkf.com). The milling machine can operate in working area of the dimensions 229x305x38 mm with a resolution of 0.25 μm. In the PCB production process the milling machine automatically selects the proper milling cutter from a 10-element set. Each tool can be exchanged during the milling process. Apart of this, the milling machine allows to hole drilling in the PCB.

Despite the high moving speed of milling head, equals to even 15 cm/s, manufacturing process of PCB using milling method is a relatively long. Circuit board with an area of 0.5 dm^2 is milled for about 15 minutes.

Figure 1. Milling machine LPKF ProtoMat S-62 with PCB connections made by milled method

Controlling of the milling machine is realized with a software LPKF CircuitCAM and BoardMaster via the USB port. This software supports the file format generated by the program Eagle, thus prepared projects can be directly implemented to milling machine. The quality of individual copper mosaic paths is controlled by a vision system and additional manual microscope that allow measuring the geometrical dimensions of the mosaic. Realization of PCB using milling method is relatively expensive, due to high price of milling cutters, which must be replaced every 2-3 PCB. Therefore, during the implementation of student projects typically the photochemical method is used.

For the photochemical method, platesetter LPKF 300-245 as well as PCB etching device (see Fig. 2) fabricated in the Department of Marine Electronics, are used. In addition, PCB can be covered using an anti-soldering mask by LPKF Promask. Holes metallization, for example for vias, can be done using LPKF MiniContac RS, shown in Fig. 3. Detailed information on the LPKF equipment can be found on the manufacturer's website (www.lpkf.com).

Equipment of manufacturing PCB, applied in the laboratory, allows to prepare circuit boards containing both THD (Trough-Hole Device) and

SMD. THDs are manually embedded and soldered using for example VTSS1 Velleman soldering station. Whereas, embedding of SMD elements, due to the small size of casing, requires specialized apparatus. In the laboratory semi-automatic SMD assembly system shown in Fig. 4, fabricated by Essemtec, is used. In turn, for reflow soldering a oven marked as RO06-Plus (the same manufacturer) is utilized.

Figure 2. The device for etching circuit boards

Figure 3. The LPKF set MiniContac RS

Figure 4. Semi-automatic SMD assembly type Expert SAFP

The SMD assembly system (Fig. 4) consists of: mounting arm, pick&place head with time-pressure dispenser, underpressure suction nozzle, armrest for operator, tray-carousel feeder for 45 types of electronic elements. It is possible to block the mounting arm in x / y / z axes. The mounting arm is automatically lowered for laying-down elements. The tray-carousel feeder is controlled by a microprocessor. Position of the head is controlled by the optical system (optical encoder, CCD camera). The optical system also contains a prism for the elimination of parallax error.

Unfortunately, the presented system requires an experienced operator, who precisely directs the mounting head over the selected area of the PCB. A computer program only indicates the stacking order of elements and enables observation of the magnified workspace screen. Detailed information of Essemtec devices can be found on their website (www.essemtec.com). One should be noted, that additional information about this laboratory are presented in (Górecki et al. 2012)

3 LABORATORY OF MEASUREMENTS OF ELECTRONIC DEVICES AND CIRCUITS

Laboratory of Measurements of Electronic Devices and Circuits is equipped with measuring devices which allow to obtain the electrical characteristics of both electronic components and circuits. Keithely's measuring equipment (www.keithley.com), e.g. 4200-SCS Semiconductor Characterization System, 2602A System Source Meter and 2410 High Voltage Source Meter, is frequently used in the laboratory.

The 4200-SCS (unit A in Fig. 5) is a high precision measuring instrument and includes hardware and software features suitable for semiconductor device characterization. The instrument is based on a PC (personal computer) architecture – it contains a motherboard with the main processor, RAM, physical hard drive, compact disk and other typical PC hardware elements, e.g. external SVGA monitor (see Fig. 5). It uses the Microsoft Windows operating system.

The 4200-SCS is modular, configurable and supports up to nine SMUs (Source-Measure Units) plugged in the back of the device chassis. Generally, SMU in the form of expansion card instrument can be used as a voltage or current source as well as a voltage or current meter. The instrument is well-suited for performing a wide range of measurements on commonly used electronic devices, including such measurements as: DC and ultra-fast pulsed current-voltage (I-V), capacitance-voltage (C-V), capacitance-frequency (C-f), drive level capacitance profiling (DLCP), four-probe resistivity (ρ, σ), and Hall voltage (V_H). Apart of this, additional manufacturer application notes describe, how to use the 4200-SCS system to make electrical measurements, for example on: PV cells or nanoscale components.

Each measurement requires the use of test fixture (unit B in Fig. 5). The test fixture is a metal case with four female triaxial connectors and a latch on the outside (see Fig. 5). Inside the test fixture is one 4-pin device holder (transistor socket), which can also be used for electronic devices with 2 or 3 leads. The test fixture greatly improves the quality of measurement by reducing leakage currents and parasitic capacitances.

The measuring system manufacturer provides an extensive set of Windows-based software tools, which greatly simplify creating test sequences and analyzing results. For instance, Keithley Interactive Test Environment (KITE) allows users to gain familiarity quickly with tasks such as managing tests and results as well as generating reports. Sophisticated and simple test sequencing and external instrument drivers simplify performing automated measurements. The KITE is designed to let users understand device behavior quickly and provides advanced test definition, parameter analysis, graphing and automation capabilities required for modern semiconductor characterization. When running a test sequence, users can view results and plots for completed tests while the sequence is still running. A multiple plots can be viewed at the same time to get a complete picture of device performance.

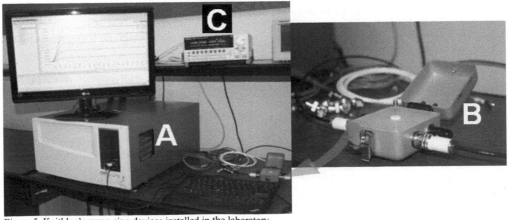

Figure 5. Keithley's measuring devices installed in the laboratory

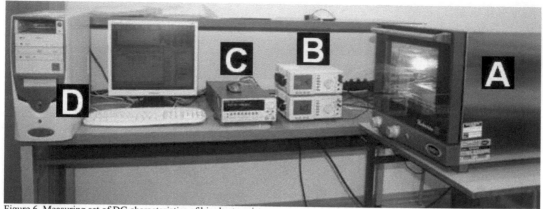

Figure 6. Measuring set of DC characteristics of bipolar transistors

Another device commonly used in the Laboratory is 2602A System Source Meter (unit C in Fig. 5). The instrument is, in fact, a stand-alone two-channel SMU device and provides high speed source-measure capability, advanced automation features and time-saving tools.

The device simplify capturing the data needed to characterize an electronic device and allow to use built-in pulsed and DC sweeps, including linear staircase, logarithmic staircase and custom sweeps. All sweep configurations can be programmed for single-event or continuous operation. The 2602A System can be programmed and controlled using the front panel or additional PC, connected by the LAN (local area network). To control the instrument operation, the TSP-Link Windows-based program is used.

In the laboratory, measurements of parameters and characteristics of modern semiconductor devices, e.g. power BJT, JFET, MESFET and MOSFET made of silicon, gallium arsenide, silicon carbide and gallium nitride, are carried out. Investigations mainly concern on the influence of selfheating phenomenon (Zarębski, 1996) on the static and dynamic characteristics of such devices. In particular, measurements of thermal parameters of semiconductor devices at different cooling conditions and at different values of the ambient temperature, have to be carried out (Zarębski, 1996).

An implementation of the research requires a number of measurements. In Fig. 6, an example of measuring set of isothermal and nonisothermal DC characteristics of power BJT, is shown.

The DUT is placed inside a heat chamber (unit A in Fig. 6), which allows to set the proper ambient temperature of the transistor. Two digital multimeters (units B in Fig. 6) along with a temperature sensors Pt-100 measure the temperature inside the chamber and the temperature of the transistor case. The main measuring device is Keithley's 2410 System Source Meter (unit C in Fig.

6) that measures the DC characteristics of the transistor. A principle of operation of the measuring device is similar to the 2602A Source Meter mentioned earlier.

For controlling the measuring source as well as for archiving and analyzing the results of measurements, PC with external monitor (unit D in Fig. 6) is used. As an example of measurement results, in Fig. 7 the output isothermal DC characteristics of BJT SiC transistor (BT-1206) are shown.

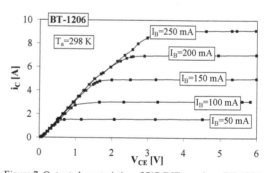

Figure 7. Output characteristics of SiC-BJT transistor BT-1206

As seen, characteristics are of the smooth shape, which confirms high quality of instruments used in the measuring set.

4 FINAL REMARKS

The paper presents the concept and implementation of new research laboratories in the Department of Marine Electronics in Gdynia Maritime University. The Laboratory of Technology and Construction of Electronic Devices allows prototyping and production of PCB, using the assembly process by through-hole and surface mount methods. The Laboratory of Measurements of Electronic Devices and Circuit is dedicated for performing a various measurements of characteristics and parameters of electronic devices and circuits. High precision and PC-controlled measuring instrumentation fabricated by Keithley is used. The described laboratories provide a complex realization of electronic circuits, ranging from the design of circuit diagram and ending on detailed measurements of constructed circuit.

The laboratories are dedicated especially for highly skilled engineers in the designing and fabrication processes of electronic equipment used in maritime industry and economy.

REFERENCES

Bibik Ł. et al. 2008. Vision of the Decision Support Model on Board of the Vessel with Use of the Shore Based IT Tools, *TransNav - International Journal on Marine Navigation and Safety of Sea Transportation*, Vol. 2, No. 3, pp. 255-258.

Dąbrowski J. & Posobkiewicz K. 2010. *Komputerowe projektowanie obwodów drukowanych*, Bydgoszcz, Wydawnictwo Tekst (in polish).

Górecki K. et al. 2012. Laboratorium projektowania i konstrukcji urządzeń elektronicznych, *Wiadomości Elektrotechniczne*, Nr 10, ss. 34-36, (in polish).

Kisiel R. 2005. *Podstawy technologii dla elektroników – Poradnik praktyczny*, Warszawa, Wydawnictwo BCT (in polish).

Wyszkowski J. & Mindykowski J. 2012. Electrical, Electronic and Control Engineering – New Mandatory Standards of Competence for Engineer Officers, Regarding Provisions of the Manila Amendments to the STCW Code, *TransNav - the International Journal on Marine Navigation and Safety of Sea Transportation*, Vol. 6, No. 2, pp. 249-253.

Zarębski J. 1996. *Modelowanie, symulacja i pomiary przebiegów elektrotermicznych w elementach półprzewodnikowych i układach elektronicznych*, Gdynia, Wydawnictwo Uczelniane WSM (in polish).

www.cadsoftusa.com – CadSoft website.
www.essemtec.com – Essemtec website.
www.keithley.com – Keithley website.
www.lpkf.com – LPKF Laser&Electronics AG website.

Maritime Education and Training (MET)
STCW, Maritime Education and Training (MET), Human Resources and Crew Manning, Maritime Policy,
Logistics and Economic Matters – Marine Navigation and Safety of Sea Transportation – Weintrit & Neumann (Eds)

Polish Activities in IMO on Electro-technical Officers (ETO) Requirements

J. Wyszkowski & J. Mindykowski
Gdynia Maritime University, Gdynia, Poland

ABSTRACT: This paper summarizes the legislative way and a role of the Gdynia Maritime University in a comprehensive review of the IMO STCW Convention and STCW Code as well as the consequences resulting from it. The legislative way covers the time - schedule of the Polish initiatives, and shortly describes the IMO Model Course Draft for Electro-Technical Officers - 2012 including the international consultation process. Additionally, basic information about the Polish proposal of Training Record Book for ETO Cadets, as a complementary documentation for IMO ETO Model Course, is presented. Final remarks are also formulated.

1 INTRODUCTION

Today and tomorrow challenges in front of maritime universities cover, among other things, a full implementation of the STCW 1978 as amended in 2010 Convention in the teaching and training programs of the universities [1].

One of the key points of the amendments to the STCW Convention and Code's Annex approved at the end of June 2010 by the Diplomatic Conference in Manila, were the standards regarding Electro-Technical Officers (ETO). These are the first in IMO history standards for electro-technical personnel and now the international community of seafarers is just in the process to create, describe and accept appropriately detailed and comprehensive documents for ETO, like IMO Model Course and Training Record Book.

For better understanding of today challenges concerning the electro-technical personnel and tomorrow consequences for onboard practice as a result, we should shortly overview a legislative way and related conditions leading to current state of the art [2,3,4,5,6].

One of the most important players for preparing and proceeding of ETO requirements, established by the Regulation III/6 and sections A-III/6 and B-III/6 of the STCW Convention and Code [6], was Poland, with significant contribution of the Gdynia Maritime University (GMU), Faculty of Marine Electrical Engineering.

2 ETO TRAINING IN POLAND

ETO training in Poland has long history and was regulated by the following documents:
– Regulation of the Ministry of Shipping and Water Economy 1958: Marine Electrician Officer – Class I, Class II, Class III,
– Regulation of the Ministry of Transport 1971: Marine Electrician Officer – Class I, Class II, Class III,
– Regulation of the Ministry of Infrastructure, 2005: Marine Electro Automation Officer,
– Regulation of the Ministry of Transport, Construction and Maritime Economy (draft), 2013: Electro-Technical Officer.

3 THE ROLE OF POLAND IN A COMPREHENSIVE REVIEW OF THE IMO STCW CONVENTION AND CODE

The first unsuccessful attempt of Poland, Norway and USA to introduce regulations for electro-technical officers to the STCW'95 Convention took place in the years 1992 - 1995.

In March 2008, Poland, France, United Kingdom and Bulgaria, submitted the definition and mandatory requirements for certification of Electro-Technical Officer (ETO) and Senior Electro-Technical Officer (SETO). It was reported in STW 39/7/12 - Appendix 1.

In 2008 the representatives of Poland in cooperation with other countries have undertaken the initiative to create the international informal working group oriented to ETO works, named ETO Forum (Table 1).

Table 1. ETO Forum members

No.	Country	First Name	Surname
1.	Bulgaria	Spaska	Georgieva
2.	France	Isabelle	Merle
		Marc	Fouliard
3.	IFSMA	Christer	Lindvall
4.	Islamic Republic	Mohammad Ali	Shahba
	of Iran	Ali Reza	Navab
5.	ITF	Mikael	Huss
		John	Bainbridge
6.	Malaysia	Adthisaya	Ganesen
7.	Poland	Janusz	Mindykowski
		Jacek	Wyszkowski
8.	Sweden	Goran	Tibblin
9.	Ukraine	Yuriy M.	Mykhaylenko
		Vadym	Zakharchenko
10.	United Kingdom	Roger	Towner
		Cleveland	Powell
11.	United States	Barry	van Vechten

This was followed by the formulation of the definition of training and certification requirements for the Electronic Officer submitted by the Islamic Republic of Iran in STW 39/7/1.

The People's Republic of China has presented its proposal on the knowledge and skill requirements for Electronic Officer in STW 39/7/44.

The delegations of Poland, France, the United Kingdom, Bulgaria and the Islamic Republic of Iran, in September 2009, consolidated from their previous documents the requirements for ETO at operation and management levels and presented them as a joint proposal issued as STW 41/7/4 – Appendix 2.

The STW 41/7/1 of January 2010 includes the view of American and German delegates that there should be two levels of training, i.e. support and operational. After an in-depth discussion it was agreed to include to the STCW Convention and the Code the requirements for Electro-Technical Officers (operational level) and Electro-Technical Ratings (support level).

The significant progress has taken place during the Conference of Parties to the International Convention on Standards of Training, Certification and Watchkeeping for Seafarers in Manila in 2010. The mandatory minimum requirements for certification of electro-technical officers were formally included in the Section A-III/6 and for certification of electro-technical ratings in A-III/4 of STCW/CONF.2/DC/2.

It is expected that the next steps within the IMO procedures will be taken in April and May 2013 when the Polish delegates will present the document STW 44/3/1: IMO Model Course Draft and Onboard Training Record Book for Electro-Technical Officer.

All the presented stages were accompanied in parallel with the International Association of Maritime Universities' (IAMU) research project taking place in FY 2012 -2012/2013.

Poland and China, under the leadership of Ukraine, take part in the activities related to the development of Model Course for Electro-Technical Officers:
- Course for Electro-Technical Officer (ETO),
- Course for Senior Electro-Technical Officer (Senior ETO).

The concept to include Course for Senior ETO in this project was originated from the Diplomatic Conference in Manila, where during the process of comprehensive review of STCW Convention and Code, delegations from large number of countries have been successfully supporting the idea of two-level standards for ETOs. These delegations proposed to establish also standards for Senior Electro-Technical Officer at the management level. In this context, the Committee of the Whole of the STCW Conference in Manila agreed to invite the Maritime Safety Committee (MSC) to consider proposal of Senior ETO with a view to establishing a new work programme item for the STCW Sub-Committee (STCW/CONF.2/CW/RD/1). In the same queue MSC invited Member Governments and international organizations to submit proposals relating Senior ETO (MSC 88/26).

4 SPECIFICATION OF MINIMUM STANDARDS OF COMPETENCE FOR ELECTRO-TECHNICAL OFFICERS

The minimum standards of competence for ETO, established in the STCW Convention and Code's Annex approved at the end of June 2010 by the Diplomatic Conference in Manila, cover 18 competences divided into 3 functions (Table 2).

Table 2. Specification of minimum standards of competence for Electro-Technical Officers

No.	Function	Competence
1.	Electrical, Electronic and Control Engineering at the Operational Level	Monitor the operation of electrical, electronic and control systems Monitor the operation of automatic control systems of propulsion and auxiliary machinery Operate generators and distribution systems Operate and maintain power systems in excess of 1,000 volts Operate computers and computer networks on ships Use English in written and oral form Use internal communication systems
2.	Maintenance and Repair at the Operational Level	Maintenance and repair of electrical and electronic equipment Maintenance and repair of automation and control systems of main propulsion and auxiliary machinery Maintenance and repair of bridge navigation equipment and ship communication systems Maintenance and repair of electrical, electronic and control systems of deck machinery and cargo-handling equipment Maintenance and repair of control and safety systems of hotel equipment
3.	Controlling the Operation of the Ship and Care for Persons on Board at the Operational Level	Ensure compliance with pollution prevention requirements Prevent, control and fight fire on board Operate life-saving appliances Apply medical first aid on board ship Application of leadership and teamworking skills

All the above standards of competence were the basis for the development of a new IMO Model Course on Electro-Technical Officer [7].

5 STW MODEL COURSE DRAFT: ELECTRO-TECHNICAL OFFICER

The content of the Model Course Draft for ETO shall be carefully designed in order to assure the full coverage of the necessary requirements.

Table 3. The content of the Model Course Draft for ETO

No.	Chapter	Sub-Chapter
1.	Introduction	
2.	Part A: Course Framework for all functions	
3.	Function 1 Electrical, Electronic and Control Engineering at the Operational Level	Part B1: Course Outline Part C1: Detailed Teaching Syllabus Part D1: Instructor's Manual
4.	Function 2 Maintenance and Repair at the Operational Level	Part B2: Course Outline Part C2: Detailed Teaching Syllabus Part D2: Instructor's Manual
5.	Function 3 Controlling the Operation of the Ship and Care for Persons on Board at the Operational Level	Part B3: Course Outline Part C3: Detailed Teaching Syllabus Part D3: Instructor's Manual
6.	Part E: Evaluation	
7.	Appendices	1. Basic Engineering Science 2. Mathematics 3. Thermodynamics 4. Mechanical Science 5. Industrial Chemistry

The document STW 44/3/1 presents a new draft Model Course for Electro-Technical Officer [7] based on material developed by the Faculty of Marine Electrical Engineering, Gdynia Maritime University, Poland in cooperation with India, Iran, Malaysia, Ukraine and European Maritime Safety Agency.

This paper consists of: introduction, 5 chapters and appendices as shown in Table 3.

6 ON BOARD TRAINING RECORD BOOK FOR CANDIDATES FOR CERTIFICATION AS ELECTRO-TECHNICAL OFFICER

The On Board Training Record Book [8], as a basis for the recognition of the officer's knowledge experience, leading to obtaining of the certificate, shall contain 5 parts: General Information, Summary record of on board training, On board training record for First, Second and Next Ship, List of training tasks and record of achievements, and Electro-Technical Operations Workbook Guidance. A detailed structure of the content of the Training Record Book, taking into account the division of the sections into related sub-sections is shown Table 4.

Table 4. The content of the Training *Record Book*

No.	Section	Sub-Section
1.	Section 1 General information	1.1 Introduction 1.2 Guidance for completing Training Record Book
2.	Section 2 Summary record of onboard training	2.1 Personal details 2.2 University/College/ Training Centre 2.3 Companies' details 2.4 Ancillary or additional training certificates achieved 2.5 Sea service record 2.6 Task Summary Chart 2.7 The on board Training Record Book review and training assessment for each ship 2.8 On board training final acceptance as required by STCW Convention
3.	Section 3 On board training record: First, Second and Next Ship	3.1 Shipboard and Safety Familiarisation 3.2 Ship's particulars 3.3 Designated Shipboard Training Officer's Reviews of Progress 3.4 Chief Engineer Officer's Monthly Reviews of Progress 3.5 Sea Service Testimonials 3.6 Specimen Signatures of Officers and other experienced staff authorised to sign off Tasks, Records and Reports
4.	Section 4 List of training tasks and record of achievements	4.1 Function: Electrical, electronic and control engineering at the operational level 4.2 Function: Maintenance and repair at the operational level 4.3 Function: Controlling the operation of the ship and care for persons on board at the operational level
5.	Section 5 Electro-Technical Operations Workbook Guidance	

7 CONCLUDING REMARKS

Due to the continuous technological development as well as new required qualifications and skills for maintenance of electrical/electronic systems, equipment and installations, there is a significant increase in employment of properly qualified Electro-Technical Officers.

Such specialists are the crew members of cruise vessels, large ferries and all kinds of special purpose vessels, and therefore their qualifications and competences had been standardised at the international level in the STCW Convention.

As in the case of other officers, there is a need to develop and approve the new IMO Model Course for Electro-Technical Officers and Onboard Training Record Book. These documents are required to establish procedures for ETO's education and training.

It seems that the next essential step in the near future will be the adoption of appropriate standards of education and training for Senior Electro-Technical Officers.

REFERENCES

[1] Mindykowski J., Charchalis A., Przybyłowski P., Weintrit A. Maritime Education and Research to Face the XXI – st Century Challenges in Gdynia Maritime University Experience, *BIT's Annual World Congress o Ocean 2012 (WCO 2012), Dalian, China, September 2012, Book of summaries, p. 183.*

[2] Wyszkowski J. & Mindykowski J., "Electrical, electronic and control engineering" – new mandatory standards of competence for engineer officers, regarding provisions of the Manila amendments to the STCW Code. In proc. *9th International Navigational Symposium on Marine Navigation and Safety of Sea Transportation, Trans-Nav 2011, Gdynia, June 2011, p. 65-69.*

[3] "Are engineers getting the electrical training they need?" In Marine Engineering Review, March 2006, p. 35-36.

[4] Wyszkowski J., Mindykowski J., Wawruch R. 2009. Novelties in the development of the qualification standards for Electro-Technical Officers under STCW Convention requirements. In proc. *8th International Navigational Symposium on Marine Navigation and Safety of Sea Transportation, Trans-Nav, Gdynia, 2009, p. 761-770.*

[5] STCW Convention. Final Act of the 1995 Conference of Parties to the International Convention on Standards of Training, Certification and Watchkeeping for Seafarers, 1978.

[6] STCW CONF.2-DC-2 - Adoption of the final act and any instruments, resolutions and recommendations resulting from the work of the conference. Draft resolution 2. Adoption of amendment to the seafarers' training, certification and watchkeeping (STCW) Code, 2010.

[7] STW 44/3/1 IMO Model Course Draft for Electro-Technical Officers, document submitted by Poland, 2012.

[8] STW 44/INF Information on the new training record book for Electro-Technical Officer submitted by Poland, 2013.

Appendix 1

INTERNATIONAL MARITIME ORGANIZATION

IMO

E

SUB-COMMITTEE ON STANDARDS OF
TRAINING AND WATCHKEEPING
39th session
Agenda item 7

STW 39/7/12
30 November 2007
Original: ENGLISH

**COMPREHENSIVE REVIEW OF THE STCW CONVENTION
AND THE STCW CODE**

Definition and mandatory requirements for certification of electro-technical officer and
senior electro-technical officer and a special training for engineering personnel managing
the operation of electrical power plant above 1000 Volts

Submitted by Bulgaria, France, Poland, and the United Kingdom

	SUMMARY
Executive summary:	This document contains proposals for amending the STCW Convention and the STCW Code in order to introduce a definition and mandatory requirements for certification of electro-technical officer and senior electro-technical officer and a special training for engineering personnel managing the operation of electrical power plant above 1,000 Volts
Action to be taken:	Paragraph 9
Related document:	STW 38/17, annex 11, paragraphs 5 and 21

Background

1 The Maritime Safety Committee (MSC), at its eighty-first session (10 to 19 May 2006), endorsed the proposal of the Sub-Committee on Standards of Training and Watchkeeping (STW) at its thirty-seventh session and decided to include in the work programme and agenda for STW 38 a high priority item on a "Comprehensive review of the STCW Convention and the STCW Code", with target completion date of 2008. The Committee instructed also the Sub-Committee to define, as a first step, the issues to be reviewed and advise it accordingly for the Committee to endorse the scope of the review and the Sub-Committee to undertake subsequently, as a second step, the authorized review.

2 The co-sponsors welcome the results of the discussions of STW 38 of the comprehensive review of the STCW Convention and the STCW Code as set out in annex 11 to the report to the Committee (STW 38/17) and the overall positive outcome of MSC 83 in this regard.

3 In accordance with paragraphs 5 and 21 of annex 11 of STW 38/17, the co-sponsors propose to include, in the Convention and the Code, a definition of the electro-technical officer and the senior electro-technical officer and mandatory requirements for certification of electro-technical officer and senior electro-technical officer. A special training for engineering personnel having management responsibilities for the operation of electrical power plant above 1,000 Volts is also proposed.

I:\STW\39\7-12.doc

Appendix 2

INTERNATIONAL MARITIME ORGANIZATION

IMO

E

SUB-COMMITTEE ON STANDARDS OF
TRAINING AND WATCHKEEPING
41st session
Agenda item 7.3

STW 41/7/4
30 September 2009
Original: ENGLISH

COMPREHENSIVE REVIEW OF THE STCW CONVENTION AND THE STCW CODE

Chapter III of the STCW Convention and Code

Note by the Secretariat

	SUMMARY
Executive summary:	This document contains the draft amended text of chapter III of the STCW Convention and Code prepared by the second *ad hoc* intersessional meeting of the STW Working Group on the comprehensive review of the STCW Convention and Code
Strategic direction:	5
High-level action:	5.2
Planned output:	5.2.2
Action to be taken:	Paragraph 2
Related document:	STW 41/7/1

1 The second *ad hoc* intersessional meeting of the STW Working Group on the comprehensive review of the STCW Convention and Code was held from 7 to 11 September 2009. The Group prepared the draft amended text of chapter III of the STCW Convention and Code as set out in annexes 1, 2 and 3.

Action requested of the Sub-Committee

2 The Sub-Committee is invited to take into account annexes 1, 2 and 3 when finalizing the draft amended text of chapter III of the STCW Convention and Code.

***

I:\STW\41\7-4.doc

Chapter 2

Drills on Board

Drills on Board
STCW, Maritime Education and Training (MET), Human Resources and Crew Manning, Maritime Policy,
Logistics and Economic Matters – Marine Navigation and Safety of Sea Transportation – Weintrit & Neumann (Eds)

Lived Experiences of Deck Cadets On Board

M.M. Magramo & L.D. Gellada
John B. Lacson Foundation Maritime University-Arevalo, the Philippines

ABSTRACT: This study aimed to find out the effects of the lived experiences on board of a Deck Cadets to their behavior and perceptions. It also attempted to determine the relation of religious affiliation, location of residence, nature of work of father/mother, and monthly income of parents as to whether it affects the ability of the Deck Cadets to cope with the shipboard training. Furthermore, it made use of a checklist type questionnaire devised by the researchers. The checklist contains examples of lived experiences which the students may or may not have experience on board. The data gathered from the questionnaires were processed through Statistical Package for Social Sciences Program(SPSS), analyzed, and then interpreted.

1 INTRODUCTION

With a predicted world – wide shortage of officers, maritime education and training will be increasingly important in the next decade. This is to ensure that the best teaching practices are used. The more effective they are, the quicker the people learn. Better training methods mean better passing rates and higher standards. Effective training is good for the shipping industry.

Shipping is an international industry. It is therefore imperative that all sea going officer share a common sense of purpose and apply rules and regulations in like manner. With a predicted world – wide shortage of officers, maritime education and training will become increasingly important in the next decade. The best teaching practices are to be used. The more effective they are, the quicker the people learn (The Nautical Institute, 1997).

Schooling (inside or outside the university) is seen to provide a foundation for young people's intellectual, physical, moral, spiritual, and aesthetic development according to the government. By providing a supportive and nurturing environment, schooling contributes to the development of students' sense of self – worth, enthusiasm for learning, and optimism for the future (Stowell, et al., 1996).

Maritime education provides career development opportunities for mariners through cutting – edge training. Such education and training is very expensive. There are scholarships that are granted to those who can qualify.

Norwegian Shipowners Association (NSA) is an employer's organization and interest group for Norwegian Shipping and offshore companies. The organization's primary fields are national and industry policies, employer issues, competence and recruitment, environmental issues and innovation in addition to safety at sea. NSA provides a cadet program in the Philippines where in its objectives are; 1) to enhance the quality of maritime education in the Philippines, 2) to promote competence among Filipino maritime officers for NIS controlled fleets; and 3) to prove that Filipino cadets can become highly competitive seafarers in the international maritime market. It has become partners to different maritime schools in the country. One of that partner schools is John B. Lacson Foundation Maritime University in Molo and Arevalo in Iloilo City. Every year, it selects 50 cadets (Nautical and Engineering). It's scholarship programs provides free upgrading courses and training supports improvement of teaching methods to faculty and staff, donation of technical equipment and material, faculty development program, cadet monitoring system, performance evaluation and assessment, and character and discipline formation. Scholars have to pass very rigid requirements (www.mtc.com.ph). Maritime teachers, trainers, and assessors have jobs that include mentoring and guiding juniors and trainees. They help the deck cadets by guiding them in the task of raising the professional standards of seafarers through education and training. During their work at sea or working ashore requires a specific detailed knowledge, a wide range of skills, and rich experiences.

Deck cadets assist and understudy the ship's deck officers in their duties. They enter into training program a large part of which will be spent on based receiving structured training and building up experience. Supported by shore – based learning, becoming a deck cadet is the first step to reaching an officer ranking on board.

Importance of safety at sea has been increased in the operations of the ships in the last decades. To achieve a safe navigation, ships are equipped according to both highly skilled individual and a high degree of team coordination. Therefore, the training of the seafarers for updated information and for better skills also becomes a crucial issue.

The purpose of this study is to explore the importance of on – the – job training experiences of the deck cadets as a means of complementing and supplementing leadership development provided in formal education programs. Their vivid positive memories of experiences will significantly affect their development as officers. Outcomes of their on – the – job experiences will appear in their personal growth and interpersonal leadership skills, knowledge and values. This will also show how the experiences will play a vital role in shaping the lives of deck cadets.

Thus, this study was conducted among the twenty-nine fourth year NSA scholars at John B. Lacson Foundation Maritime University, who had just completed from their one year training on seafaring abroad to find out how their experiences have shaped their lives now, how their behavior changed to more active and how they will improve their abilities.

1.1 *Theoretical Framework*

This study is anchored on the following theories: Erickson's psychosocial development theory, Vygotsky's theory of social development, and Dewey's experiential learning theory.

Erickson's psychosocial development theory deals with students' experiences of transition wherein they undergo identity and role confusion. It is during this stage where teachers and parents have roles to play by providing the students with opportunities to develop initiative. Building their self esteem (Eceles et al, 1983) it is also seen as a means of helping to alleviate transitional difficulties. Teachers need to look at ways to build students' self esteem and provide experiences for students to find themselves in their new environment (Walter, 2002).

Vygotsky (1978) was a social constructivist who is responsible for the theory of social development. Cognitive functions originate as products of social interactions and learning as more than accumulation of new knowledge. This theory helps one to understand that cognitive and learning is a social collaborative process with others (Rogoff, 1990).

Experiential learning theory of Dewey states that experience is the source of learning and development. It offers a fundamentally different view of learning process from that of the behavioral theories of learning based on an empirical epistemology. From this perspective emerge some very different prescriptions for the conduct of education, proper relationships among learning, work, and other life activities, and the creation of knowledge itself. This experiential learning theory suggests a holistic integrative perspective on learning that combines experience, perception, cognitive and behavior. Experiences compromise knowledge or skills or observation of something or event gained through involvement in or exposure to that thing or event. After On – the – Job experience is important in complimenting and supplementing leadership development. The experiences also improve their communication (listening, speaking, and writing) skills, sensitivity to and respect for others, self – confidence, and decision making (http://vocserve.berkely.edu)

Experiences of students during trainings may pose serious problems for them. They may face social, curriculum, and peer challenges (La Rue, 2003) such as fear of being bullied (Akos, 2002) and friendship (Green, 1997). Students sometimes neither fear being lost in the new environment (Kaplan, 1996) nor find the new challenges too vigorous (Hatton, 1995).

Experiences can be investigated by seeking insights from student themselves. It is anticipated that understanding what students experience, schools, teachers, and policy makers will be better equipped to facilitate students' transition during this time.

Transition is an important process whereby students experience the problems of re identifying themselves, the uncertainty of the new environment. Transition is a challenging issue confronting the students.

Experiences while it is tried should be investigated so that the meaning of experience can be discovered.

Transition of experiences from classroom into the workplace seems to be based mainly on academic and practitioners' views on student development. All students experience transition. This transition is a very significant phase (Legters and Kerr, 2001) because it coincides with developmental stages. It has major effects on students' sense as they undergo cognitive, emotional, physical, social and psychological developmental changes at this time.

Transition is a process of moving from the known to the unknown (Green, 1997). Students go through transitions when they start school, when they leave this primary school to go to high school, when they leave school to go to tertiary institution and when they go to their work places. Transition is therefore, experienced by all students.

1.2 Objectives of the study

This study tried to find out the lived experiences of deck cadets while on board. What are the lived experiences of the deck cadets while on board? This investigation was conducted at the John B. Lacson Foundation Maritime University during the second semester of the school year 2011 – 2012.

Transitions between home and primary school, and experienced early school success, tend to maintain higher levels of social competence and achievement.

2 METHOD

The descriptive method of research was employed in this study since its main purpose is to find out the lived experiences of deck cadets while on board who are the fourth year Norwegian Shipowners Association (NSA) scholars enrolled at John B. Lacson Foundation Maritime University during the first semester of the school year 2011 – 2012.

2.1 The Respondents

The respondents of the study were the twenty nine (29) fourth year NSA scholars enrolled at John B. Lacson Foundation Maritime University, Arevalo, Iloilo City during the first semester of the school year 2011 – 2012. They were classified according to their religious affiliation, location of residence, nature of work of parents, and monthly income of parents.

3 RESULTS

Based on the responses of the twenty – nine respondents, the following is the first ten popular lived experiences.

"It is their time to be away from their family and home." "I experienced deteriorating condition of the weather at sea." "I worked for long hours." "I communicated with my family through text, calls, chat, and e-mails." "I developed camaraderie with colleagues." "I watched TV, read books, and listened to music during my free time." "I always feel very homesick." "I felt seasickness during varied weather condition." "I enjoyed the comfort in my own cabin." "I enjoyed playing sports using high tech gadgets."

Most of the respondents are first timers in being away from their families and they may be feeling lost. There are a lot of changes in their routine activities from the time they wake up in the morning until time they go to sleep. This is the time when they have to adjust themselves to the new environment they are in. They are at their own - without the caring guidance of parents. They are now deprived of the kisses, cuddles, and passionate reminders of parents, sibling and friends.

Being homesick is common among youth. A feeling of anxiety is also experienced. Sometimes, it is associated with sadness and hopelessness.

Getting along with colleagues, dealing with new work pressures, managing own finances, meeting academic demands – all while being away from home for the first time and without sources of support – can leave them with feelings of being confused and stressed out.

Changes in weather conditions are felt by these students at the most when they are at sea. Unlike, on land during bad weather conditions, they just stay inside their homes. But during this period, students have to experience aside from heavy rains, thunder and storms, there is still the swaying of the ship due to large waves. They are experiencing a greater danger compare than that in their homes. They have to keep their spirit right through prayers, positive thinking and making themselves brave and prepared for any eventuality.

Long, tedious, and tiring hours of skill work on any part of the ship is such an experience because they cannot do otherwise. They have to do their work to the fullest because it is expected of them. Their commitment to their work should prevail even though they are on the verge of surrendering.

Their constant communication with their love ones through text, phone calls, emails, and chats are their consolation amidst the homesickness and long work. They find consolation to exchanging news with parents of what is happening from both ends. During communications, tears are flowing without letting each other knew, hearts are shouting but are prevented from being heard. Students make themselves understand that being homesick is being part to becoming mature and it is a normal thing. Encouragements given by parents also console the students. Constant communication is a source of comfort, perspective and inspiration. Befriending and enjoying the company of colleagues is also one way of forgetting sickness and tiredness. It even improved their personality, communication skills, and their attitudes especially if colleagues are of different nationalities. The presence of these colleagues keeps their minds being interested to these people. Ethics are sensitive issues which they are now applying to cross international attitudinal borders.

The amenities and comforts that are provided by the vessels to the students are worthwhile experiences for them and the feeling of being superior in terms of having enjoyed the luxury which they have not experienced before. It is already an accomplishment to them. The latest high tech gadget they are using can also be rewarding and consoling to their being away from home.

4 FINDINGS

Based on the investigation, the following conclusions were drawn.

The top five lived experiences of the deck cadets were:

- It is their time to be away from my family and home.
- I experienced a deteriorating weather at sea.
- I worked for long hours.
- I communicated with my family through text, calls, chat, and e mails.
- I developed camaraderie with colleagues.
- The last five were the folowing :
- I witnessed the change in time zones.
- I had a difficult time communicating with co – workers of different nationalities.
- I acted as a ship captain during one of my duties.
- I served some passengers during the sea accident.
- I participated in responding to a vessel on fire.
- residence, nature of work of parents, and monthly income of parents.

5 CONCLUSIONS

Cadets experience mostly homesickness when they are away from home for the first time.

The independent variables used in the study have no great/direct relationship with the lived experiences of deck cadets.

There may be other factors which may have direct relationship with the dependent variable.

6 RECOMMENDATIONS

Parents and teachers should support, comfort, and console cadets while on board so that they will experience vivid and exciting memories on deck.

Cadets should always bear in mind that being away from home is part of growing up leading to maturity and coupled with responsibility. Communications to parents and vice versa should be maintained.

Proper mind set, faith in God, and love for the family should be the guiding posts of cadets while on board.

REFERENCES

Akos, P. (2002). Students Perception of Transition from elementary to Middle School. Professional School Counseling.

Dockett, S. and Perry, B. (2003). The Transition to School: What's Important. Educational Leadership.

Eccles, JS (1983). Experiences, Values, Academics, Behaviors. San Francisco, Freeman.

Green, P. (1997). Moving from the World of the known to the Unknown: The transition from Primary to Secondary School. Melbourne Studies in Education.

Hatton, E. (1995). Middle School Students' Perceptions of School organization. Unknown.

Kaplan, L. S. (1996). Where's your focus. High academic Standards versus personal social development. National Association of Secondary School Principals Bulletin.

Kvale, D. (1996). Interviews. London: Sage Publications.

Legters, N. and Kerr, K. (2001). The Efects of Transition to High School: An investigation of reform practices to promote ninth grade success. Campbridge, Massachusetts, John Hopkins University

Mc Namara, C. (1999). General Guidelines for Conducting Interviews. Minnesota

Pinnel, P (1998). Middle Schooling: A Practitioners Perspective. Victoria, Autralia.

Rogoff, B. (1990). Apprenticeship in Theory and Cognitive Development in Social Context . New York: Oxford University Press.

Stowell, LP, Rios, FA, Mc Daniel, J. (1996). Working with Middle School Students. Highett Victoria: Hawken, Brownlow Education

The Nautical Institute, 1997.

www.cdc.gov/health youth. retrieved January 29, 2012

www. mtc.com.ph.

www.nmp.gov.ph

www.gearbulk.com.

www.http://vocserve.berkerly.edu.

www.http://www.who.int. World Health Organization. Interviews.

Drills on Board
STCW, Maritime Education and Training (MET), Human Resources and Crew Manning, Maritime Policy,
Logistics and Economic Matters – Marine Navigation and Safety of Sea Transportation – Weintrit & Neumann (Eds)

The Research of Self-evaluation on Fire-fighting Drill on Board

W. Jianjun & W. Shengchun
Shanghai Maritime University, Shanghai, China

ABSTRACT: The paper talks about measures for self-evaluation of fire-fighting drills on board responding to PSC inspections. A brief background about the PSC inspection on fire-fighting drills on board is introduced in the beginning. Then the paper will state the significance of fire-fighting drill self-evaluation. Following that, legal ground is discussed on International Conventions, regulation of related regional group, national maritime laws and rules. Furthermore, the paper discusses some PSC states and PSC inspection, and gives some examples of detention of ships due to fire-fighting drills. Moreover, the paper translates the procedure for fire-fighting drills on board, organization, personnel and implement will be involved. In the following importantly, introducing the systematics theory, the paper lists self-evaluate principle and the method, focuses on the preparation, performance and rehabilitate of drill and develops self-evaluation criterion. Finally, some suggestions are raised to carry out self-evaluation in order to responding to the PSC inspection on fire-fighting drills.

1 INTRODUCTION

In order to put out a fire on board in time and successfully, and also to enhance the emergency response capabilities of ship's master and crews, every ship must comply with fire-fighting drills to meet the requirements of the SOLAS Convention. As an important part of ship management, self-evaluation on fire-fighting drills should be carried out on board. The self-evaluation should be implemented under the leadership of the master and related officers to learn a lesson from the former fire-fighting drills and further improve operations as much as possible in the next drill, in order to respond to a fire in an emergency.

Something undesirable could be found in the self-evaluation based on the author's experience on board: many officers don't understand who should be in charge of self-evaluation, what should be self-evaluated and how to self-evaluate. New regulations impose more requirements on fire safety, which put greater pressure on ships to respond to PSC inspections, especially fire-fighting drills in some Port States. Therefore, it is more advantageous to pass the PSC inspection successfully if ships carry out self-evaluation efficiently in advance.

2 THE LEGAL GROUND FOR PSC INSPECTION OF FIRE-FIGHTING DRILL

2.1 *The related regulations of the SOLAS convention* [1]

Specific regulations concerning fire-fighting drills have been developed in the International Convention for The Safety of Life at Sea (the SOLAS convention). According to the provisions of Regulation II-2/15 "Instructions, on-board training and drills", fire-fighting drills shall be conducted and recorded in accordance with the provisions of regulations III/19.3 and III/19.5 as general requirement 2.2.5 provided. For passenger ships, "in addition to the requirement of paragraph 2.2.3, fire-fighting drills shall be conducted in accordance with the provisions of regulation III/30, having due regard to notification of passengers and movement of passengers to assembly stations and embarkation decks" as additional requirements 3.1 provided.

Furthermore, the SOLAS convention has also developed special requirements for helicopter facilities on some related ships, such as Regulation II-2/18.8 "Operation manual and fire-fighting arrangements", which provides "fire-fighting personnel consisting of at least two persons trained for rescue and fire-fighting duties, and fire-fighting equipments shall be immediately available at all times when helicopter operations are expected" on the basis of provision II-2/18.8.3; it also provides that "on-board refresher training shall be carried out and additional supplies of fire-fighting media shall be provided for training and testing of the equipment "as regulation II-2/18.8.5 provided. In the Regulation II-2/30 "drills" for all passenger ships, "an abandon ship drill and fire-fighting drill shall take place weekly. The entire crew need not be involved in every drill, but each crew member must participate in an abandon ship drill and a fire-fighting drill each month as required in regulation II-2/19.3.2", and "all passengers shall be strongly encouraged to attend these drills".

2.2 *The related regulation of the STCW code* [2]

Chapter II of the STCW code has developed standards regarding the master and deck department as section A-II/1 and section A-II/3 provided. chapter III has developed standards regarding engine department as section A-III/1 provided. These provisions focus on the "Prevent, control and fight fires on board" and provide criteria for evaluating competence in table A-II/1, A-II/3 and A-III/1. Firstly, the type and scale of the problem is promptly identified and initial actions conform to the emergency procedure and contingency plans for the ship. Secondly, "evacuation, emergency shut-down and isolation procedures are appropriate to the nature of the emergency and are implemented promptly". Thirdly, "the order of priority, and the levels and timescales of making reports and informing personnel on board, are relevant to the nature of the emergency and reflect the urgency of the problem".

In chapter V, we can find standards regarding special training requirements for personnel on certain types of ships, such as mandatory minimum requirements for the training and qualifications of masters, officers and ratings on oil and chemical tanker and liquefied gas tankers in section A-V/1-1 and A-V/1-2 respectively. Minimum standard of competence in basic training for oil and chemical tanker cargo operations and that for liquefied gas tanker cargo operations are listed in table A-V/1-1-1 and table A-V/1-1-2. Criteria for evaluating competence on carrying out fire-fighting operations, including that initial actions and follow-up actions on becoming aware of an emergency, must conform with established practices and procedures, actions taken on identifying muster signals must be appropriate to the indicated emergency and comply with established procedures, clothing and equipments must also be appropriate to the nature of the fire-fighting operations, similarly, the timing and sequence of individual actions must be appropriate to the prevailing circumstances and conditions, and

extinguishment of fire should be achieved using appropriate procedures, techniques and fire-fighting agents.

Additionally, chapter VI, table A-VI/1-2 "Specification of minimum standard of competence in fire prevention and fire-fighting" develops criteria for evaluating competence on "minimizing the risk of fire and maintaining a state of readiness to respond to emergency situations involving fire", that is "initial actions on becoming aware of an emergency conform with accepted practices and procedures" and "actions taken on identifying muster signals are appropriate to the indicated emergency and comply with established procedures"; and this table also lists details of criteria for evaluating competence on fighting and extinguishing fires, which focus on clothing and equipments, timing and sequence of individual actions, extinguishment of fire, breathing apparatus procedures and techniques. Table A-VI/1-4 "specification of minimum standard of competence in personal safety and social responsibilities" provides the criteria for evaluating competence with emergency procedures of "the master and deck department in charge of a navigational watch on ships of 500 gross tonnage or more". The criteria include "initial actions on becoming aware of an emergency conform to established emergency response procedures" and "information given on raising alarm is prompt, accurate, complete and clear". Table A-VI/3 shows minimum standard of competence in advanced fire-fighting, in the table, criteria for evaluating competence on controlling fire-fighting operations aboard ships are related to actions taken to control fires, the order of priority, timing and sequence of actions, transmission of information and personal safety during fire control activities. In the aspect of organizing and training fire parties, composition and organization of fire control parties must ensure a prompt and effective implementation of emergency plans and procedures. Operational effectiveness of all fire systems and equipments must be checked out during inspection and maintenance. Causes of fire are identified and the effectiveness of countermeasures must be evaluated when investigating and compiling reports on fire.

2.3 *The related regulation in some regional PSC agreements*

Port State Control involves enforcement activities controlled over the vessels and carried out by the government of the foreign port in which the vessels operate to ensure compliance with applicable domestic and international requirements to ensure safety of the port, environment and personnel [3]. The Port State, upon identifying a substandard vessel, was authorized to detain the vessel until corrections to eliminate hazards to the port and return to a seaworthy condition were achieved, and ultimately, to eliminate substandard ships from the waters of participating countries. The regulations of Port State Control were adopted in the amendment to the SOLAS in 1994, which confirmed the legal position of PSC inspection on international vessels.

Besides, the International Maritime Organization has also encouraged some regional PSC meetings to conduct PSC inspections by the same standard. In January 1982, 14 European States agreed to establish a system of control, which resulted in the signing of the Paris Memorandum of Understanding (MoU) on Port State Control ("Paris MoU" in short). In the early 1990s the Asia-Pacific (Tokyo) MoU came into being in the Far East, which largely involves western Asia-Pacific rim States. Nearly at the same time, the South American States, along with Mexico and Cuba, formed the Viña del Mar (Latin American) Agreement. The end of the 1990s has also seen the establishment of regional MoUs in the Caribbean, the Mediterranean and Indian Ocean. The West and Central African (Abuja) MOU and the Black Sea MoU have also recently been established. A further MOU is being planned to cover the Arabian Gulf region.

Recently, from September 1st to November 30th 2012, the Paris MoU and Tokyo MoU on Port State Control have launched a joint Concentrated Inspection Campaign (CIC) on Fire Safety Systems to verify the fire safety arrangements, maintenance records and other applicable documentations, which was to ensure compliance with SOLAS Chapter II-2/ Construction - fire protection, fire detection and fire extinction arrangements on board ships [4]. The Indian Ocean, Black Sea and Viña del Mar MoUs on PSC have also been conducting a CIC covering compliance with the same regulatory requirements during the same period. In addition to the physical inspection and testing, the crew may be required to demonstrate their familiarity with the fire-fighting systems and appliances, and may also be called upon to conduct a fire-fighting drill for the benefit of the PSCOs [5], according to the 12 items in the provided checklist, which is "where a fire-fighting drill was witnessed; was it found to be satisfactory?" [6]

2.4 Some national maritime laws and rules on fire-fighting drill

Unlike other major maritime countries, the United States has not participated in any regional MOUs grouping. Under the US Coast Guard's Port State Control (PSC) Program, it undertakes control measures on a unilateral basis. According to US Coast Guard, Navigation and Vessel Inspection Circular (NVIC) was adopted to provide guidance on the enforcement of the 1995 Amendments to the STCW Convention and Code. Coast Guard Port State Control Officers (PSCOs) used the provided guidance and procedures to conduct PSC inspection in ensuring compliance with the requirements of the STCW Convention and Code. [7]

Related to the fire-fighting drill inspection, the PSCOs will examine the specific new crewmember familiarization procedures in the first stage, which is the general examination. During the general examination, if the PSCOs find a situation which indicates that the fire-fighting drill can not meet the requirements due to the inability of crewmember(s) to perform their assigned duties during fire-fighting drills, or the drill poses a danger to persons, property or the environment, the examination will be transformed into an expanded examination which focuses on correcting the apparent deficiencies. The PSCOs will provide an opportunity and permit the master to correct the deficiencies at that time; for instance, they allow the crew to receive instructions and repeat a fire-fighting drill after failing the first time or to replace one crew member by another who is qualified to hold that position. If the situation can not be corrected, ships may be detained under the STCW Convention and Code.

3 THE PROCEDURE OF PSC INSPECTION OF FIRE-FIGHTING DRILL

3.1 The procedure of PSC inspection under the Tokyo MoU

During the Asia-Pacific PSC inspection within the Tokyo MoU [8], the PSCOs may witness a fire-fighting drill carried out by the crew assigned to these duties on the muster list.

In the first stage, the PSCOs will consult with the master and then select one or more specific locations of the ship for a simulated fire and assign a crewmember to the location(s) to activate a fire alarm system or use other means to sound alarm.

In the next stage, at the simulated fire location the PSCOs describe the fire indication to the crewmember and observe how the report of fire is transferred to the bridge, which is generally the control center.

In the following stage, ship will sound the fire alarm to summon the fire-fighting team to their stations. The PSCOs will observe the fire-fighting team arriving on the scene, properly donning their protective equipments and using fire-fighting equipments to fight the simulated fire. The team leaders should be giving orders appropriate to their members and passing the words back to the bridge at the same time.

Moreover, crews should response to personnel injuries by communication and using of stretcher and medical teams. The PSCOs will monitor them handling the stretcher and wounded properly through narrow passageways, doors and stairways, which are difficult and takes practice.

Furthermore, those crewmembers assigned to close manually operated fire doors and dampers should pay more attention to the duties in the areas of the simulated fire(s) during the drill. Those crews assigned to other duties related to emergency equipments will be asked to explain their duties and demonstrate their familiarity possibly in the drill.

3.2 *The procedure of PSC inspection under the Indian Ocean MoU*

Under the Port State Control of Indian Ocean [9], the PSCOs will determine whether the crew members are familiar with the duties assigned to them and locations where duties are performed in fire-fighting drill on the muster list. Generally, after consultation with the Master, the PSCOs require a fire-fighting drill to ascertain the awareness and promptness of the crewmembers in an emergency. During this drill, the PSCOs will ask the crew members to explain their duties including the procedure. For instance, if a crew is assigned to operate the fixed fire-fighting equipments, which cannot be operated during the simulated drill, the PSCOs will evaluate whether the person required to operate the system can explain the complete procedure.

3.3 *The procedure of PSC inspection in some state maritime administration*

According to ST. Vincent and the Grenadines Maritime Administration as an example, during the fire-fighting drill inspection on ferries and passenger ships, "injured/unconscious" persons should be placed in the identified suitable area, the PSCOs observe the performance of search-rescue team, and the fire-fighting operations of fire-fighting team. During the drill, in addition to some general items, the PSCOs will observe evacuation of passengers from the identified area by assigned crew members, ashore organization informed by a designated person, report process of the fire-fighting team leader to the bridge and control process of the Master in the entire drill. Under the program of ST. Vincent PSC inspection, lack of communication and control, incorrect use of fire-fighting equipments are common deficiencies. [10]

3.4 *The procedure of PSC CIC*

In response to the requirements and practice of a joint Concentrated Inspection Campaign (CIC) launched by Paris MoU and Tokyo MoU on Port State Control, every vessel should confirm the following items before surveys.

The fire fighters' outfits including personal equipments should comply with the SOLAS Ch II-2 /R10.10 & R14.2.2. The Emergency Escape Breathing Devices (EEBD) must comply with the SOLAS Ch II-2/R13.3.4 & 13.4.3. The portable extinguishers ready for use in locations as per fire plan must meet SOLAS Ch II-2/R10.3.2.4. The test of automatic audible alarm should sound prior to the release of a fixed gas fire-extinguishing medium into spaces in which personnel normally work according to SOLAS Ch II-2/R10.5. The fire protection systems and fire-fighting systems and appliances must be maintained ready for use to meet the SOLAS Ch II-2/R14.2.1. The crew must be familiar with the location and operation of fire-fighting systems and appliances so that they may be called upon to use them in accordance with the SOLAS Ch II-2/R15.2.2'.

The test of the sprinkler system should trigger an automatic visual and audible alarm for the section on the basis of the SOLAS Ch II-2/Reg 10.6.The activation of any detector or manually operated call point can initiate a visual and audible fire signal at the control panel on the bridge or control station to meet the SOLAS Ch II-2/R7.4.2.The Emergency Fire pump must be capable of producing at least two jets of water in the drill and emergency to comply with the SOLAS Ch II-2/R 10.2.2.3.1& R2.2.4, and the isolating valves of the fire main must be marked, maintained and easily operable in accordance with the SOLAS Ch II-2/R10.2.1.4.

All the above-mentioned should be demonstrated in the fire-fighting drill, which should satisfy the PSCOs when witnessing in the PSC inspection. Once the satisfaction has been achieved, the detention of ship would never happen as a result of the CIC.

3.5 *The statistical analysis of detention deficiencies of fire-fighting drill*

A report on detention deficiencies per category group found on DNV ships from 2006 to 2011 illustrates clearly that the item of fire safety measures plays a leading role in the deficiencies, as detailed in figure 1 [11,12]. In the six years, DNV ships were found to have 1547 deficiencies on fire safety measures, among which 512 cases, accounting for 1/3, were detainable deficiencies, which reached 18 percent of total detention deficiencies. [13]

According to a report of detainable and ISM Related Deficiencies noted in PSC inspection reports of Class NK & NK-SMC ships during Jan-Aug 2012 [14], 115 ships in total were detained because of their serious deficiencies. 7 detained ships in port in Japan, USA and Spain were related to fire-fighting drills. The most frequent cause was that crews were unable to demonstrate proficiency in fire-fighting drills.

Another report on detention of Hong Kong Ships due to fire-fighting drills in 2007 shows that two cases of detention were because of fire-fighting drills, which were inspected by USCG in USA, while 46 of them were owing to their serious deficiencies [15]. Another two ships inspected in Italy and Spain under Paris MoU were instructed to rectify deficiencies before departure because fire-fighting drills were hardly sufficient or crew members lack knowledge of fire-fighting drills. Through these cases, some points can be gained that the PSC inspection on fire-fighting drill under USCG is very strict. If the fire-fighting drill does not satisfy the PSCOs of the USCG, they will give deficiencies action code 30 indicating grounds for detention, which is the most serious result in the PSC inspection.

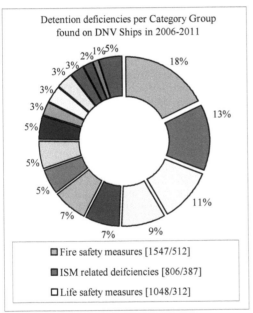

Detention deficiencies per Category Group found on DNV Ships in 2006-2011

18%
13%
11%
9%
7%
7%
5%
5%
5%
3%
3%
3%
3%
2% 1% 5%

☐ Fire safety measures [1547/512]

■ ISM related deifciencies [806/387]

☐ Life safety measures [1048/312]

Figure 1. Detention deficiencies per category group found on DNV ships from 2006 to 2011. The first number in square brackets stands for total deficiencies per category, while the last number stands for quantity of detainable deficiencies.

4 THE SELF-EVALUATION OF FIRE-FIGHTING DRILL ON BOARD

4.1 Definition

Within the system engineering theory, evaluation is to determine the property of research objects in accordance with some specific purposes, and turn the property into denominated objective quantity or subjective utility. In this research, self-evaluation is defined as a series of activities via which ships evaluate their own fire-fighting drills, i.e. evaluation

of crew performance in fire-fighting drills [16]. It is the continuous process of evaluating all items involved in the preparation, performance, and rehabilitation of the drill.

4.2 Principle

In accordance with SOLAS 74/CII-2/R15.2.2.3, crew performance of fire-fighting duties should be evaluated periodically by conducting on-board drills and identifying areas in need of improvement. The self-evaluation of fire-fighting drill is to ensure the operational readiness of the fire-fighting organization and that competency in fire-fighting skills is maintained. Self-evaluation should meet the most important principle that is "drill must ensure the safety of involved crews", which is the key factor in the success of the fire-fighting drills.

In recent years, some shipping companies have experienced a number of serious accidents during fire-fighting drills. Many guiding documents from the shipping industry provide advices and instructions to help prevent accidents during fire-fighting drills. Fire-fighting drills are covered by the provisions on occupational safety and health on board, which should be planned, organized and performed so that they are safe in every respect. Before the drill, possible hazards to crewmembers in the execution of the drill should be identified and respective risk should be mitigated [17]. Additionally, fire-fighting drills prescribed by national laws and regulations and by international instruments should be conducted in a manner that ensures the disturbance of the rest period is minimized and does not induce fatigue. [18]

4.3 Points of self-evaluation

4.3.1 Preparation stage

Firstly, crew participation should be maximized. Personnel arrangements should include every crew member except on-duty crews. In particular, those crew members assigned to other duties related to fire-fighting drills, such as the manning of the emergency generators, the CO2 room, the sprinkler and emergency fire pumps, should also be involved in the drill. In the second, all related equipments must be in good order and readily available in an emergency. The persons in charge, such as third officer and engineer, should maintain the fire-fighting equipments and devices regularly. Another key point is that the type and position of the fire scenario should be varied in a well conceived sequence, which covers most parts of the ship and all types of fire-fighting. For the purpose of a fire drill, the alarm should be activated and the requisite actions be taken in accordance with the ship's safety and health policy.

4.3.2 Performance stage

The first step: time and place of fire-fighting drill should be kept as a secret until the noticed crew activates the fire alarm that delivers the signal of fire location, which is helpful to improve the awareness and promptness of the crew members during an emergency. The best time and location should be determined through coordinating with the Master and/or ship's safety officer. The master should hold drills considering locations where the ship is most likely to experience a fire, where most recent drills have been held, and while minimizing disruptions to cargo operations [19].

Point 2 is planning a realistic emergency [20]. The drill should, as far as practicable, be conducted as if there were an actual emergency. All involved crews must participate in the fire-fighting drill with a positive attitude and be fully confident of a realistic fire-fighting other than a simulation. Available resources (such as smoke-generating machine) should be utilized to make drill as realistic as possible [19]. For example, in order to prevent unnecessary risk and gain realistic effectiveness, the recovery of a hoist stretcher should be carried out without persons on the stretcher, where a similar load can be used instead [21]. And fire-fighting drills should be planned in such a way that regular practice depends on the type of ships and cargo in various emergencies.

The third point is to keep a timeline of events. Certain emergency situations require timely response, without any delay during the whole drill. As an example of that, when hearing the warning of fire doors remote released, which is announced on the public address system, crew should run away from the fire area as far as possible because the remote controlled release of fire doors can also involve a risk of personal injury [22].

Moreover, checking the necessary arrangements for subsequent abandoning of the ship must be implemented according to muster list. Many ships generally neglect abandon ship drills or conduct fire-fighting drills several minutes after abandon ship drill. This situation should be avoided.

4.3.3 Rehabilitation stage

In the rehabilitation stage, equipments used during drills should be immediately brought back to its fully operational condition and any faults or defects discovered during the drills be remedied as soon as possible. The effectiveness of the drill will be evaluated. The master should hold a meeting or discussion to get feedback together and conduct a thorough debriefing of the lessons learned. This is the most important element of the drill or exercise as it allows crewmembers to identify those areas in which they can improve their respective response efforts [23]. For instance, the master can give some comments on incapacitation of crewmember and malfunctions of simulated equipments including radios, fire-fighting equipments, and lighting.

4.4 The criterion of self-evaluation

The criterion of self-evaluation on fire-fighting drill on board would be found in the following table 1. The master can apply this reference table to self-evaluation, and obtain a score of the whole drill after having graded every item in the checklist. The checklist designs 25 kinds of items belonging to 7 stages. The highest score will run up to 125 points if 25 points have been performed perfectly. If the score is lower than 100, this drill would be not satisfactory, and the whole drill must be redone. The crews require supplemental training and corrective action should be taken in the shortest possible time to respond to the seriously defective drill. If the drill gains a higher score with only several items not at good level, the corresponding knowledge and skill must be checked carefully and the related roles must be well disciplined.

Table 1. The criterion of self-evaluation on fire-fighting drill on board

No.	Items	Self-evaluation Elements	Self-evaluation Criteria	Score
1	Reporting	Content of report: the location, size, kind and time of the fire	No Miss=5, Other=0	
2	Bridge of fire	The initial caller should be most likely connected to the designated fire source in his job role	Good=5, Average=3, Poor=0	
3	Initial actions	Fire-Signal is correct with the Muster list	No Miss=5, Other=0	
4		Ship's alarm announcement contains the location, size, kind and time of the fire	No Miss=5, Other=0	
5		Proper extinguisher has been used and Local ventilation has been shut	No Miss=5, Other=0	
6	Mustering	Assembling condition after sound emergency alarm:	Good=5, Average=3, Poor=0	
7	and preparation	Each crew wearing condition: the crews should have taken out their personal effects	Good=5, Average=3, Poor=0	
8		Wearing time of fireman's outfit	Less 3 min=5, Less 5 min=3, Over 5 min=0	
9		Wearing condition of fireman's outfit	No Miss=5, Other=0	
10		Starting time of Emergency Fire Pump	Less 5 min=5, Less 10 min=3, Over 10 min=0	
11		Condition of Emergency Pump Presser with two hoses at the same time.	Over 12 min=5, Less 12 min=0	
12		Starting time of Electrical supplies and lighting	Less 5 min=5, Less 10 min=3,	

13		Time of listed damper close	Over 10 min=0 Less 5 min=5, Less 10 min=3, Over 10 min=0
14	Entry into	Each team reporting level during drill	Good=5, Average=3, Poor=0
15	the fire zone	The condition of check and report about the fire team to the bridge	No Miss=5, Other=0
	and fire-fighting	(BA pressure/Time of entry/Name/Mask secure and Safety Lanterns are in all good order)	
16		The condition of the fire hose(proper length, boundary cooling)	No Miss=5, Other=0
17	Rescue	The report of the injured (Name/location/status)	No Miss=5, Other=0
18	operation	The condition of transference of the injured person to a safe location	Good=5, Average=3, Poor=0
19		The condition of monitor and first aid the injured person(s)	Good=5, Average=3, Poor=0
20	Use of a Fixed	All ventilation and opening been closed	No Miss=5, Other=0
21	Fire Extinguish-	The Roll call has been reported to the Bridge	No Miss=5, Other=0
22	ing System	Any crew can explain the manual operation method	Good=5, Average=3, Poor=0
23	Master's	Condition of each team action to be taken:	Good=5, Average=3, Poor=0
24	Comments	Each crew attitude for drill	Good=5, Average=3, Poor=0
25		Whole performance	Good=5, Average=3, Poor=0
1	**Suggestion**	**Crews required supplemental training**	**Yes / No**
2		**Corrective Action**	**Yes / No**

5 CONCLUSION

Some suggestions are raised to carry out self-evaluation in order to respond to PSC inspections of fire-fighting drills. Firstly, sufficient knowledge and understanding of the legal background about the PSC inspection on fire-fighting drills on board are needed. More research should be conducted by ship owners, ship operators, ship class association and some related maritime research institutions, focusing on the International Conventions such as the SOLAS convention, STCW convention and code, regulation of related regional PSC agreements, some national maritime laws and rules on fire-fighting drill. The master and crews should be proficient in the procedure of the PSC inspection on fire-fighting drill, whatever PSC agreements are, including main maritime administration, Paris MoU, Tokyo MoU, Indian Ocean MoU and USCG. Most importantly, every ship had better study and analyze the cases of detention deficiencies in recent years whatever regional PSC agreements are, and discipline its crew members regularly and efficiently. Of course, the related pre-post training should be carried out on land to comply with the requirements of PSC inspections. In order to pass the PSC inspection on fire-fighting drills, what have been mentioned above should not be neglected or omitted.

REFERENCES

1. International Maritime Organization. 2004. International Convention for The Safety of Life at Sea.
2. International Maritime Organization. 2012. The Manila Amendments to the Seafarers' Training, Certification and Watchkeeping (STCW) Code.
3. U.S Coast Guard. (n.d.). Port state control (PSC). Retrieved from http://www.uscg.mil/d13/dep/news/port_state_control.htm
4. Paris and Tokyo MoUs on PSC will hold joint CIC on Fire Safety Systems. 2012. Paris MoU on Port State Control. Retrieved from http://www.parisMoU.org/Publications/Press_releases/2012.06.01/Paris_and_Tokyo_MoUs_on_PSC_will_hold_joint_CIC_on_Fire_Safety_Systems.htm
5. Port State Control - Indian Ocean, Black Sea and Viña del Mar MOUs - Concentrated Inspection Campaign on Fire Safety Systems. 2012. Retrieved from http://www.westpandi.com/Publications/News/Port-State-Control-Indian-Ocean-and-Black-Sea-MOUs-Concentrated-Inspection-Campaign-on-Fire-Safety-Systems/
6. The American Club. 2012. Member Alert.
7. United States Coast Guard. (n.d.). Navigation and vessel inspection circular No. 3-98.
8. Tokyo MOU Secretariat. 2008. Asia-Pacific Port State Control Manual, Revision 2/2008.
9. IOMOU Secretariat. 2001. Manual for PSC officers.
10. ST. Vincent and the Grenadines Maritime Administration. 2005. Guidance for Operational Control on Ferries and Passenger Ships.
11. DNV. 2009. Port State Control - Top detention items. Reduce the risk of detentions.
12. DNV. 2012. Port State Control - Top detention items. Be prepared and reduce the risk of detentions.
13. DNV. 2012. Port State Control- Synopsis of Frequent Findings and Detention Items.
14. NIPPON KAIJI KYOKAI (Class NK). 2012. Detainable and ISM Related Deficiencies noted in PSC inspection reports of Class NK & NK-SMC ships in Jan-Aug 2012. Retrieved from http://www.classnk.or.jp/hp/zh/index.html
15. HongKong Marine Department. 2008. Port State Control Inspections (To: Shipowners / Ship Managers and Classification Societies).
16. Germanischer Lloyd. 2012. PSC Information Manual.
17. International Labour Organization. 1997. Accident prevention on board ship at sea and in port. Geneva.
18. International Labour Organization. 2009. Guidelines for port State control officers carrying out inspections under the Maritime Labour Convention, 2006. Geneva.
19. U.S. Coast Guard Sector. 2008. Auxiliary Assistant Port State Control Examiner Performance Qualification Standard
20. Fairfield-Maxwell Services, Ltd. 2010. SMS Highlights & News. FMSL Safety Alert: P-2.
21. Danish Maritime Authority. 2002. Guidance on safety during abandon ship drills and fire-fighting drills on board ships.

22. Germanischer Lloyd. 2006. International Code for Fire Safety Systems, GL consolidated up to Res. MSC.206 (81) adopted 2006.
23. Kwiatkowska K. & Kalucka P. 2010. Application of Thermal Analysis and Trough Test for Determination of the Fire Safety of Some Fertilizers Containing Nitrates. International Journal on Marine Navigation and Safety of Sea Transportation 4(4): 441-445.

Chapter 3

Human Resources and Crew Manning

Human Resources and Crew Manning
STCW, Maritime Education and Training (MET), Human Resources and Crew Manning, Maritime Policy,
Logistics and Economic Matters – Marine Navigation and Safety of Sea Transportation – Weintrit & Neumann (Eds)

The Problem of the Human Resource Supply Chain on the Social Network at Maritime Education and Training

T. Takimoto
Graduate School of Maritime Sciences, Kobe University, Kobe, Japan

K. Hirono, M. Rooks & M. Furusho
Kobe University, Kobe, Japan

ABSTRACT: This study analyzes the social network in the processes of maritime education and training. The objective of this study is to focus on the shortage of seafarers in the maritime global transportation as mentioned by the BIMCO. The authors divide the processes of maritime education and training into three categories. These processes are "Family", "Maritime education institute" and "Maritime Company". They are not systematically connected, but are found in the processes between social network. There is the "Maritime education institute" process between "Family" process and "Maritime company" process; it has the social network to both. Therefore, the teaching staff members, in their roles as job advisors (hereinafter "teaching staff") in the "Maritime education institute" use the social network in conjunction with "Family" and "Maritime companies" too. The teaching staffs communicate with students using these processes. The teaching staffs are the carriers in regards to how these processes are related. The authors surveyed aspects of "personality" and "social network" pertaining to teaching staffs, and quantitatively analyzed these process regarding social network.

1 INTRODUCTION

BIMCO and ISF research supply and demand of the seafarer human resources once every five years. The research report "MANPOWER UPDATE" proved the most comprehensive assessment global supply of and demand for merchant seafarers available. They have two purposes. First, one is to describe the current worldwide supply and demand situation of seafarers. Secondly, one is to make predictions of the likely situation 5 – 10 years in the future, in order to assist the industry to anticipate changes and to take appropriate action. The study was updated since 1995 after evidence was found a significant shortage of officers.

The supply side of maritime educational institutes has a social mission to supply maritime experts to consumers (i.e. maritime companies) side. The supplier and the consumer need to exchange information for the employment of maritime experts. However, this communication requires a relationship (social network) as the basis between the supplier and the consumer. Investigating of status of these social network in maritime educational institutes is very important in regards to the supply of human resources. Also, it is very meaningful for thinking of future human resource strategies in maritime education training and maritime industry.

The object of this study is to quantify the current status of social network used by maritime educational institutions for communication with companies. Also, the investigation explores the factors that affect social network.

2 SOCIAL NETWORK

2.1 Social Network

The human has relationship by some role: the job, the family, the friends and so on. These relationships are ties of each other. And it is the social network that show it by ideally and virtually. Also this concept can apply to the organizations and the groups. Putnam(2000) made several important statements on social network. He defined the relationship into three main factors. They are "Network", "Trust" and "Precept".

Granovetter studied the problems of social network in 1973. The study made it clear that a function of social networking is to search for jobs. The public and private outplacement services as well as "introduction" and "personal connections" affected the job change. Granovetter revealed that successful job change by "strong relationship" of family, relatives and close friend in U.S.. (Granovetter 1973)

Burt analyzed the networktructure of the brokers. Brokers (mediator) have two activities. First, they find the gap in the network that do not exist in the relationship. Secondly, they control the flow of information and resources as a mediator. Burt(1992) argued, about the benefits of connecting weak relationships and no relationships.

To know the social value of the organization is to know the actual conditions of a social network.

2.2 *Process of human resource supply chain in MET*

Supply Chain Management (SCM) is concept for product supply that is used in production management of industry. SCM is a set of process flow from procurement of raw materials, making, shipping to sales. Another definition is provided by the APICS Dictionary when it defines SCM as the "designing, planning, executing, controlling, and monitoring of supply chain activities with the objective of creating net value, building a competitive infrastructure, leveraging worldwide logistics, synchronizing supply with demand and measuring performance globally."
There is the production process of not only the general products but of the human resources. Therefore, Maritime Education and Training (MET) can be divided into several processes using the concept of the SCM.

MET can be classified into three main processes. These are "development of personality ", "learn specialized skills" and "social practices".

Personality and basic education is required for human to live in the society. G. Allport said that life philosophy and self-consciousness are some of the elements that make up the personality. Also, A. Maslow said that self-actualization is one of the basic human needs in Maslow's hierarchy of needs.

These theories have meaning that good home discipline and the elementary knowledge develops in basic education institutions. These can learn as base of family. From these, "development of personality" can be defined "Family" Process.

The specialized knowledge and the skill need for getting social advantage. The students need learning the expertise before working in the society. Maritime educational training is one of them. The students go on to specialized university or technical school and will learn the mentality, basic specialized subjects, and practical training as the specialist. From these educational contents develop the identity for the field of maritime industry.

This "learn specialized skills" process is that education develops the basics special subject and the

professional identity. Author redefined it to "Maritime education institute".

The graduating students can get the working field by entering employment to the company. The worker is given duties and roles from the company. They need to learn new specialized knowledge and skills, for understanding and making efficient. It is a continuous learning when work belonging to the company. These duty and role are always changing by promotion, career change and so on. Author defined as "maritime companies" to region corresponding to "learn specialized skills".

The author will use the term "Human Resource Supply Chain (HRSC)" to refer to one of these processes as the flow of human resource development. Figure 1 illustrates the structure of HRSC.

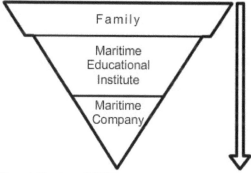

Figure 1. Structure of HRSC

2.3 *Gap between a process and a process*

The HRSC concept looks at the macroscopic human resource educational process for maritime experts. However, it does not have strict integrity; these processes are made by individual organization. Therefore, there is a gap between each process.

The cooperation with each process is necessary for smooth supply of human resources(i.e. job placement assistance). And it needs to know the interests of the other party to the cooperation process. Because, the organization, group, human and so on in each process are not able to build cooperative relationship to the unknown partners.

Figure 2 illustrates the cooperative relationship of each process in the HRSC.

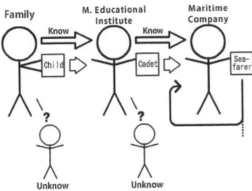

Figure 2. Cooperative relationship of each process in HRSC

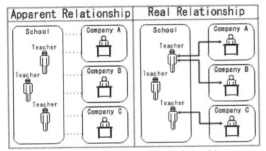

Figure 3. Apparent relationship and Real relationship

2.4 Personal social network in a process

To communicate is necessary the relationship that is either already having a social network or connecting a new relationship. It's similarly in communication with individuals, organizations and processes. Individuals and individual members are working in these. To social network of the organization replaces personal social network. The figure 3 indicates that image of relation between a school to some companies. It is assume that there is already a relationship between a school and company. It is easy to say that connection between organizations or groups is "apparent" relationship. This relationship is a personal relation between the persons of the school and the company. It is "real relationship". All members of organizations have their own personal social network. Therefore, it is possible to know social network and the social attractiveness of an organization by knowing the personal social network in an organization.

3 SURVEY

3.1 Defining keyword

To survey the social network that "maritime education institutes" have between the "maritime company" in HRSC requires several steps.

First, the definition and verification of survey items is required. In this study, we define a personal social network as social network of an organization. Additionally, two factors that affect personal social network are reviewed. These are the inside factor and the outside factor.

3.1.1 Inside factor as personality of teaching staff

Teaching staffs have a wide variety of backgrounds as proportional with age. This background is the career experience of working. Age effects life experience. The career can be classified into two main groups: an Educational career at an academic institution, and a Job career in a company. The author refers to the "Inside factor" as age and career.

3.1.2 Outside factor as the size of the educational institution

Teaching staffs interacted with other teaching staff and students every day. This relationship can be divided into horizontal and vertical relationships. The horizontal relationship denotes being able to build a new relationship as a mediator with other teaching staffs. The vertical relationship denotes being able to build a new hierarchical relationship. For example, it is the relation between seniors and juniors. The horizontal and vertical relationships are analyzed by investigating to the number of teaching staffs and students. The author uses the "outside factor" as the number of teaching staffs and students.

3.2 Research methods

This study was made using a questionnaire. There are two reasons for doing so. First, it needed constant quality of survey responses. Secondly, survey targets were available in several different countries.. The questionnaire was composed of 25 questions concerning outside and inside factors. Author divided these factors to four group questions. These are "personality", "organization", "recruitment situation" and "reaction of students". In the "personality" asked the teaching staff personality as age, major, career (Job and Educational) and number of knowing company. In the "organization" asked as number of teaching staff and student. In the "recruitment situation" asked recruitment situation of the company and the employment situation of the student. In the "reaction of students" asked student interest and reaction of career education. Also, to asked qualitatively and quantitatively these elements.

It was administered to 224 teaching staff members at random via e-mail in maritime educational institutions in 31 countries and regions.

3.3 Result

The author was able to obtain some data over a two-month period. Data was obtained from 39 people of 19 countries. The response rate was 17%. The following is a list of regional groups that submitted answers.(regions were grouped using the informal setting of the United Nations.): Western European and Others accounted for about 54% of responses. Asia-Pacific area group about 31%.Eastern European area group about 8%. Africa area group about 8%. The author could not get a response from the Latin American and Caribbean group. Figure 4 indicates the area ratio of respondents.

Figure 4. Proportion of respondents by Area

4 ANALYSIS OF SOCIAL NETWORK

4.1 Data analysis

The data acquired during this research was analyzed using statistical analysis methods (using STATA for Windows program, 12th version). The following data methods for analyzing quantitative descriptive statistics, were used; cross tabulation, correlation analysis using the linear regression method. The significance level was that set at 5% in this analysis. All missing values in the data were excluded before analysis.

4.2 Situation of Social network

In order to know the status of the social network for teaching staff, 25 questions were asked. Number of companies as personal relationship asked from between 0 and 50 by 5 steps. The result is that personal social network connected to an average of 20.8 companies. The median value was 16 companies. Social network in data tended towards smaller network.

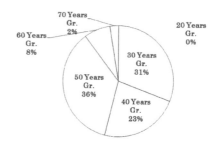

Figure 5. Age distribution of the respondents

4.3 Inside factor as personality

4.3.1 Inside factor as age

Time is given opportunities that do many things. To put it plainly, person is given experience with age. This time element was able to replace to the age of the teaching staff. The size of the social network by age was analyzing. The age value had an average value of 47.8 years old, and the median value was 45 years old. There was no group of respondents in their 20's. Figure 5 indicates the age structure of the respondents.

Table 1 shows the result of the relation between age and the social network. The p-value was 6.7% , which is bigger than a significance level at 5%. According to these findings, age is therefore not significant in regards to social network.

Table 1. Relation between Age and Social network

	Social network
Age	0.517
	(1.89)
_cons	-3.674
	(-0.28)
N	36

t statistics in parentheses
* $p<0.05$, ** $p<0.01$, *** $p<0.001$

4.3.2 Inside factor as educational career

The survey results were plotted Figure 6 by diamonds in order to analyze the relationship between social network and the educational career of the respondents.

The educational career of respondents averaged 15.7 years, with a median 14 of years. Table 2 is result of a detailed analysis of education careers. The p-value was 26%, the adjusted R-squared was 0.068. The p-value was bigger than the significance level at 5%. The significance was therefore not confirmed from education career standpoint.

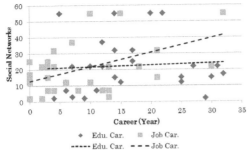

Figure 6. Result of survey at Educational career and Job career

Table 2. Relation between Educational career and Social network

	Social network
Educational Career	0.115
	(0.31)
_cons	20.51**
	(3.17)
N	31

t statistics in parentheses
* p<0.05, ** p<0.01, *** p<0.001

4.3.3 Inside factor as Job career

Survey results were plotted Figure 5 by squares. The job career of the respondents averaged 8.4 years with a median 4 years. Table 3 illustrates the detailed results of job careers and social network. The p-value of the job career was 0.0%. The adjusted R -squared of the coefficient was determined to be 0.23. Significance was confirmed from the job career stand point. The p-value was smaller than significance level at 5%. The significance has been confirmed from job career.

Table 3. Relation between Job career and Social network

	Social network
Job Career	0.898**
	(3.38)
_cons	12.72***
	(3.74)
N	36

t statistics in parentheses
* p<0.05, ** p<0.01, *** p<0.001

The survey results of total career were plotted Figure 7. Table 4 illustrates the detailed results. The p-value was 0.2%. Adjusted R-squared of the coefficient of determination was 0.261. The significance was also confirmed from the total career.

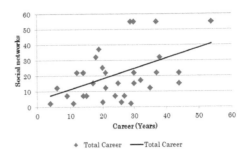

Figure 7. Relation of total career and personal social network

Table 4. Relation between Total career and Social network

	Social network
Total Career	0.792**
	(3.41)
_cons	3.381
	(0.55)
N	31

t statistics in parentheses
* p<0.05, ** p<0.01, *** p<0.001

4.4 Outside factor as the size of the educational institution

4.4.1 Outside factor as number of the teaching staff

The survey results were plotted Figure 8 by diamonds in order to analyze the number of teaching staffs. The number of teaching staff averaged 54.7 persons, with a median 49 of persons. Table 5 is result of detailed analysis of the number of teaching staff. The adjusted R-squared were -0.043. The significance was therefore not confirmed from number of teaching staff.

Table 5. Relation between number of teaching staff and Social network

	Social network
Number of teaching staff	0.00601
	(0.05)
_cons	24.09**
	(3.51)
N	25

t statistics in parentheses
* p<0.05, ** p<0.01, *** p<0.001

4.4.2 Outside factor as the number of students

The survey results were plotted Figure 8 by squares in order to analyze the number of students. The number of student averaged 196 persons, with a median 100 persons. Table 6 is result of detailed analysis of the number of student. The Adjusted R-squared of the coefficient of determination was 0.125. The p-value was smaller than significance level at 5%. The significance has been confirmed from the number of students

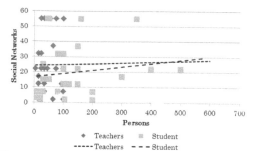

Figure 8. Relation between number of teaching staffs and students

Table 6. Result of relation between number of students and Social network

	Social network
Number of students	0.0293*
	(2.39)
_cons	15.88***
	(4.68)
N	34

t statistics in parentheses
* p<0.05, ** p<0.01, *** p<0.001

4.5 *Consideration*

For present, these may be useful to look more closely at some of the more important features of about the critical factor that has affect the social network. Factors of social network were some groups these are inside and outside factors. The significance could be confirmed some factors. These are the job career and the total career of teaching staffs in inside factor, and the number of students in outside factor. In order to calculate the Pearson's product-moment correlation coefficient analyzing these factors.

Table 7 is result of the correlation coefficient as job career, number of students and social network (job career group). The Table 8 is result of correlation coefficient as total career, number of students and social network (total career group).

The job career group confirmed that the job career has the correlation coefficient at 0.5011, and number of student has the correlation coefficient at 0.3896. The total career group confirmed that the total career has correlation coefficient at 0.5347, and number of student has correlation coefficient at 0.3896. The job career group and the total career group have the significance level at 5%, and these have the significant to factors.

This result will make clear the social network of between teaching staffs and companies. It confirmed the factors of the job career of teaching staffs and the number of students. The job career builds the social network. It is a strong effect by the career in the company. It seems reasonable to suppose that the company expand the personal network by the job

and the role with the external. The number of students also builds the social network. This effect is weaker than job career. The teaching staff can make the relationship with students, recruiting companies and the graduates of different generations, if there is many students in the school. However, it is a debatable point. These details are required the panel survey to wide and long-term.

Table 7. Result of correlation coefficient of Job career, number of student and Social network

	Social net.	Job career	No. of students
Social network	1.0000		
Job Career	0.5011*	1.0000	
Number of students	0.3896*	0.1906*	1.0000

* p<0.05

Table 8. Result of correlation coefficient of Total career, number of student and Social network

	Social net.	Total Car.	No. of students
Social network	1.0000		
Total Career	0.5347*	1.0000	
Number of students	0.3896*	0.0047	1.0000

* p<0.05

5 CONCLUSIONS

The authors studied to quantify outline of a social network of maritime education institution.

It was observed in the preceding chapter that the personal social network of the teaching staff in "Maritime educational institute" are affected by the job career and the number of students. As author said at the outset, Granovetter and Burt discussed the question of the relationship between social network and employment. However, changing of the capacity of number of student is difficult. It is important the social performance of the teaching staff for that keep or increase the social network under the condition.

In maritime educational institute, it is increasing the students who do not become seafarer officer. It is known by survey results of BIMCO. The supply human resources need the new social network which was connects to the various quarters that is not only to the seafarer. These connections will be the infrastructure of generate a new identity of the maritime education institutions.

REFERENCES

BIMCO & ISF (2000) BIMCO/ISF Manpower Update: SUMMARY, BIMCO
BIMCO & ISF (2005) BIMCO/ISF Manpower Update: SUMMARY, BIMCO
BIMCO & ISF (2010) BIMCO/ISF Manpower Update: MainReport, BIMCO

Burt, Ronald S.(1992) Structural Holes: The Social Structure of Competition., Harvard University Press

Burt, Ronald S. (2010) Foundations of Social Theory Cambridge: Belknap Press of Harvard University Press

Granovetter, Mark S. (1973) the Strensth of Weak Ties. American Journal of Sociology 78(6):1360,Grootaerrt, C., ed.

N Lin & RS Burt (1988) Social Capital in the Creation of Human Capital. The American Journal of Sociology 94:95-120

Putnam,Robert (2000) Bowling Alone : the Collapseand Revivalof America Community, NewYork: Simon&Schuster.

Ronald B. Adler & Russell F.(2006) II Proctor, Looking Out Looking In. 12th, Wadsworth Pub Co

Human Resources and Crew Manning
STCW, Maritime Education and Training (MET), Human Resources and Crew Manning, Maritime Policy,
Logistics and Economic Matters – Marine Navigation and Safety of Sea Transportation – Weintrit & Neumann (Eds)

A Code of Conduct for Shipmasters. Intermediate Report on a Research Project in Progress

T. Pawlik & W. Wittig
Bremen University of Applied Sciences, Centre of Maritime Studies, Bremen, Germany

ABSTRACT: In the aftermath of the recent "Costa Concordia" disaster several media reports raised the question whether Shipmasters are bound by some kind of a "Code of Honour" or "Code of Conduct". In other industries and professions such codes have been in existence for a long time. Based on a pre-study, selected experts were asked to express their views on the idea of a Code of Conduct for Shipmasters within the framework of a Delphi study. This paper outlines further steps in the on-going process to develop a Code of Conduct for Shipmasters.

1 INTRODUCTION

In order to stimulate the discussion about the necessity of a Code of Conduct for Shipmasters, but also to collect the professional views of shipmasters on the subject matter, in April 2012 an online-survey was initiated by the Centre of Maritime Studies at Bremen University of Applied Sciences, together with IFSMA, the International Federation of Shipmasters' Associations.

As a result, 87 % of the respondents regarded a Code of Conduct for Shipmasters as important or even very important. 76 % expressed the opinion that a Code of Conduct can be trained in MET. In particular, the following benefits were expected to derive from a Code of Conduct:
- Minimum standard of expected behaviour
- Additional guidelines to standards such as STCW
- Professional behaviour in regard to safety and environmental protection, especially in cases of emergency
- Improved public perception and reputation of shipmasters ("Respect")
- Role model for good leadership on board ship
- Cornerstone for a professional culture
- Priority of professional duties (esp. in potential conflicts with company requirements)
- Protection against criminalisation.

In total there were 279 proposals for specific issues to be dealt with in a Code of Conduct. The authors grouped these responses as follows:
- Ethical Behaviour
- Leadership
- Personality
- Cross-Cultural Competency
- Public Image
- Respect
- Skills & Knowledge
- Legal and regulatory obligations

In order to further progress the development of a Code of Conduct for Shipmasters, it was decided to consult more experts by means of a Delphi study. The concept of a Delphi study will be introduced in section 3 of this paper after an outline of general functions and selected examples of Code of Conducts in section 2. Section 4 exemplifies further steps of the on-going research project.

2 PROFESSIONAL CODES OF CONDUCT: FUNCTIONS AND SELECTED EXAMPLES [1]

Professional codes of conduct or codes of ethics have been in existence for a long time. Most probably the ancient Hippocratic Oath can be regarded as the core root for all modern professional codes of conduct. The Hippocratic Oath still requires physicians "to prescribe only beneficial treatments, according to his abilities and judgment; to refrain from causing harm or hurt; and to live an exemplary personal and professional life." [2]

Today, especially professions with a close interaction to their respective customers or clients tend to apply codes of conduct, sometimes only on the corporate level but in many cases also or exclusively on the level of their professional associations. For example, the American "National Association of the Deaf (NAD) and the Registry of Interpreters for the Deaf, Inc. (RID)" delivers to its members a very detailed code of conduct in line with the following definition: "A code of professional conduct is a necessary component to any profession to maintain standards for the individuals within that profession to adhere." [3] Table 1 delivers an overview on selected examples for Codes of Conduct.

Table 1. Examples for professional codes of conduct.

Title	Branch	Content	Date of Introduction
WMA Declaration of Geneva (based on Hippocratic Oath)	Medical Science	"[…] to declare their commitment to assume the responsibilities and obligations of the medical profession."[4]	1948, last revision in 2006
Aviators Model Code of Conduct	Aviation	"Innovative tools advancing aviation safety and offering a vision of excellence for aviators."[5]	2003, frequently updated
Ethical Principles of Pscho-logists and Code of Conduct	Psycho-logists	Aims to protect psychologists' clients and educates the members and young academics related to ethical standards [6]	2003, occasionally amended
WMA Declaration of Helsinki	Physi-cians and other partici-pants in medical research	"[…] ethical principles for medical research involving human subjects including research on identifiable human material and data."[7]	1964, last revision in 2008
Rule of St. Benedict	Bene-dictine monks	Organizes the life of the monastic community [8]	Probably from the early 6th century
People in Aid Code of Good Practice	Humani-tarian aid and de-velop-ment agencies	"[…]to improve agencies' support and management of their staff and volunteers." [9] "[…]to improve standards, accountability and transparency amid the challenges of disaster, conflict and poverty." [10]	1997, revision in 2003

Examples for professional codes of conduct can also be found in maritime industry related professions. The Federation of National Associations of Ship Brokers and Agents (FONASBA) is one case in point: FONASBA has issued a general code of conduct which has been adopted by a number of its member associations. In chapter 3 of that general code, the following rules of the professional code of conduct are stipulated:
"Members will:
1 *ensure that all activities are carried out honestly within the highest standards of professional integrity,*
2 *by proper management control, create and maintain a high standard of confidence that all duties will be performed in a conscientious and diligent manner,*
3 *observe all national and international laws and any local regulations appertaining to the shipping industry,*
4 *operate from a permanent address with all the necessary facilities and equipment to conduct business in an efficient and timely manner,*
5 *take great care to avoid any misrepresentation and ensure that all activities are subject to the principles of honesty and fair dealing,*
6 *ensure that for all dealings, the necessary authority is held from the proper party and that no action will be taken which knowingly exceeds that authority,*
7 *ensure that brokers, acting for an owner, shall only offer firm a vessel for any one cargo at any one time,*
8 *ensure that charterers' brokers will only make firm bids of a cargo or cargoes to one vessel or one shipowners' broker at any one time,*
9 *ensure that a vessel or cargo will not, in any circumstance, be quoted unless duly authorised by a principal,*
10 *ensure that all business enquiries are bona fide by making all reasonable enquiries before placing them on the market."*[11]

A recent and prominent example for the introduction of a new professional code of conduct is the MBA Oath which should give orientation and guidance to MBA (Master of Business Administration) graduates and has been sworn by almost 7,000 MBAs since its introduction in 2009 [12]. The MBA Oath is interpreted as "part of a larger effort to turn management from a trade into a profession" [13]. Indeed, it has been argued that codes of conduct are constitutive elements of a profession. [14] "A profession's code of ethics is perhaps its most visible and explicit enunciation of its professional norms. A code embodies the collective conscience of a profession and is testimony to the group's recognition of its moral dimension." [15] Frankel has categorized various functions of professional codes of conduct, out of which the following four were used in our survey: 1) Guidance: a code of conduct should be like a compass and provide direction for correct professional behaviour; 2) relation to the public: a code should be a basis for the public's expectations and evaluation of professional performance; 3) professional socialization: a code should help to foster pride in the profession and strengthen professional identity and allegiance and 4), reputation: a code of conduct should help to gain the public's trust and enhance its status. From a conceptual point of view three types of professional codes of conduct can be distinguished: 1) Aspirational codes (what are the ideals to strive for?), 2) educational codes, which comment and interpret in depth the norms stipulated in the codes and 3) regulatory codes with detailed governing rules.[16]

3 DELPHI STUDY ON CODE OF CONDUCT FOR SHIPMASTERS

"The Delphi method is an iterative process used to collect and distil the judgments of experts using a series of questionnaires interspersed with feedback. The questionnaires are designed to focus on problems, opportunities, solutions, or forecasts. Each subsequent questionnaire is developed based on the results of the previous questionnaire. The process stops when the research question is answered: for example, when consensus is reached, theoretical saturation is achieved, or when sufficient information has been exchanged." [17]

Figure 1 outlines the typical process of a Delphi study and table 2 summarizes the main advantages but also the disadvantages of this research technique.

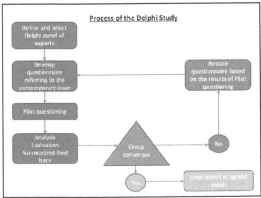

Figure 1: Process of the Delphi Study [18]

Table 2: Advantages and Disadvantages of Delphi Studies

Advantages	Disadvantages
democratic procedure [19]	time-consuming, especially for the organizers [20]
more experts can participate instead of debating personally	high commitment of expert panel is required
preserving anonymity to avoid mutual influence [21]	high operating costs [22]
to solve dissension between project collaborators [23]	tendency to eliminate extreme positions [24]
dread of unconventional thoughts will be reduced due to anonymity [25]	median score represents majority opinion only [26]
prejudices decrease [27]	decision-making process can be influenced by the organizer [28]
flexible as Delphi technique is individually adaptable [29]	traveling costs may occur
useful if issue is manifold [30]	
different experts = different points of view [31]	

Meanwhile more and more Delphi studies are organised via the internet, especially in cross-border research. This approach was also chosen for the purposes of this research as the advantages outweigh the disadvantages: Anonymity is kept [32], processing is fast which supports higher rates of commitment and enthusiasm, the results are already digitalized [33] and the online approach avoids travel and paper costs.

In January 2013 the first round of the Delphi study on a Code of Conduct for Shipmasters was started. It was decided to address high-level maritime decision-makers who have a master mariners background as relevant experts for the given research topic. In order to safeguard a fair geographical distribution of the survey four experts from each continent were chosen (see table 3). At the time of writing the first Delphi round is in progress. In order to find the experts' consensus it is intended to conduct three Delphi rounds.

Table 3: Geographical distribution of the expert panel

Africa	Asia	Europe	North-America	South-America	Australia/Oceania
Ghana	India	France	Canada	Argentina	Australia
Madagascar	Iran	Ireland	Jamaica	Brazil	Indonesia
Nigeria	Japan	Sweden	Mexico	Chile	New Zealand
South Africa	Pakistan	Turkey	USA	Ecuador	Philippines

The experts were asked to contribute their professional views on the following topics:
- The importance of a professional code of conduct for shipmasters.
- Functions of a professional code of conduct for shipmasters.
- Prioritisation of the crucial elements to be incorporated in a professional code of conduct for

shipmasters as proposed by participants of the pre-study.
- Proposals for passing on the contents and spirit of a professional code of conduct for shipmasters to students of maritime studies.

4 FURTHER STEPS OF THE RESEARCH PROJECT

The results of the Delphi study on a Code of Conduct for Shipmasters will be presented and discussed during TRANSNAV 13 in Gdynia.

The next step will be to set up a first draft for the intended code which will then be further elaborated with the expert panel of the Delphi study.

Based on this the final draft will be developed. This draft will be presented to IFSMA AGA 2014, since IFSMA has been chosen as the umbrella organisation due to the fact that the majority of the respondents of the pre-study were of the opinion that an international professional organisation would be the right body to further propose and safeguard the introduction of a Code of Conduct for Shipmasters. As a Code of Conduct for Shipmasters has to be integrated into MET in a globally harmonized manner, a toolbox for MET institutions will be developed and piloted.

ACKNOWLEDGMENTS:

The authors wish to thank the members of the expert panel for their valuable input as well as Mrs. Roberta Howlett from the IFSMA Secretariat for proof reading the paper. Thank you also to Annette Grimm, Franziska Schmeichel and Nicole Valsamakis for their research support.

REFERENCES

[1] Parts of this chapter are taken from Pawlik/Wittig: A Code of Conduct for Shipmasters, IAMU-AGA 13 proceedings, St. Johns, 2012
[2] Britannica Online Encyclopedia, "Hippocratic oath (ethical code)"
[3] Registry of Interpreters for the Deaf, Inc. (RID), "Code of Professional Conduct", accessed April 25, 2012, http://www.rid.org/ethics/code/index.cfm
[4] http://www.ncbi.nlm.nih.gov/pmc/articles/PMC1121898, assessed January 7, 2013
[5] http://www.secureav.com, assessed January 7, 2013
[6] http://www.apa.org/ethics/code/index.aspx, assessed January 7, 2013
[7] http://www.wma.net/en/30publications/10policies/b3, assessed January 7,2013
[8] http://www.middle-ages.org.uk/benedictine-rule.htm, assessed January 7, 2013
[9] http://www.peopleinaid.org/pool/files/code/code-en.pdf, assessed January 7, 2013
[10] http://www.peopleinaid.org/code, assessed January 7, 2013
[11] Norsk Skipsmeglerforbund: Code of Conduct of The Federation of National Associations of Ship Brokers and Agents (FONASBA), Chapter 3, accessed March 24, 2012, http://shipbroker.no/tjenester/fonasba-quality-standard/kriterier/code-of-conduct/
[12] The MBA Oath, "Signers", accessed April 25, 2012, http://mbaoath.org/list-of-oath-signers/
[13] The Economist (4 June 2009): A Hippocratic Oath for managers forswearing greed. MBA students lead a campaign to turn management into a formal profession.
[14] Carr, David: Education, Profession and Culture: Some Conceptual Questions, British Journal of Educational Studies , Vol. 48, No. 3 (Sep., 2000), pp. 248-268
[15] Frankel, Mark S.: Professional Codes: Why, How, and with What Impact? Journal of Business Ethics 8 (1989), p. 110
[16] Frankel, Mark S.: Professional Codes: Why, How, and with What Impact? Journal of Business Ethics 8 (1989), p. 110
[17] http://www.fepto.eu/storage/files/articole/Delphi%20method%20for%20Graduiate%20research.pdf, assessed January 7, 2013
[18] Adapted from http://phd-thesis.wikispaces.com/file/view/Delphi_process.JPG
[19] http://www.fepto.eu/storage/files/articole/Delphi%20method%20for%20Graduiate%20research.pdf, assessed January 2, 2013
[20] „Projektmanagement", F. Bea/S. Scheurer/S. Hasselmann, Lucius & Lucius Verlagsgesellschaft, 2008, S.150, ISBN 978-3-8252-2388-5
[21] „Projektmanagement", F. Bea/S. Scheurer/S. Hasselmann, Lucius & Lucius Verlagsgesellschaft, 2008, S.150, ISBN 978-3-8252-2388-5
[22] http://www.dse.vic.gov.au/effective-engagement/toolkit/tool-delphi-study, assessed January 2, 2013
[23] „Organisation", Klaus Olfert, 16.Auflage, NWB Verlag, 2012, ISBN 978-3-470-51376-8, S.382
[24] http://pareonline.net/pdf/v12n4.pdf, assessed January 5, 2013
[25] „Projektmanagement", Manfred Burghardt, Siemens, 4. Auflage, Publicis MCD Verlag, 1997, S.545 ISBN: 3-89578-069-3
[26] „Project Management – The Managerial Process 4e",C.F. Gray, E.W. Larson, 2008 ,S. 123, ISBN 978-0-07-128751-7
[27] http://www.sofind.de/glossar_pm/l_370_Delphi-Methode.html, January 7, 2013
[28] http://www.robertsevaluation.com.au/index.php?option=com_content&task=view&id=49, assessed January 2, 2013
[29] http://www.dse.vic.gov.au/effective-engagement/toolkit/tool-delphi-study, January 2, 2013
[30] http://www.dse.vic.gov.au/effective-engagement/toolkit/tool-delphi-study, January 2, 2013
[31] http://www.robertsevaluation.com.au/index.php?option=com_content&task=view&id=49, assessed January 2, 2013
[32] Bea, F., Scheurer,S., Hasselmann: Projektmanagement, 2008, p.150
[33] http://www.fepto.eu/storage/files/articole/Delphi%20method%20for%20Graduiate%20research.pdf, assessed January 2, 2013

Human Resources and Crew Manning
STCW, Maritime Education and Training (MET), Human Resources and Crew Manning, Maritime Policy,
Logistics and Economic Matters – Marine Navigation and Safety of Sea Transportation – Weintrit & Neumann (Eds)

Human Factors and Safety Culture in Maritime Safety

H.P. Berg
Bundesamt für Strahlenschutz, Salzgitter, Germany

ABSTRACT: As in every industry at risk, the human and organizational factors constitute the main stakes for maritime safety. Furthermore, several events at sea have been used to develop appropriate risk models. The investigation on maritime accidents is, nowadays, a very important tool to identify the problems related to human factor and can support accident prevention and the improvement of maritime safety. Operation of ships is full of regulations, instructions and guidelines also addressing human factors and safety culture to enhance safety. However, even though the roots of a safety culture have been established, there are still serious barriers to the breakthrough of the safety management. One of the most common deficiencies in the case of maritime transport is the respective monitoring and documentation usually lacking of adequacy and excellence. Nonetheless, the maritime area can be exemplified from other industries where activities are ongoing to foster and enhance safety culture.

1 INTRODUCTION

The strengthening of safety culture in an organization has become an increasingly important issue for all high risk industries. A high level of safety performance is essential for business success in intensely competitive global environment. The most important objective is to protect individuals, society and the environment by establishing and maintaining an effective protection against the respective hazards. This is achieved through the use of reliable structures, systems and components as well as adequate clear procedures, and acting people which are committed to a strong safety culture.

The term 'safety culture' first appeared in the International Atomic Energy Agency's initial report following the Chernobyl disaster. In the early investigation of this accident, the initial emphasis was focussed on plant deficiencies. However, more thorough analyses also identified organizational, cultural, and managerial issues and showed a lack of an adequate safety culture.

In the nuclear industry, international organizations such as the International Atomic Energy Agency (IAEA) recognized the important role that all regulators should play in monitoring safety performance in the nuclear industry. Following the Chernobyl accident, the IAEA published two guides on safety culture (IAEA 1991 and 2002).

More recently, the IAEA developed the following definition for safety culture (IAEA 2006): "The assembly of characteristics and attitudes in organizations and individuals which establishes that, as an overriding priority, protection and safety issues receive the attention warranted by their significance."

In the meantime, inquiries into many major accidents such as the King's Cross fire, Piper Alpha and the Herald of Free Enterprise and more recently the accidents in the Mexican Gulf (Deepwater Horizon blow-out) in 2010 have found faults in the organisational structures, safety management systems and the prevailing cultures, throwing the importance of safety culture into the spotlight.

The most recent event in the nuclear industry was in March 2011 when a magnitude 9.0 earthquake off the coast of Japan and the resulting tsunami caused extensive damage at the Fukushima Daiichi nuclear power plant. A full understanding has yet to emerge of what Japanese authorities and the nuclear industry have learned about safety culture implications at the Daiichi plant, as facility and equipment damage from the earthquake and resulting tsunami has been the initial focus of event studies. As with all major events, getting to the underlying safety culture issues requires more time and further analysis before all root causes can be identified, but a recent report by the Japanese Government strongly points to safety culture issues (NAIIC 2012).

Although there is a wealth of information, articles and reports relating to safety culture, there is still no universal definition or model. Safety culture has been defined in a variety of ways including (Berg 2011):
- The ideas and beliefs that all members of the organisation share about risk, accidents, and ill-health,
- A set of attitudes, beliefs or norms,
- A constructed system of meaning through which the hazards of the world are understood,
- A safety ethic.

The assessment of safety culture is key to identifying a companies current level of safety culture (known as its maturity or development level) in order to identify how to learn and improve. There are a number of different assessment methods including:
- Safety attitude surveys (using questionnaire to elicit workforce attitudes),
- Safety management audits (using an audit process and trained auditor to examine the presence and effectiveness of safety management systems)
- Safety culture workshops (involving a cross-section of the workforce to consider perceptions of the safety culture and elicit improvement ideas)
- Safety performance indicators (analysing data on indicators such as the number of safety tours performed or near miss data).

A safety culture assessment allows an organisation to better understand how its people perceive safety and the company's approach to safety management. It allows the organisation to identify both strengths and weaknesses that then enable it to continuously monitor and improve its approach to safety.

As in every industry at risk, the human and organizational factors constitute the main stakes for maritime safety. Furthermore, several events at sea have been used to develop appropriate risk models. Operation of ships is full of regulations, instructions and guidelines also addressing human factors and safety culture to enhance safety. However, management ashore and on board need not only to ensure that the formal skills are in place but also ensure, encourage and inspire the necessary attitudes to achieve the safety objectives.

2 HUMAN FACTORS IN MARITIME SAFETY

2.1 Role of human factors

Regulations and systems have not achieved the desired effects in averting marine accidents which are a result of human errors and account for 80% of those occurring worldwide.

Studies have shown (Rothblum 2000) that human error contributes to:
- 84-88% of tanker accidents,
- 79% of towing vessel groundings,
- 89-96% of collisions,
- 75% of fires and explosions.

These estimations are still valid. Thus, the maritime transport system is 25 times riskier than the air transport system according to the accounts for deaths for every 100 km. Intensification of sea trade for last ten years causes the increasing of potential risk to the ship safety.

The implementation of the International Safety Management Code (IMO 2008) has played a significant role in addressing this issue through training and education of crew members but to some extent casualties can be prevented by eliminating other indirect causes including hardware, such as equipment systems.

It must be noted that if the possible cause of an accident is human error, finding and eliminating the root cause of such errors is vital for preventing recurrence - whether it is related to human element, hardware factors, organizations and management factors.

However investigation in human factors, main cause of such accidents, is increased nowadays and the methodologies to carry out such an investigation are being developed by several institutions. These methodologies, adopted from the investigation on risk analysis are frequently based on the estimation of risk levels, whose values, in the case of human factor investigation are not always clear.

In any case, a comprehensive risk assessment consists of:
1 Identifying the hazard in the system
2 Evaluating the frequency of each type of accident
3 Estimating accident consequences
4 Calculating various measures of risk, such as death or injuries in the system per year, individual risks or frequency of accidents of a particular kind.

For improvements in operability and working environment it is necessary to ensure that the operability is not poor or inconvenient or is encountering obstacles during operations. Since it heightens the risk of an accident, it is important to pay attention to the arrangement and layout of equipment. Hence it is important that operators work in congenial and safe surroundings.

It is clear that total safety over ships operation can not be achieved, but it is possible to obtain a high degree on it. Research on the influence of human factors over maritime accidents is, also, very difficult. On the one hand we find that an accident involves the interaction of individuals, equipment and environment, as well as unforeseen factors

(Caridis, 1999), and on the other hand, human factors comprise operative human errors –derived from personnel own qualifications, or from their physical, mental and personal conditions- and situational errors– derived from work environment design, management problems, or human-machine interface, amongst others

Being aware that risk is an inherent factor of maritime activity which can not be totally removed and that errors are part of human experience, it is expected that elements such as good management policies, effective training and having suitable qualifications and experience, can reduce the occurrence of human errors.

The practical application of this kind of analysis seems clear: obtaining the cause parameters, both direct and indirect parameters, from the studied factor, one can better understand the root of the presence of such a factor, and one can take specific and direct corrective actions to try to minimize the accident risk. The main weakness of this method lies in the lack or shortage of data related to accidents and incidents on maritime domains.

Even though the roots of a safety culture have been established, there are still serious barriers to the breakthrough of the safety management.

The poor reporting practises cause further problems. The information about the non-conformities, accidents and hazardous occurrences does not cumulate at any level of the maritime industry. The personnel of the other ships cannot learn from the experiences of the other vessels. There are no possibilities to interchange information about incidents between the vessels. The company cannot utilize the cumulative information when improving its safety performance. Companies do not have the opportunity to learn from other companies' mistakes. The national maritime administrations are powerless in their attempts to develop the maritime safety.

The fundamental philosophy of the IMS Code (IMO 2008) is the philosophy of continuous improvement. The procedures for reporting the incidents and performing the corrective actions are the essential features of the continuous improvement. If this information is not provided the successful cycle of continuous improvement cannot function.

Operation of ships is full of regulations, instructions and guidelines which officers and crew are expected to know and adhere to. A culture of safety may perhaps be achieved through written instructions, but in the end it is a question of a common mind-set throughout the organisation. Management ashore and on board need not only ensure that the formal skills are in place but also ensure, encourage and inspire the necessary attitudes to achieve the safety objectives. Statistics prove beyond doubt that investing in a good safety culture provides results and pays off in the long term.

The effort of allocating various forms of human error as verified accident causes is surely not a trivial task. Moreover, this difficulty is augmented in the case of maritime transport, since the respective monitoring and documentation is usually lacking of adequacy and excellence. Nonetheless, marine industry can be exemplified from other sectors of industry (e.g. civil aviation, nuclear plants), where considerable load of attention is already given in pinpointing and revealing various involved aspects of human element extracted from comprehensive databases of safety relevant events.

Human behaviour and performance can be the prevailing factors that prescribe the level of safety for numerous maritime transport procedures and practices of management. This means that they can also influence, in a considerable degree, the protection of marine and coastal environment. Thus, a feasible way to reduce the frequency and severity of naval accidents is, by identifying the contributing factors to the so-called human error, and by investigating for methods, which will either eliminate or mitigate these mistakes.

Over the last 40 years or so, the shipping industry has focused on improving ship structure and the reliability of ship systems in order to reduce casualties and increase efficiency and productivity. We've seen improvements in hull design, stability systems, propulsion systems, and navigational equipment. Today's ship systems are technologically advanced and highly reliable.

Yet, the maritime casualty rate is still high. Why, with all these improvements, was it not possible to significantly reduce the risk of accidents? It is because ship structure and system reliability are a relatively small part of the safety equation. The maritime system is a people system, and human errors figure prominently in casualty situations.

Human reliability also influences the overall system reliability in automatic systems. This influence can both be negative (e.g. human working error) or positive (e.g. controlling system breakdowns or system problems).Human performance could be defined as the human being's execution of an action with the purpose of accomplishing a given task.

2.2 Example of an accident

The example provided in the following (Gard 2012) deals with the accident grounding. A vessel is under way on an ocean crossing with course set out from start to end. The course is set out and the voyage planned on a small scale planning chart. The course is set to pass some small groups of mid-ocean islands and the CPA (Closest Point of Approach) is considered and thought to be well on the safe side.

On a nice tropical night with calm seas and good visibility, the vessel makes its approach to pass one group of islands well on the port side some time after midnight.

The chief officer observes during the last two hours of his 1600-2000 hrs watch that a slight breeze and current are working together to set the vessel slightly off course and towards the islands ahead. He therefore makes a correction to the course to compensate for the drift and setting to keep the vessel on its intended course. When handing over the watch at 2000 hrs, the chief officer makes the second officer aware of this.

The second officer continues to plot the positions throughout his watch and observes that the vessel is still drifting somewhat off course to the effect of making the CPA to the islands ahead less safe than planned. He therefore makes some minor course adjustments to compensate for drift and setting. At midnight the watch is handed over to the first officer, who is also made aware of the drift and the course adjustments. At 0040 hrs the vessel runs aground at full speed on the beach of a small low atoll. The beach is mainly sand and pebbles and slopes at a low angle into the sea so the vessel suffers minor damage but can not be re-floated with its own power. A costly salvage operation follows.

The human aspect of this accident is discussed in the following. The positions were plotted in the same small scale planning chart covering the entire ocean where the voyage was planned and the course set out. In a small scale chart it is difficult to accurately measure small distances and observe small deviations from the course between hourly plots. The reason for using a small scale chart was probably that it was not considered necessary to conduct "millimetre" navigation when crossing the ocean. The island on which the vessel grounded was marked on the chart in use, but only as a small dot and the course was set to pass at what seemed to be a safe distance.

The drift and current, however, worked together to set the vessel off course towards the island and it is painfully obvious that the corrective actions taken by the navigation officers were not adequate.

It can be concluded that the grounding would not have happened if:
- a large scale chart had been used for position plotting since it would then have become apparent that the course was heading gradually towards the island, and/or
- a much wider passing had been planned in the first place, and/or
- a considerable safety margin had been applied when the corrections were made to compensate for drift and setting.

It is also possible that proper look-out and use of radar could have been an issue. On the other hand, the island was very low and it is arguable that it could not have been spotted visually in time in the dark tropical night. It is unclear whether and why the island was not seen on the radar, but it is a known fact that radars are subject to a lot of interference in tropical waters and it could be that both the rain and sea clutter settings had been adjusted to deal with that, thus at the same time removing or diminishing the radar image of the island.

3 MARITIME SAFETY CULTURE

3.1 *General aspects*

Safety culture can be viewed from many angles (Berg 2011). Typically, the environment close to safety managers of the organizations provides most of the research material, and consequently the middle management view dominates. Similarly, employee perspective is strong in internal material of the organizations, typically work instructions and safety management documentation. From the top management viewpoint, lesser amount of practical information is available.

However, in shipping, and especially on board ships the organization is hierarchic, due to tradition and the need for clarity in emergency operations. Therefore, safety considerations depend strongly on the actions of the masters and the officers of the ships, and the interactions of the land-based organization (Räisänen 2009).

One typical feature of shipping is that ships are manned with crews of multiple nationalities, and the much of it is carried out in international setting, outside national legislations. These issues complicate the communication and interactions within the ships, between them, and with the land-based stakeholders.

Moreover one has toemphasize the effects of national culture, which is less prominent in related safety discussions of other fields (Håvold 2005, Håvold 2007).

The prevailing goals, principles and procedures in an organisation, which can safeguard against errors and when errors are encountered through which it is possible to react with subsequent changes in practises before serious incident or accident occurs. Accident investigation is part of the maritime safety culture - a reactive one - but an excellent observer point.

In the Baltic Sea the maritime traffic is rapidly growing which leads to a growing risk of maritime accidents. Particularly in the Gulf of Finland, the high volume of traffic causes a high risk of maritime accidents. The growing risks give us good reasons for implementing the research project concerning maritime safety and the effectiveness of the safety measures, such as the safety management systems.

In order to reduce maritime safety risks, the safety management systems should be further developed.

The purpose of the METKU Project (Development of Maritime Safety Culture) which started 2008 was to study how the ISM Code has influenced the safety culture in the maritime industry, to evaluate the development of safety culture in maritime industry and to examine the weaknesses found in the safety management systems of shipping companies. The main results found were that maritime safety culture has developed in the right direction after the launch of the ISM Code in the 1990´s (Heijari & Tapainen 2010).

In this study it has been discovered that safety culture has emerged and it is developing in the maritime industry. Even though the roots of the safety culture have been established there are still serious barriers to the breakthrough of the safety management. These barriers could be envisaged as cultural factors preventing the safety process. Even though the ISM Code has been effective over a decade, the old-established behaviour which is based on the old day's maritime culture still occurs, e.g., there are still serious barriers to the breakthrough of the safety management. These barriers could be envisaged as cultural factors preventing the safety process.

However, experience has shown that there are perceived gaps between the desirable leadership qualities, and what is currently being delivered. These primarily concern:
– Clear two-way communication,
– "Tough empathy",
– Openness to criticism,
– Empathy towards different cultures,
– Ability to create motivation and a sense of community,
– Knowing the crew's limitations,
– Being a team player.

Moreover, there are other important explicit barriers to effective safety leadership that relate to the current structure of the industry, standards, practices and economic pressures. These barriers would need to be addressed irrespective of the personal qualities and skills of the Master.

Operation of ships is full of regulations, instructions and guidelines which officers and crew are expected to know and adhere to. The ISM Code has to a large extent codified what is known as good seamanship.

A culture of safety may perhaps be achieved through written instructions, but in the end it is a question of a common mind-set throughout the organisation. Management ashore and on board need not only ensure that the formal skills are in place but also ensure, encourage and inspire the necessary attitudes to achieve the safety objectives.

3.2 Examples of safety culture approaches

The extent of good safety leadership (and more broadly good safety management arrangements) appears to be highly variable across companies. Safety management arrangements are generally most highly developed in the tanker sector, and least highly developed in the dry cargo sector.

However, the research results (Little 2004) confirm that a good safety performance can be achieved with a committed leader who has the key qualities described above, without necessarily having the most sophisticated management arrangements.

For the maritime area, one safety culture approach can be illustrated by a pyramid as shown in Figure 1, accompanied by the elements "lessons learned" and "safety as a value" which are important for the entire organisation to succeed in a sound safety culture (Drouin 2010).

A further safety culture approach is illustrated in Figure 2.

Figure 1. Safety culture pyramid for the maritime area according to Drouin (2010).

The central premise of this model, discussed in more detail in (ABS 2012) is that improvements in organizational safety culture can lead to enhanced safety performance.

The first step is an assessment of the existing safety culture to identify areas of strength, weaknesses of defences, and opportunities for improvement against operational incidents, personal injuries, etc.

This model also incorporates a process for identifying an organization's potential leading indicators of safety. There are two ways of conducting this process:

1) By the identification of objective leading indicators. This is done by correlating safety culture metrics with safety performance data. This is the preferred approach because of its objectivity; because it utilizes metrics that the organization has

collected; and it does not require a survey of the workforce which can be time-consuming.

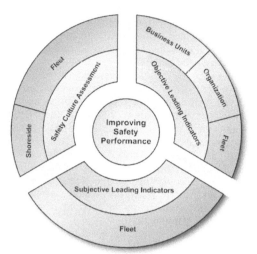

Figure 2. The ABS safety culture and leading indicators model according to ABS (2012).

This can be done at three levels:
- At the Organizational level,
- Across Business Units,
- Across the Fleet.

2) By the identification of subjective leading indicators from a safety culture survey. These indicators are based on the values, attitudes, and observations of employees. This method may identify potentially beneficial safety culture metrics not yet tracked by the organization. This approach may be used when the organization lacks sufficient metrics to use the objective leading indicators process. There are a number of criteria for undertaking a leading indicators programme, and for each type of assessment.

4 CONCLUDING REMARKS

In general, safety culture has been found to be important across a wide variety of organizations and industries. While initial studies of safety culture took place in jobs that have traditionally been considered high-risk, organizations in other areas are increasingly exploring how safety culture is expressed in their fields. Overwhelmingly, the evidence suggests that while safety culture may not be the only determinant of safety in organizations, it plays a substantial role in encouraging people to behave safely.

The essence of safety culture is the ability and willingness of the organization to understand safety, hazards and means of preventing them, as well as ability and willingness to act safely, prevent hazards

from actualising and promote safety. Safety culture refers to a dynamic and adaptive state. It can be viewed as a multilevel phenomenon of social processes organizational dimensions, and psychological states of the personnel.

In the nuclear field safety culture is still seen as an important task (CNSC 2012), and also the German activities are ongoing (Berg 2008, Berg 2010, Kopisch & Berg 2012).

In February 2011, i. e. one month before the Fukushima accident, a further international activity in the nuclear field on safety culture started. The general objective is to establish a common opinion on how regulatory oversight of safety culture can be developed to foster safety culture. It is intended that the output of the meeting will form the basis for a Safety Report Series document providing guidance on how regulators and licensees can deal with the safety culture components in order to continuously foster a positive safety culture. Moreover, the IAEA is also working on a document how to perform safety culture self-assessments.

In the meantime, there are many demanding aspects of seafaring such as the inability of employees to leave the worksite, extreme weather conditions, long periods away from home, and motion of the workplace. Some of these are unchangeable and are a reflection of the nature of the domain.

However, it is possible to modify, supplement, and introduce new strategies or interventions to potentially reduce the impact these factors have on the health and welfare of the individual seafarer (Parker et al. 2002).

There are many human factors influencing safety in this domain as have been presented in this review: fatigue, automation, situation awareness, communication, decision making, team work, and health and stress. These issues were reviewed within a framework that proposed that these individual factors can be contributory causes in accident causation, however the safety climate on ship will also influence whether or not an individual engages in safe behaviours or not. The review also considered the current status of attempts to address these human factors issues prevalent in the maritime industry. The review demonstrated that there are many "gaps" in the maritime literature, and a number of methodological problems with the studies undertaken to date.

Maritime transportation is characterized by a level of safety of the order of 10^{-5} per movements which is inferior to that of air transportation (10^{-6}), however it is comparable to the level of safety of rail transportation and much higher than the level of safety of road transportation. Thus, for passenger transportation in Europe, the risk of fatality is estimated at 1.1 for road transportation (for 10^8

person kilometre) and at 0.33 for ferry transportation (Mackay, 2000).

Within this context, the risk of accident and - more precisely - the place of the human factor in this risks, are central issues. The human factor, indeed, seems to be the main cause of incidents at sea (Hetherington, Flin & Mearns 2006) describing the factors that contribute to incidents and accidents: factors which cause a decrease in performance (fatigue, stress, and health problems), insufficient technical and cognitive capacities, insufficient interpersonal competencies (communication difficulties, difficulties mastering a common language), organizational aspects (safety training, team management, safety culture).

A closer look at the questions of human-machine cooperation and at the role of automation in maritime accidents is taken in (Lützhöft & Dekker 2002).

In the case of a crew or team working together, the shared mental representation is one of the elements at the heart of the performance. The methods developed in cognitive psychology to analyze this mental structure can be used to evaluate the impact of Bridge Resource Management (BRM) on the work of a crew (Chauvin 2011). A study of this type was carried out some years ago (Brun et al. 2005).

However (see Salas et al. 2006), these studies remain marginal and recent in the maritime field, even though they are numerous and have been developed for several years in the field of air transportation.

Because human error (and usually multiple errors made by multiple people) contributes to the vast majority of marine casualties it is necessary to prevent human error of paramount importance in order to reduce the number and severity of maritime accidents. Many types of human errors were described, the majority of which were shown not to be the "fault" of the human operator. Rather, most of these errors tend to occur as a result of technologies, work environments, and organizational factors which do not sufficiently consider the abilities and limitations of the people who must interact with them, thus "setting up" the human operator for failure.

This general problem is also discussed for CRG casualties (Kobylinski 2009).

Human errors can be reduced significantly. Other industries have shown that human error can be controlled through human-centered design. By keeping the human operator uppermost in our minds, we can design technologies, work environments, and organizations which support the human operator and foster improved performance and fewer accidents.

There is often a delay between the development of weaknesses in safety culture and the occurrence of an event involving a significant safety consequence. The weaknesses can interact to create a potentially unstable safety state that makes an organization vulnerable to safety incidents. Within the nuclear industry, there have been a number of high profile cases in different parts of the world that have been linked to a weakened safety culture.

By being alert to the early warning signs, corrective action can be taken in sufficient time to avoid adverse safety consequences. Both the organization (which could be a specific plant or utility) and its regulators must pay attention to signs of potential weakness.

Some organizations that have encountered difficulties with their safety culture have previously been regarded as good performers by their industry peers. Good past performance is sometimes the first stage in the process of decline.

The investigation on maritime accidents is, nowadays, a very important tool to identify the problems related to human factor that, studied with attention can be one mainstay to accident prevention and to the improvement of maritime safety.

The long-term positive trend in ship safety, with year-on-year improvements, has now been reversed (Madsen 2011). This is worrying. It's time to take a new look at the maritime industry's safety culture. A stronger focus on safety culture, safety training and competence assessment is needed.

Statistics show that the shipping industry's accident frequency rate has started to rise from a historically low level. Technology, rules and compliance will never achieve the expected level of safety unless there is a greater focus on the human element.

Historically, the safety focus in shipping has been on technical improvements. Most shipping company employees dealing with the operation of vessels have a technical background. Audits and inspections pay great attention to technical compliance. This technical focus has resulted in major improvements to ship safety. But now it is time to focus more on the soft issues. To improve maritime safety one needs to adopt a threefold approach:

1 Improved safety culture,
2 Improved training schemes,
3 A formal competence assessment programme.

The last two aspects are also addressed in (Chirea-Ungureanu & Rosenhave 2012) recommending a cross cultural training to deal with the specific situation onboard.

An organisation that decides to improve its safety culture should follow a systematic, closed-loop process. A typical enhancement process is presented in Figure 3.

The first step consists of defining what safety culture is and understanding what is meant by safety culture in the respective management organisation. This requires identifying the characteristics of safety culture to look at, and their sub-components. These first two steps are important because to measure safety culture effectively, an organisation must define and describe what it is attempting to measure.

The next (3rd) step of the process enters the assessment stage, where the organisation carries out or commissions a survey to measure its own safety culture. Surveys and other techniques contribute to the identification of strengths and weaknesses of the safety culture (4th step). On the basis of this assessment an action plan is developed (5th step).

The actions are effected to improve safety culture (6th step).

After a reasonable period (e.g. at least two years), safety culture can be assessed again iteratively to determine if the situation has improved.

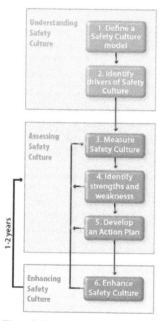

Figure 3. Safety culture enhancement process

The iteration timeframe depends on the time required to carry out the assessment, define the plan and put in place all planned actions and mature the enhancement. The iteration should not occur too quickly, as safety culture takes time to change.

Results of a recent study (Oltedal 2011) indicate several deficiencies in all parts of a traditional safety management system defined as:

– reporting and collection of experience data from the vessel;
– data processing, summarizing, and analysis;
– development of safety measures and
– implementation.

The underreporting of experience data is found to be a problem, resulting in limitations related to the data-processing process. Regarding the development of safety measures, it is found that the industry

emphasizes the development of standardized safety measures in the form of procedures and checklists. Organizational root causes related to company policies (e.g., crewing policy) is to a lesser degree identified and addressed. The most prominently identified organizational influential factors are the shipping companies crewing policy, which includes rotation systems, crew stability, and contract conditions, and shipboard management. The companies' orientation toward local management, which includes leadership training, educational, and other managerial support, are also essential. The shore part of the organization is identified as the driving force for development and change in the shipboard safety culture. Moreover, management ashore and on board need not only to ensure that the formal skills are in place but also ensure, encourage and inspire the necessary attitudes to achieve the safety objectives.

REFERENCES

American Bureau of Shipping (ABS) 2012. Guidance notes on safety culture and leading indicators of safety, January 2012.
Berg, H.P. 2008. Safety management and safety culture assessment in Germany. Proceedings of the ESREL Conference 2008, Safety Reliability and Risk Analysis: Theory, Methods and Applications, Vol. 2, Taylor & Francis Group, London, 1439 – 1446.
Berg, H.P. 2010. Risk based safety management to enhance technical safety and safety culture. Transactions ENC 2010 – European Nuclear Conference, May/June 2010.
Berg, H.P. 2011. Maritime safety culture. Proceedings of the XV International Scientific and Technical Conference on Marine Traffic Engineering, 12 – 14 October 2011, Akademia Morska, Szczecin 2011, 49 – 59.
Brun, W., Eid, J., Johnsen, B.H., Labertg, J.-C., Ekornas, B. & Kobbeltvedt, T. 2005. Bridge resource management training: enhancing shared mental models and task performance? In H. Montgomery, R. Lipshitz & B. Brehmer (Eds), How professionals make decisions, 183–193), Mahwah: LEA.
Caridis, P. 1999 CASMET. Casualty analysis methodology for maritime operations. National Technical University of Athens.
Canadian Nuclear Safety Commission 2012. Safety culture for nuclear licensees, Discussion paper DIS-12-07, August 2012.
Chirea-Ungureanu, C. & Rosenhave, P.-E. 2012. A door opener: teaching cross cultural competence to seafarers. International Journal on Marine Navigation and Safety of Sea Transportation, Vol. 6, Number 4, December 2012.
Chauvin, C. 2011. Human factors and maritime safety. The Journal of Navigation 64: 625–632.
Drouin, P. 2010. The building blocks of a safety culture. Seaways, October 2010, 4-7.
Gard 2012. Safety culture - incidents resulting from human error. Gard News 207, August/October 2012.
Håvold, J.I. 2005. Safety culture in a Norwegian shipping company, Journal of Safety Research, Vol. 36, 441-458.
Håvold, J.I. (2007) National cultures and safety orientation: A study of seafarers working for Norwegian shipping companies, Work & Stress, 21 (2):73-195.

Heijari, J., & Tapainen, U. 2010. Efficiency of the ISM Code in Finnish shipping companies, Report A 52, Centre for Maritime Studies, Turku.

Hetherington, C., Flin, R. & Mearns, K. 2006. Safety in shipping: The human element. Journal of Safety Research, 37, 401–411.

International Atomic Energy Agency 1991. Safety culture, A report by the International Nuclear Safety Advisory Group, Safety Series, No. 75-INSAG-4, IAEA, Vienna, Austria.

International Atomic Energy Agency 2002. Key practical issues in strengthening safety culture, A report by the International Nuclear Safety Advisory Group, INSAG-15, IAEA, Vienna, Austria.

International Atomic Energy Agency 2006. The management system for facilities and activities. GS-R-3 Safety Standards Series – Safety Requirements, IAEA, Vienna, Austria.

International Maritime Organization (IMO). 2008. International Safety Management (ISM) Code.

Kobylinski, L. 2009. Risk analysis and human factor in prevention of CRG casualties. International Journal on Marine Navigation and Safety of Sea Transportation, Vol. 3, Number 4, December 2009.

Kopisch, C. & Berg, H.P. 2012. The role of the regulator in the field of safety culture. Proceedings of the SSRAOC Workshop, Antwerp, Belgium, January 2012.

Little, A. 2004. Driving safety culture, identification of leadership qualities for effective safety management. Final Report to Maritime and Coastguard Agency, Part 1, Cambridge, October 2004.

Lützhöft, M. H. & Dekker, S. W. A. 2002. On your watch: automation on the bridge. Journal of Navigation, 55(1), 83–96.

Mackay, M. 2000. Safer transport in Europe: tools for decision-making. European Transport Safety Council Lecture.

Madsen, O.M. 2011. A new look at safety culture. DNV Forum 2011 No. 2.

Oltedal, H.A. 2011. Safety culture and safety management within the Norwegian-controlled shipping industry, state of art, interrelationships, and influencing factors. PhD Thesis, University of Stavanger.

Parker, A. W., Hubinger, L. M., Green, S., Sargent, L., & Boyd, R. 2002. Health stress and fatigue in shipping. Australian Maritime Safety Agency.

Räisänen, P. 2009. Influence of corporate top management to safety culture, A literature survey. Turku University of Applied Sciences, Reports 88, Turku.

Rothblum A. 2000. Human error and marine safety. Maritime Human Factors Conference, Linthicum, MD, March 13-14, 2000.

Salas, E., Wilson, K.A., Burke, C.S., & Wightman, D.C. 2006. Does crew resource management training work? an update, an extension, and some critical needs. Human Factors 48 (2), 392–412.

The National Diet of Japan 2012. The official report of the Fukushima Nuclear Accident Independent Investigation Commission - Executive Summary. Available at: http://naiic.go.jp/wp-content/uploads/2012/08/NAIIC_report_lo_res5.pdf.

Human Resources and Crew Manning
STCW, Maritime Education and Training (MET), Human Resources and Crew Manning, Maritime Policy,
Logistics and Economic Matters – Marine Navigation and Safety of Sea Transportation – Weintrit & Neumann (Eds)

The Role of Intellectual Organizations in the Formation of an Employee of the Marine Industry

N.A. Kostrikova & A.Ya. Yafasov
Baltic Fishing Fleet State Academy, Kaliningrad, Russia

ABSTRACT: The role of intellectual organizations in the formation of professional competence, social adaptability and volitional qualities of workers of maritime industry is actual in terms of growth of natural and man-made emergency situations, unstable socio-economic environment. In the marine industry, except for the education system and research, such organizations have prospects of development as the most competitive in the shipbuilding, fishing industry, port management, the logistics of fishing in the ocean and coastal fisheries, transport and storage prior to further processing. In this article we propose a holistic resource approach to the formation of intellectual organization as a major maritime educational complex that integrates all stages of maritime specialist training, from a seaman to a master of a large ship including active research and intense maritime practice.

1 INTRODUCTION

With the entry of the world economy to the turbulent phase [1,2], a specific role of improved sustainability and the competitiveness of the Russian economy have been acquired, as well as ensuring economic and food security, the country's attractiveness for investment, for residence and intellectual creativity. New realities require the organization of the modern system of general and continuing professional education that is in a dynamic research and innovation environment, integrated into the world educational area, but kept the best traditions and experience of domestic education.

According to Academician Engelhardt, "The life is a unity of three processes: substance, energy and information." Any new information is the product of intellectual work of human activity, reflects the ideology or environmental phenomena. Labor, human activity associated with energy supply and nutrition. Thus the fundamental problems of humanity could be divided into energy, food, world outlook and associated with the habitat. Development of Russia in the paradigm of solving these problems can be provided in a following ways:
- consecutive transition to the production of the science intensive products, with minimization of impact on environment and expenses of natural resources, development of technological innovative entrepreneurship;
- maximum mobilization of the intellectual capital and people spiritual potential, ensure continuity of generations, increases of scientific potential and culture;
- organization of investment funds, tools and mechanisms of development of infrastructure and innovative manufacturing to ensure the growth of domestic investments within the next 10-15 years not less than 20% to previous year;
- creation of choice possibility of the foreign investments, answering to problems of formation of information - industrial society with kernels in the form of the intellectual organizations and intellectual clusters;
- creation of rapidly developed infrastructure - communications, telecommunications, GIS, transportation, roads, banks, hotels in the framework of the public-private partnership;
- organization of modern system of repatriation of compatriots and the thought-out migratory policy in compliance with the principles of humanity and economic reasonability;
- organization of a continuous system of training and retraining of personnel, providing constantly updated intellectual resources for the solution of post-industrial economy problems.

Even against crisis the leading countries are engaged in building-up of investments in technological updating and by that they accurately designate the main line of competitiveness in the future world competition and what it gives them in perspective.

2 INTELLECTUAL CLUSTERS

Considering the problems of education, recreation and continuous updating of intellectual resources and their role in national economy, it is necessary to pay attention to one very important point: the direction and nature of evolution of any society is defined not only by the amount of the gross domestic product (GDP) volume, but also how this gross domestic product is reached. If it is a way of intellectualization of work, it develops society and promotes growth of its competitiveness. If it is a way of traditional primitive extensive technologies, the motivation to development weakens, and the state loses its positions in the international markets, its economy degrades. The extreme case with rich natural resources is the «Dutch disease» of the economy.

Today we live in a century of the world economic, political and cultural integration and unification, as consequence of which we do have world labor division, labor migration and productive resources, standardization and unification of the legislation, economic and engineering processes, the convergence of different nation's cultures. These processes cause an intense competition for intellectual resources in the world market for a simple reason – the maximum surplus value in any product belongs to the intellectual component the share of which in the world economy in comparison with material and productive resources increases every year.

Within the next few decades of the modern alternative globalization 10-15 regional economics - "free trade areas" can be founded. They will protect themselves from negative influence of global economy by means of protectionism, gold, "raw" or other standard as basis for the establishment of foreign exchange rates.

The successful solution of problems in economy of knowledge of the future is connected with development of the intellectual organizations, "intellectual clusters" [3] which activity is based on the holistic views and resource approach, and moreover, social resources have not smaller value in comparison with natural, productive and intellectual resources as we could see from the events of the last years in Iraq, Libya and Syria. It is necessary to pay attention to the new phenomenon of the social environment in a number of developed countries in Europe - growing of social fragmentation and decrease of stability development in connection with this phenomenon. Therefore the economy

intellectualization is represented as an objective regularity which is based on the formation of the intellectual organizations, the intellectual clusters. We consider the small innovative enterprises (SIE) as a quantum of those intelligent clusters in production.

The cluster means a network of the independent productive and/or service firms, suppliers and consumers, the innovative organizations – the research institutes (RI), the design engineering bureaus (DEB), universities, the scientific and technical complexes (STC), SIMs and other market institutes (brokers, consulting, engineering and other firms) concentrated geographically and interacting with each other within a uniform field and redistribution of a surplus value.

The main features of the intellectual cluster are:
- the organized intellectual ability to expect future changes, forks, their reasons and consequences;
- integrating new knowledge with the existing in cluster organizational, institutional, individual and collective intellectual capital, ability to create new knowledge and competences;
- ability to adhere to the development purposes, effectively using and reproducing various resources, first of all - intellectual.

The realization of the general principles of clustering policy of the intellectual organizations assumes:
- combination of the general and specific stimulant means of clustering processes according to fractal model of intellectual clusters, using the principle "from below to up" and "from top to down";
- ensuring the fundamental principles of a complementarity and compliance: local, municipal, regional and national clusters - making part of the international clusters;
- communication processes of the intellectual clustering with scientific, technological and educational infrastructures at the active state-private partnership (SPP) with the leading role of the state in this process;
- history-, culture-, traditions- and experience-based policy of the intellectual clustering.

It is necessary to emphasize fractality of the intellectual clusters and their non-contradiction to the fundamental principles of the complementarity and compliance therefore providing the continuity and indissolubility of economy development of the certain countries and regions with harmonious entry into the future design of the world economy.

Today the level of the innovative enterprises in Russia is less than 15%, and production share in the international market – less than 0,5%. The Kaliningrad region has no immunity from this particular disease. Nevertheless, there are the real preconditions for transition from production of low repartition to development of the full-scale difficult productions based on new technologies for which highly qualified specialists on highly paid workplaces will be demanded. It would allow to reduce one of the main shortcomings of social and economic development, both of the region, and the country as a whole – to begin the systematic and steady growth of still thin layer of organized and socially active middle class on the basis of development of technological innovative entrepreneurships.

3 INNOVATIONS LOGISTICS IN BFFSA

In the paradigm of the human capital development and creation of the intellectual organizations, a new structure and logistics of innovation, education, research and preproduction is being developed and continuously improved in the Baltic Fishing Fleet State Academy since 2007 [4]. It is shown in Figure 1 and 2 as of December, 2012.

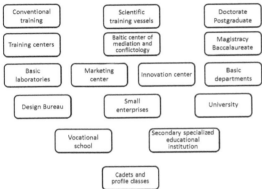

Figure 1. The structure of Maritime Entrepreneurship University

Structure and logistics under consideration take into account the vertically and horizontally integrated educational complex with following features:

1 specific character of the fishing industry - a basic component of food security and the maritime industry in general, have specific requirements for the personnel training, the most important elements of which are in addition to training is high moral and volitional qualities, knowledge of modern information and GIS technologies and the ability to use them, high general knowledge, good knowledge of foreign languages, a mandatory practice on board the vessels, the ability to make quick decisions in extreme cases;

2 the need for development of research and innovation environment, technology and simulator framework reflecting the current level of world shipbuilding, shipping and fishing industry at the universities, which allow us to train the specialists meeting the requirements of the world's leading maritime Power of the XXI century, as the Concept for long-term development for the period up to 2020 has positioned Russia;

3 the development of the educational system of maritime industry and accelerated modernization of the technological base of the shipbuilding industry, new technologies of shipbuilding, construction of a new fishing fleet and the development of modern logistics of production, transportation, processing and delivery of fish and other marine products to the world and the Russian domestic market must take place simultaneously, generating a synergistic effect of the industry;

4 necessity of further development of large multi-level vertically and horizontally integrated maritime university complexes in the industry, combining fishing Higher and Secondary educational institutions, research institutes and design bureaus, small enterprises at universities, basic departments and laboratories at the largest enterprises of the coastal regions, small educational-scientific-industrial complexes (SESIC).

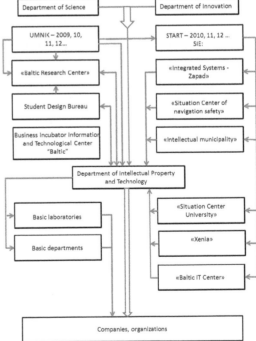

Figure 2. Logistics of Maritime Entrepreneurship University

Infrastructure of innovation support is, on the one hand, a component of scientific and innovation management in general, a platform of technological entrepreneurship, on the other hand - an intermediary between suppliers of new products -

structural units and employees of the university, engaging in research and innovation, and consumers - organizations and enterprises willing to pay for intellectual products. In the Academic infrastructure of innovation's support the Quality management system provides structural units which are responsible for innovation service (preparation of documents, mediation within R & D, legal support of contracts, technological consulting, etc.).

One of the primary goals of scientific and innovative activity is a fostering system for young talented researchers and teachers-innovators, and release of scientific departments engaged in research from performing non-core functions, as well as grouping of such functions in specific subdivisions. The above structure presents a kind of technological business incubators, greenhouses for young or novice innovators. Such companies as the Baltic Research Center, Student Design Bureau operate in this way.

4 CONCLUSION

Such an approach consistently increase scientific and innovative potential of the Academy, through which cadets and students, young scientists and teachers have the opportunity to participate actively in the programme of the Foundation for Assistance to Small Innovative Enterprises in Scientific Research and Technical Sphere, which is called "Member of Youth Scientific Innovation Contest " - "M.Y.S.I.C." This programme is a launching pad for young researchers. Since 2007 more than 1000 young scientists have participated in intellectual competitions. More than 50 candidates were prepared by various departments of the Academy and reached the final of the annual competition. 14 persons gained two year grants in the amount of 400 thousand rubles to finance the implementation of R&D projects.

As a result, academy currently has a "belt" of young innovators SIE`s where they realize projects for innovative products for the marine industry and coastal areas. In particular, in 2012 two new enterprises "Baltic IT Center" and "Ksenia" topped up to the innovative structure of BFFSA which had listed 5 SIEs before. Those companies run their business in the field of manufacturing and information and communication technologies. They received grants from Foundation for Assistance to Small Innovative Enterprises in Scientific and Research Technical sphere within the program "M.Y.S.I.C to START" of 1 million rubles each for the implementation of R & D projects and launching new products this year. The total amount made in the last years by small innovative enterprises exceeded 20 million rubles. Thus, the intellectual organizations form cells of new economy, creating a basis for new methods of control over social and production systems with use of creative technologies, and, the intellectual organizations develop not only in the production sphere, but also in the public and municipal administration sphere.

BIBLIOGRAPHY

[1] M. Ershov. World economy: prospects and obstacles for recovery. Moscow, Economy questions, № 12, 2012, p.61-83.
[2] Global Trends 2030: Alternative Worlds. A publication of the National Intelligence Council. December 2012.
[3] Krasnyansky I.Yu., Merkulov A.A., Yafasov A.Ya. Model of intellectual system of monitoring and management of municipality: "Intellectual municipality". Collection of papers "Modernization: innovative solutions in economy and management". / North Western academy of public service; under the editorship of. A.Ya.Yafasov. - St. Petersburg, 2010, p. 11-40.
[4] Volkogon V.A. New approaches in training for the fishing industry in Russia. St. Petersburg, Economics and Management № 1/4 (40), 2009, p.90-96.

Human Resources and Crew Manning
STCW, Maritime Education and Training (MET), Human Resources and Crew Manning, Maritime Policy,
Logistics and Economic Matters – Marine Navigation and Safety of Sea Transportation – Weintrit & Neumann (Eds)

Wasted Time and Flag State Worries: Impediments for Recruitment and Retention of Swedish Seafarers

C. Hult & J. Snöberg

Kalmar Maritime Academy, Linnaeus University, Sweden

ABSTRACT: This study focuses on impediments for seafarers' motivation at work for the specific shipping company (organizational commitment), and seafarers' motivation for working in their particular occupation (occupational commitment). The study takes its departure in the lingering difficulties to recruit and retain qualified senior seafarers in the Swedish shipping sector. Statistical analyses are employed, using a survey material of 1,309 Swedish seafarers collected in 2010. The results show that the main negative effects on seafarers' commitment at work primarily have to do with invested – or wasted – time, and flag state worries. Flagging-out imposes a significant decline in organizational commitment for all seafarers due to the perception of social composition onboard. The oldest seafarers demonstrate diminished occupational commitment under a foreign flag due to degree of satisfaction with the social security structure. The youngest seafarers display a decline in occupational commitment related to time served on the same ship, partly due to a decline in satisfaction with work content for each year on the same ship. In the concluding discussion, the findings are discussed in more details and recommendations are put forward.

1 INTRODUCTION

1.1 *The attempt*

This study aims to investigate Swedish seafarer's attitudes to their work and occupation. The study is undertaken in response to the needs of the shipping industry to recruit and to improve retention of onboard crew. The two main areas of enquiry are seafarers' motivation at work for the specific shipping company (organizational commitment), and seafarers' motivation for working in their particular occupation (occupational commitment). The study is based on a sample of 1,309 seafarers taken year 2010 from the Swedish Register of Seafarers.

2 BACKGROUND

2.1 *More foreign flags on Swedish ships*

In 2010, the year of the seafarer, the shipping industry had been hit hard by the ongoing financial crisis. In Sweden, worries were exacerbated by a wave of flagging-out of ships. Compared to 2009, the Swedish-registered merchant navy had declined by 12,5 percent, calculated as ships of 300 gross tonnes and above (Lighthouse, 2010). The situation

was probably felt most keenly by Swedish ratings because opportunities to compete with other nationalities were further constrained by the foreign registrations. It was argued that among Swedish seafarers only senior officers could feel relatively secure in light of current developments. (Swedish Employment Service Maritime 2010: 21.)

2.2 *Recruitment and retention of seafarers*

In the Swedish shipping sector there is large proportion of employees who are approaching the end of their active careers. The average age of onboard personnel rose by 11 percent between 2006 and 2009 (Swedish Employment Service Maritime, 2010: 8). The difficulty has to do with recruiting and retaining qualified and experienced onboard personnel (Swedish Maritime Administration, 2010: 4). One common estimate is that the average time a Swedish ship's officer remains in the occupation is only eight years (Swedish Employment Service Maritime, 2010: 10; Swedish Maritime Administration, 2010: 17).

Thus, we find two rather disparate trends in the Swedish shipping sector; one that is making the labour market for Swedish seafarers smaller, and one that makes it more difficult for ship-owners to recruit and retain educated seafarers. The question we need to ask is how the seafarers' commitment to their work is affected in this development? Does the current situation comprise any serious impediments for recruitment and retention in the shipping sector?

3 RESEARCH AND THEORETICAL CONSIDERATIONS

3.1 *Earlier research, job satisfaction and commitment*

Quantitative research on seafarers' attitudes to work is sparsely occurring. There is a handful of studies from differentt parts of the world (e.g. Guo et al., 2005; Turker & Er, 2007; Guo et al., 2010; Sencila et al., 2010; Pan et al., 2011). In Sweden, studies have been mainly based on interviews and observations, with focus on subjects like seafaring life, culture, stress, fatigue, and safety. There are, however, three Swedish studies with quantitative approaches; two focusing job satisfaction onboard merchant ships (Werthén, 1976; Olofsson, 1995) and one focusing commitment to work (Hult, 2012).

Job satisfaction and commitment are related phenomena, but the terms are far from synonymous. By definition, commitment encompasses a certain measure of motivation and dedication, which is not necessarily the case with job satisfaction (c.f. Steers & Porter, 1987: 29; Steers, 1984: 132). Job satisfaction is customarily regarded more as an emotional response to a work situation (Steers, 1984: 428-444). Accordingly, job satisfaction can be regarded as one of several underlying factors to work-related commitments (Steers, 1984: 442; Hult 2005).

3.2 *Organizational commitment*

Organizational commitment has to do with loyalty and dedication in the specific job and to the specific organization. Research has indicated important relationship between this type of commitment and employee turnover (e.g. Steers, 1977). Organizational commitment is driven mainly by what people feel they get out of the job and how far this aligns with individual preferences. If work-related preferences and perceived conditions agree, the individual is expected to express a high level of dedication and loyalty (Hult, 2005).

Earlier research has shown that perceived non-financial gains have a stronger positive impact on loyalty and dedication at work than perceived financial gains (Hult, 2005). But rather than differentiating between financial and non-financial factors, the distinction in this research tradition is between *internal* and *external* factors. Internal factors are rewards one receives *within* the work –

such as a sense of well-being and pride from the feeling that one is doing good and important work – and external factors are rewards one receives *for* work performed – pay and other benefits (Hackman and Lawler, 1971; Lincoln & Kalleberg, 1990: 98). This does not mean external factors are less important, but their positive effects on commitment tend to be relatively short-lived (Herzberg et al., 1993: 70-83).

3.3 *Occupational commitment*

If organizational commitment has to do with the specifics of working life, occupational commitment has to do with the generalities for an occupation. Like in organizational commitment there is a more qualitative and emotional driver, which is the aspect that has been given the most attention in earlier research (c.f. Lee et al., 2000). It is, for example, primarily within an occupation that people can develop a sense of status and identity. Earlier research shows that the duration of education, age, and years invested in the occupation have positive effect on occupational commitment (Nogueras, 2006).

It has been reported that perceptions of social quality and leadership in the workplace influence occupational commitment (Van der Heijden et al., 2009). A positive correlation between occupational commitment and perceived autonomy at work has also been reported (Giffords, 2009), as well as a strong correlation with organizational commitment (Lee et al., 2000). In other words, if people have a long, thorough professional education, are motivated in their day-to-day work and happy with their specific jobs, they can be expected to demonstrate a strong emotional relationship to their occupations.

Occupational commitment differs from organizational commitment in that satisfaction with a specific workplace may be low even while identification with the occupation is high. It is hard to imagine that the opposite situation would be particularly common, although it is entirely possible that higher satisfaction with the occupation also strengthens commitment to the specific workplace. It is also, most likely, easier to switch workplace and remain in the same occupation than the reverse, which might contribute to a tendency to direct dissatisfaction towards the particular job rather than the occupation. Research has in any case shown that strong occupational commitment restrains decisions to leave a job (Nogueras, 2006).

3.4 *What to expect from this study?*

We may expect differences in commitment depending on age. Older people generally express greater loyalty towards their work than do younger people (Mathieu & Zajac, 1990; Guo et al., 2005;

Hult, 2012). Here we need to consider that age is intertwined with the time accumulated in certain work situations which theoretically can have both positive and negative effects on attitudes depending on individual experiences in the job.

With respect to the foreign registration of flag, it has been argued that seafarers are "vulnerable to exploitation and abuse" as employees in what is called "the world's first genuinely global industry" (ILO, 2012). Seafarers have been advised to always be aware of what flag the "ship is flying and where necessary, ask for assistance to find out what are the laws of that flag State" (ITF, 2012). It has been indicated that flagging-out can pose significant damage to seafarers commitment to their work (Hult 2012), but the mechanisms at work here has not yet been studied in detail.

Hypothetically, the youngest seafarers, and the oldest may be most sensitive to foreign registration. When it comes to the youngest age group, it has been questioned whether we could expect young people to put their faith in a sector that continuously seeking the most convenient flag (SBF, 2010). When it comes to the oldest age group, it has been pointed out that social security and especially the forthcoming retirement income can become a disappointment under foreign flag (Sjömannen, 2011: 12-17). Thus, we may expect flag state worries to be a major impediment for organizational and occupational commitment in the youngest and oldest age groups. Based on earlier research we may expect the strength of flag state effect to differ for young seafarers' depending on trade area and type of ship (Guo et al., 2010), and that no effect of flag state would have anything to do with working in nationally mixed crews (Guo et al., 2005).

4 METHOD

4.1 The sample

This study is based on a sample taken from the Swedish Register of Seafarers using unrestricted random selection of deck and engineering personnel for the men, and of catering personnel for both men and women. As women are strongly underrepresented among deck and engineering personnel, all women from these departments were drawn into the framework. Data were collected via postal surveys during the period of 8 March to 8 September 2010.

The final material consists of 1,309 respondents with an answering rate of 54%. After control of different aspects, such as gender, age, onboard position, trade area, and type of ship, the material was found representative for Swedish seafarers on Swedish-controlled ships. However, it is still (as always) difficult to estimate the likeliness of non-

response effects on the *attitudinal* representativity. An educated guess would be that people who take great interest in their work may be more likely than others to complete this type of questionnaire and therefore be over-represented in the sample. If so, the attitudinal patterns in the analysis would still be correct, but the levels of commitment would be slightly overestimated (Hult & Svallfors, 2002).

The questionnaire was based on pre-existing questionnaires from the International Social Survey Programme, Work Orientations III study (ISSP, 2005). Additional items were all developed with theoretical connections to earlier research on work related commitment. The items were adapted to the specifics of the seafaring occupation.

4.2 The analysis

The Statistical Package Social Science (SPSS) is used throughout the analysis. Dependent and explanatory indices are constructed using Principal Component Analysis (PCA). In order to control for competing variables, multiple regression analysis (OLS regression), allowing adjusted effects, are used in several steps of the analysis.

5 RESULTS

5.1 Dimensions of commitment

The dependent variables are based on attitude questions expressed as statements on which respondents were asked to take a position by selecting a fixed option on a five-point Lickert Scale, from *strongly agree* to *strongly disagree*. After mapping the pattern of latent factors underlying a number of indicators, using PCA, the appropriate indicators for occupational and organizational commitment came out as shown in Table 1.

Table 1. Dimensions of commitment

Occupational Commitment, indicators:
There are qualities to the seafaring occupation that I would miss in another occupation.
The seafaring occupation is part of my identity.
The seafaring occupation is not just a job, it is a lifestyle.
I feel proud of my occupation as a seafarer.
I would prefer to remain in the seafaring occupation even if I were offered a job with higher pay on land.

Organizational Commitment, indicators:
I am willing to work harder than I have to in order to help the shipping company I work for succeed.
I am proud to be working for my shipping company.
I would turn down another job that offered quite a bit more pay in order to stay with this shipping company.

All indicators were then recoded so that 0 denotes the option that entails the lowest commitment and 4 the highest, and then summarized within particular index. To facilitate interpretation of the results, each index is then divided by its maximum values and multiplied by 100. Each index is thus permitted to vary between 0 and 100. Table 2 shows the mean value and standard deviation for each index. The high mean value and the low standard deviation for occupational commitment indicates that seafarers are quite united in their high commitment to, and identification with, the seafaring occupation. Cronbach's Alpha is a test of the internal correlation among the indicators in each index – the higher the value (between 0 and 1), the more reliable the index. Both indices turn out sufficiently stable.

Table 2. Work attitudes index – Swedish seafarers in 2010

	Occupational commitment	Organizational commitment
Mean value (0-100)	71.7	50.3
Standard deviation	16.78	20.81
Cronbach's Alpha	0.82	0.73

5.2 *Age and invested time*

In Figure 1 we can see how organizational and occupational commitments are distributed among four age categories 19-30, 31-42, 43-54, and 55+. There is a weak increase in occupational commitment (which is also the consistently stronger commitment type) between the two youngest categories and between the two oldest. We find a considerably stronger and pronounced linear increase in organizational commitment with age (the consistently weaker commitment type).

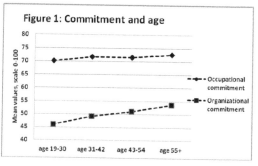

Figure 1: Commitment and age

As previously mentioned, the measure of age is often intertwined with different aspects of investment in time, in our case illustrated with how many years seafarers have worked at sea, how long they have worked in their present positions, how long they have served on their present ships, and how long they have been working for their specific shipping companies. In order to control for undesired levels of collinearity between these variables, a Variance Inflation Factor (VIF) test was made for models in Figures 2-4. In no case the VIF value reached over 2,7 indicating fully acceptable levels of collinearity.

Figures 2-3 show that the significant effects for both types of commitment – when effects from all types of time investments are mutually controlled for – are positive in direction. Dark bars indicate statistical significance. Positive effects ascend from the 0 line and negative effects descend. Each bar shows the average effect for one year. The chart's 0-line thus represents the average position at one year's less invested time. Because the effects represented by the chart refer to an average value change per year, the effects are also small – but even small effects become significant if the tendency is sufficiently stable.

Figure 2: Occupational commitment and invested time

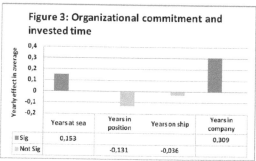

Figure 3: Organizational commitment and invested time

After control for each age group (not shown here) it became clear that only occupational commitment is vitiated by any negative age-related effect from invested time. Figure 4 reveals a surprisingly strong negative effect per year invested on the ship in the 19-30 age group (other aspects of time are controlled for, not shown).

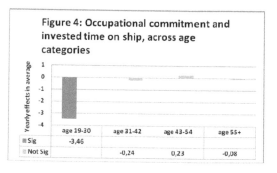

Figure 4: Occupational commitment and invested time on ship, across age categories

5.3 Age, crew nationality and flag state

As touched upon in the introduction, there is a strong international element to shipping. Accordingly, factors including nationally, mixed crews and flag state need to be investigated with mutually control.

Figure 5 shows that organizational commitment is strongly and negatively affected by a foreign flag, when nationality of crew is controlled for. Thus, loyalty and commitment to the employer are considerably higher among seafarers on Swedish-flagged ships. No effects, other than that of flag, reach significance.

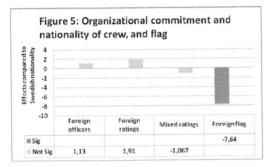

Figure 5: Organizational commitment and nationality of crew, and flag

When it comes to occupational commitment, it became clear that crew nationality and flag state lack general impact (not shown here). However, investigation of each age group revealed interesting figures for the oldest age group. Figure 6 shows a large negative effect of foreign flag in the 55+ age category. Clearly, older seafarers demonstrate less interest in the seafaring life if their ship is flagged-out.

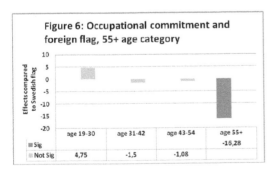

Figure 6: Occupational commitment and foreign flag, 55+ age category

5.4 Explaining the impediments

When trying to explain differences in commitment, it is a good idea to start with controlling for respondents' workplace position. This is because positions often entail differences in perception of the work content. Figure 7 shows those onboard-positions where the decline in occupational commitment due to invested time on ship, in the age category 19-30, is most pronounced. It is junior deck officers and catering ratings that display the most pronounced decline related to time invested on ship. However, multivariate statistical tests (not shown here) revealed that time on ship impose a significant effect on occupational commitment in its own right, i. e. independently of not only onboard-position, but also of type of ship and trade-area.

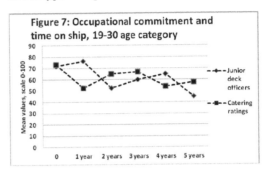

Figure 7: Occupational commitment and time on ship, 19-30 age category

The same procedure was conducted concerning the finding that foreign flag pose a significant and negative effect on organisational commitment for all ages. Figure 8 shows that senior deck officers, junior deck officers, junior engine officers, and deck ratings are those onboard positions that display the most pronounced decline in commitment with foreign flag. However, multivariate tests (not shown) made clear that also the effect of flag state remain significant and independent of onboard position.

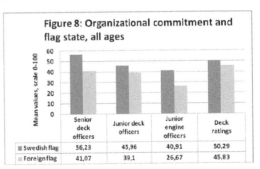

Figure 8: Organizational commitment and flag state, all ages

The finding that foreign flag has a significant and negative effect on occupational commitment for the age category 55+, was also tested with onboard position. Figure 9 tells us that the strongest effects are found among senior deck officers, junior engine officers and deck ratings. Multivariate tests (not shown) made clear that also the initial effect of flag state remain significant and independent of onboard position concerning the age category 55+.

Thus, in order to properly explain the observed impediments for organizational and occupational commitment, we need to turn our attention to differences in the degree to which seafarers valuate different aspects of their job related to the extent to which the aspects are perceived satisfied at work.

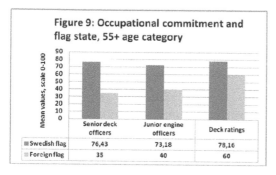

Figure 9: Occupational commitment and flag state, 55+ age category

	Senior deck officers	Junior engine officers	Deck ratings
Swedish flag	76,43	73,18	78,16
Foreign flag	35	40	60

The material contains several statements concerning the importance and perceived satisfaction with various aspects of the job. After mapping the pattern of latent factors three appropriate dimensions (each containing importance and satisfaction with aspects) came out as presented in Table 3.

Table 3. Cronbach's Alpha, Explanatory dimensions

	Importance of the aspects	How well the aspects are satisfied
Social security structure:	0.65	0.59
Job security		
Being useful to society		
Good leadership		
A union that looks out for members' interests		
Good relations with the land organization		
Social composition onboard:	0.81	0.71
Carefully considered gender composition		
Carefully considered age composition		
Carefully considered ethnic composition		
Job content onboard:	0.70	0.75
Good career opportunity		
Interesting tasks		
Autonomy		
Skills development		

All indicators were summarized within separate indices and divided by their maximum values and multiplied by 20 so that the explanatory variables can vary between 0 *not important* (valuations) and *not satisfied* (perceptions), to 20 *very important* (valuations) and *completely satisfied* (perceptions).

In Table 4-6 these variables are used as control on previous results. In Table 4, Step I, the effect of foreign flag on organizational commitment is, as expected, negative and significant. Moving from Step I to Step II, the variables of social composition are entered, both showing positive and significant effects while the effect of flag is declining and becomes insignificant. The flag effect is thus explained by the valuation and perception of the *social composition onboard*. We can see that organizational commitment is increasing with 1,84

(on the scale of 0-100) for every positive change in perception of the social composition (on the scale 0-20). Thus, the result tells us that the perception of ill considered social composition onboard flagged-out ships impairs organizational commitment among Swedish seafarers.

The analytical principles described above are the same for Table 5-6. In Table 5, the impaired occupational commitment under a foreign flag for the oldest age group is explained by the degree of satisfaction with the *social security structure*. However, the decline in occupational commitment related to time served on the same ship for the youngest age group is difficult to explain statistically (Table 6). However, *perceived job content* is part of the explanation, indicating that the perceived value of job content declines somewhat with time on ship. It is worth noting that the initial effect of time on ship is smaller here than in Figure 4, due to no time collinearity in Table 6.

Table 4. Organizational commitment, flag state and main explanatory dimension, all ages

	Organizational commitment	
	Step I	Step II
Foreign flag (comp. with Swedish)	**-7.16***	-4.27
Carefully considered social composition is important (0-20)	-	**0.54***
Having a carefully considered social composition (0-20)	-	**1.84***
Proportion of explained variance (%)	*0.6*	*9.7*
Number of respondents	*1121*	*1063*

Significance levels: bold and *** = 0.001 level, ** = 0.01 level, * = 0.05 level.

Table 5. Occupational commitment, flag state and main explanatory dimension, 55+ age category

	Occupational commitment	
	Step I	Step II
Foreign flag (comp. with Swedish)	**-15.86****	-9.49
Social security structure is important (0-20)	-	**1.12***
Having a good social security structure (0-20)	-	**1.70***
Proportion of explained variance (%)	*2,1*	*16,2*
Number of respondents	*350*	*298*

Significance levels: bold and *** = 0.001 level, ** = 0.01 level, * = 0.05 level.

Table 6. Occupational commitment, time on ship and main explanatory dimension, 19-30 age category

	Occupational commitment	
	Step I	Step II
Years on ship	**-2.21***	**-1.99***
Job content is important (0-20)	-	.21
Having a good job content (0-20)	-	**1.42***
Proportion of explained variance (%)	*6,5*	*17,8*
Number of respondents	*220*	*210*

Significance levels: bold and *** = 0.001 level, ** = 0.01 level, * = 0.05 level.

6 CONCLUDING DISCUSSION

6.1 *The findings*

The initial questions in this study concerned impediments for seafarers' motivation at work for the specific shipping company (organizational commitment) and impediments for seafarers' motivation to work in their particular occupation (occupational commitment). It was hypothesised that the youngest and the oldest seafarers would be the most sensitive for the current development due to flag state worries, and that this would lead to a decline in their organizational and occupational commitment.

The hypothesis has only partly been supported by this study. The result shows that the main negative effects on seafarers' commitment at work primarily have to do with invested time and flag state. The youngest age category (19-30 years of age) display a decline in occupational commitment related to time served on the same ship, partly due to a decline in satisfaction with work content for each year on the same ship. The oldest age category (age 55+) demonstrates a significantly diminished occupational commitment under foreign flag compared to a Swedish. This effect is explained by satisfaction with the social security structure in current employment.

Another important finding from this study is that flagging-out impairs seafarers' organizational commitment in general, independently of age. This effect is explained by the degree to which the social composition onboard is perceived as carefully considered.

6.2 *Discussion and recommendations*

If the younger seafarers drop in occupational commitment related to time on ship is only partly explained by a decline in satisfaction with work content, we may ask for the remaining reason for this loss in commitment. The remaining part of the effect might possibly be explained by a mismatch between an idea about the adventure and freedom of the seafaring life and the reality for young seafarers walking the same deck for too long. From the individuals' perspective the time spent as a seafarer may therefore be felt like a poor investment described as *wasted* time rather than invested. For a young officer who has spent several years in maritime education and training prior to the introduction to the first ship, the waste becomes substantial, covering the perhaps most important period in life.

Based on these findings we strongly recommend the shipping companies to incorporate sincere career coaching in their HRM-activities directed to their young employees. The results also call for a development of individual rotating systems for young employees with regards to trade area and ships. For small companies with one or only few vessels, it may be of mutual advantage to make cross-company arrangements to share and lend young employees within some sort of informal crew pool system.

The older seafarers, on the other hand, demonstrate diminished occupational commitment under a foreign flag due to dissatisfaction with the social security structure. The oldest category of seafarers are thus more committed to the seafaring occupation if they sail under a Swedish flag because they perceive better leadership, better relations with the land organization, have a strong Swedish union, better job security and feel they have ties to, and can contribute to home society.

Based on these findings we strongly recommend the shipping companies to uphold a certain sense of homeland belongingness for onboard employees on flagged-out ships. It is important to keep management policies and routines close to expectations related to important motherland institutions. It is also important to retain a dialogue with the union, and to facilitate the development of sufficient social security agreements for employees under foreign flag.

Flagging-out also impairs seafarers' organizational commitment in general, independently of age, due to dissatisfaction with the degree to which the social composition onboard is perceived as carefully considered. Here it is important to underline that this does not have anything to do with national/cultural conflicts onboard since crew nationalities has been controlled for (Figure 5). Instead it only has to do with the degree to which the social composition onboard is *carefully considered*. The finding tells us, however, that the social composition onboard is something very important for the seafarers.

Based on these findings we strongly recommend the shipping companies to carefully consider the social composition onboard their ships. Other than that, we can only conclude that flag state is important for the seafarers and that flag state and seafarers' perception of the social composition onboard are indeed correlated phenomena. Intensified research on the question of exactly why a change of flag comes with a change of the degree to which seafarers view the social composition onboard to be *carefully considered* is definitely needed.

Finally, if the current difficulties to recruit and retain experienced and qualified seafarers would be settled within a lasting trend, then the shipping companies will face a gigantic challenge trying to crew their future vessels. Most likely, this challenge does not apply only for Swedish controlled shipping. Therefore, internationally comparative research on organizational and occupational commitment is needed. Everyone with interests in the shipping sector should welcome initiatives to find means to keep seafarers' commitment at a high level. For now, prudent recommendations from this study can hopefully eliminate flag state worries and make the invested time to a success for seafarers.

ACKNOWLEDGEMENT

We like to thank the Swedish Mercantile Marine Foundation for funding the data collection and for in many ways being helpful throughout the whole procedure. We also like to thank one anonymous reviewer for valuable comments.

REFERENCES

Giffords, E. D. (2009). 'An Examination of Organizational Commitment and Professional Commitment and the Relationship to Work Environment, Demographic and Organizational Factors' *Journal of Social Work*, Vol. 9. 386-404.

Guo, J. L., Liang, G. S. & Ye, K. D. (2005). 'Impacts of Seafaring Diversity: Taiwanese Ship-Officers' Perseption' *Journal of the Eastern Asia Society for Transportation Studies*. Vol. 6, 4176-4191.

Guo, J. L., Ting, S. C. & Lirn, T. C. (2010). 'The Impacts of Intership at Sea on Navigation Students' Seafaring Commitment' *Maritime Quarterly*. Vol. 19(4). 77-107.

Hackman, J. R. & Lawler III, E. E. (1971). 'Employee reactions to job characteristics' *Journal of Applied Psychology Monograph*. Vol. 55. 259-286.

Herzberg, F., Mausner, B. & Snyderman, B. B. (1993). *The motivation to work* (New ed.). New Brunswick, New Jersey: Transaction.

Hult, C. (2005). 'Organizational Commitment and Person-Environment Fit in Six Western Countries' *Organization Studies*, Vol. 26(2). 249-270.

Hult, C. (2012). 'Sjömäns motivation i arbete och yrke' in Hult, C. (2012) (ed) *Sjömän och sjömansyrke 2010 – en studie i attityder till arbete och yrke under olika skeden i sjömanslivet*. Kalmar Maritime Academy, Linnaeus University.

Hult, C., & Svallfors, S. (2002). 'Production Regimes and Work Orientations: A Comparison of six Western Countries' *European Sociological Review*, Vol. 18. 315-331.

ILO (2012). International Labour Organization. Seafarers. Retrieved 2013-01-06: http://www.ilo.org/global/standards/subjects-covered-by-international-labour-standards/seafarers/lang--en/index.htm

ISSP (2005). International Social Survey Programme 2005: Work Orientation III. GESIS Data Archive, Cologne, Germany, ZA4350, NSDstat. Retrieved 2010-01-29: http://zacat.gesis.org/webview/index.jsp?object=http://zacat.gesis.org/obj/fStudy/ZA4350

ITF (2012). International Transport Workers' Federation, Seafarers. Retrieved 2013-01-06: http://www.itfseafarers.org /your_legal_rights.cfm

Lee, K., Carswell, J. J., & Allen, N. J. (2000). 'A Meta-Analytic Review of Occupational Commitment: Relations with Person- and Work-Related Variables' *Journal of Applied Psychology*, Vol. 85(5), 799-811.

Lighthouse (2010). Svensk Sjöfart: Nyckeltal januari 2010. Gothenburg. Retrieved 2010-05-10: http://www.sweship.se/Files/100608Nyckeltal2010.pdf

Lincoln, J. & Kalleberg, A. L. (1990). *Culture, Control and Commitment - A Study of Work Organization and Work Attitudes in the United States and Japan*. Cambridge: Cambridge University Press.

Mathieu, J. E. & Zajac, D. M. (1990). 'A Review and Meta-Analysis of the Antecedents, Correlates, and Consequences of Organizational Commitment' *Psychological Bulletin*, Vol. 108. 171–94.

Nogueras, D. J. (2006). 'Occupational Commitment, Education, and Experience as a Predictor of Intent to Leave the Nursing Profession' *Nursing Economics*, Vol. 24. 86-93.

Olofsson, M. (1995). *The work situation for seamen on merchant ships in a Swedish environment. Gothenburg*. Chalmers University of Technology.

Pan, Y. H., Lin, C.C. & Yang, C. S. (2011). 'The Effects of Perceived Organizational Support, Job Satisfaction and Organizational Commitment on Job Performance in Bulk Shipping' *Maritime Quarterly*. Vol 20(4), 83-110.

SBF (2010) Sjöbefälsföreningen. Retrieved 2013-01-06: http://www.sjobefalsforeningen.se/index.php/hem/informati on/arkivet/tidskriften-sjoebefael/10-information-information/137-sjoebefael-32010

Sencila, V., Bartuseviciene, I., Rupsiene, L. & Kalvaitiene, G. (2010). 'The Economical Emigration Aspect of East and Central European Seafarers: Motivation for Employment in Foreign Fleet' *International Journal on Marine Navigation and Safety of Sea Transportation*. Vol 4(3). 337-342.

Sjömannen (2011). 'Utflaggning' SEKO sjöfolk, Breakwater Publishing, Sweden. Retrieved 2013-01-06: http://www.sjomannen.se/tidningar/2011_3/files/assets/download s /publication.

Steers, R. M. (1977). 'Antecedents and Outcomes of Organizational Commitment' *Administrative Science Quarterly*. Vol. 22, 46-56.

Steers, R. M. (1984). *Introduction to Organizational Behavior* (2nd ed.). Glenview, Illinois: Scott, Foresman and Company.

Steers, R. M. & Porter, L. W. (1987). *Motivation and Behavior*. New York: McGraw-Hill Book Company.

Swedish Employment Service Maritime (2010). Prognos Arbetsmarknad Sjöfart 2010 och 2011. Stockholm. Retrieved 2010-09-15: http://www2.hik.se/dokument/ .%5CSjo/ prognos_sjofart _2011_v10.pdf

Swedish Maritime Administration (2010). Handlingsplan för ökad rekrytering av personal till sjöfartssektorn. Norrköping, Sweden. Retrieved 2011-02-24: http://sjofartsverket.se/pages/25929/08-3114 Rekryteringsuppdraget.pdf

Turker, F. & Er, I. D. (2007). 'Investigation the Root Causes of Seafarers' Turnover and its Impact on the Safe Operation of the Ship' *International Journal on Marine Navigation and Safety of Sea Transportation*. Vol 1(4), 435-440.

Van der Heijden, B., van Dam, K., & Hasselhorn, H. M. (2009). 'Intention to leave nursing - The importance of interpersonal work context, work-home interface, and job satisfaction beyond the effect of occupational commitment' *Career Development International*, Vol.14. 616-635.

Werthen, H.-E. (1976). *Sjömannen och hans yrke: en socialpsykologisk undersökning av trivsel och arbetsförhållanden på svenska handelsfartyg*. Doctoral thesis, Gothenburg: University of Gothenburg.

Human Resources and Crew Manning
STCW, Maritime Education and Training (MET), Human Resources and Crew Manning, Maritime Policy,
Logistics and Economic Matters – Marine Navigation and Safety of Sea Transportation – Weintrit & Neumann (Eds)

Analysis of Compensation Management System for Seafarers in Ship Management Companies: An Application of Turkish Ship Management Companies

E. Çakir & S. Nas
Dokuz Eylul University Maritime Faculty, Turkey

ABSTRACT: Compensation management is an important activity of the HRM. It runs through complete process of the HRM and it plays very important part in the activities of HRM. The role of compensation management in companies is to motivate employee, keep them in organization by enhancing loyalty to organization, attract and retain qualified employees and increase productivity of employees. To determine appropriate wage levels is the most important factor in compensation management system. Wage is a management tool that has to be used effectively in order to accurate objectives of companies and put costs at expected levels. On the part of employee, wage constitutes income and source of living and on the other part it is a cost for employer.

In this study, compensation management system in Turkish ship management companies was focused on. Also, the factors, which are affecting to the compensation management system, were explored. The main purpose of this study is to reveal how the compensation package, proposed for seafarers, is formed by ship management company. For this purpose, a structured interview form was developed considering to the HRM literature and applied to the Turkish ship management companies' human resource managers in 2011. Considering to the collected interview data and the reviews of academics, questionnaire was developed. The developed questionnaire was applied in 2012, during the 17th Career Days of Dokuz Eylul University Maritime Faculty to the Turkish ship management companies' human resource managers.

1 INTRODUCTION

Compensation is one of the most important human resource functions in an organization. For many organizations, compensation is the biggest single cost of doing business (Bhatia 2010). Sherman and Bohlander (1992) suggest that, indeed, compensation's share is 20% of total costs in manufacture industry and 80% of total costs in service industry.

The term, compensation, refers to all forms of financial returns and tangible benefits that an employee receives as a part of employment relationship (Bhatia 2010). Compensation can be classified as direct compensation, indirect compensation and non-monetary compensation. Examples of direct compensation are wages, salaries, overtime pay, commissions and bonuses. Indirect compensation includes benefits that may consist of life, accident, health insurance, pensions and pay for vacation or illness. And non-monetary compensation can include any benefits an employee receives from an employer. This includes career and social rewards such as job security, flexible hours and opportunity for growth, friendships, praise and recognition.

The compensation gains a social – economical identity after stepping into to industrial society. Because, it is the mean of living of worker and social state understanding adopts the idea of earning revenue for living humanely. This understanding brings legal regulations related to compensation. Unions become an important actor in this area with organizing workers

Compensation management is seen as a strategic field by organizations. Koss (2008) suggests that more than any other are in HR, ignoring pay and performance systems can be devastating. It is a very expensive and laborious process to hire new employees, buy back trust of current employees and renew the organization's energy and motivation level.

HRM departments of the shipping companies should dwell on to the turnover rates and factors that are affecting it. Its importance not only comes from the heavy cost of training and employing a new crew but rather losing an dependent, safe worker who knows, understands and obeys the regulation and company policies on safety and operation of the ship (Turker&Er 2007).

Many organizations include compensation into their programs determined to reach their objectives. Compensation management is an effort to constitute pay systems which contribute to attain of organization's objectives. Generally it constitutes of job analysis, job definitions, job evaluation, compensation surveys, building pay structure, determining compensation policies, payment and feedback stages (Gonuldas 2006).

Compensation interests many parts like employee, employer and government. For the employee, compensation is fundamental to his standard of living and is a measure of the value of his services or performance. For the employers, compensation is a significant part of their cost. And for the government, it affects the aspects of macroeconomics, stability such as employment, inflation, purchasing power and socio-economic development in general.

In this study, compensation management system in ship management companies was focused on. Also, the factors, which ones are affecting to the compensation management system, were explored. The main aim of this study is to reveal how ship management companies determine seafarers' wage and to examine the attractiveness of the wage level to retain qualified seafarers. For this purpose, a structured interview form was developed from the HRM literature and applied to the Turkish ship management companies' human resource managers to gather the data to use in this study. And obtained data was analyzed with content analysis method.

2 OBJECTIVES OF COMPENSATION MANAGEMENT

According to Rynes (1987), compensation management particularly important as a recruitment tool because (1) it is a vehicle for satisfying a wide array of human needs, (2) salary offers are expressed in clear and comparable terms, (3) starting salaries have implications for future salary progression, and (4) pay systems '' communicate so much about an organization's philosophy, values and practices.''

The objectives of compensation management are numerous and might include the following (Gonuldas 2006):
– To acquire qualified personnel. Compensation needs to be high enough to attract applicants. Pay levels must respond to supply and demand of workers in the labor market since employers compete for workers. Premium wages are sometimes needed to attract applicants who are already working for others.

– To retain present employees. Employees may quit when compensation levels are not competitive, resulting in higher turnover.
– To ensure equity. Compensation management strives for internal and external equity. Internal equity requires that pay be related to the relative worth of jobs, so that similar jobs get similar pay. External equity means paying workers what comparable workers at other firms in the labor market pay.
– To motivate employees for better performance.
– To ensure equal pay for equal work, that is, each individual's pay is fair in comparison to that of another person doing a similar job.
– To reinforce reward desired behavior. Pay should reinforce desired behaviors and act as an incentive for those behaviors to occur in the future. Effective compensation plans reward performance, loyalty, experience, responsibilities and other behaviors.
– To control costs. A rational compensation system helps the organization obtain and retain workers at a reasonable cost. Without effective compensation management, workers could be over or underpaid.
– To support, communicate and reinforce an organization's culture, value and competitive strategy.
– To comply with legal regulations. A sound wage and salary system considers the legal challenges imposed by government and ensures the employer's compliance.
– To facilitate understanding. The compensation management system should be easily understood by human resource specialists, operating managers and employees.
– To supply further administrative efficiency. Wage and salary programs should be designed to be managed efficiently, making optimal use of the human resource information system, although this objective should be a secondary consideration compared with other objectives.

3 METHODOLOGY

3.1 *The aim of the study*

The main aim of this study is to reveal how the compensation package, proposed for seafarers, is formed by Turkish ship management company.

3.2 *Survey*

In 2011, during the 16th Career Days of Dokuz Eylul University Maritime Faculty, interview studies were made with 19 ship management companies about their compensation management applications. For this interview "structured interview form", developed considering the compensation literature, was used. Considering the collected interview data and the reviews of academics, questionnaire was developed. Developed questionnaire was applied in 2012 during the 17th Career Days of Dokuz Eylul University Maritime Faculty.

The questionnaire consists of 2 main parts related to compensation management. In the first part of the questionnaire, "the effects on determining the compensation of seafarers" was asked to the respondents. The Likert scale was used in this part to evaluate each variables [(1) strongly disagree, (2) disagree, (3) neutral, (4) agree and (5) strongly agree].

In the second part of the questionnaire, "compensation policy" was asked to the respondent. The frequency scale was used in this part to evaluate each variable [(1) never, (2) rarely, (3) sometimes, (4) very often and (5) always].

3.3 Sample size

In December 2012, 25 ship management companies were invited to attend the 17th Winter Career Days of Dokuz Eylul University Maritime Faculty. The developed questionnaire was applied in these days. Participant ship management companies are operating 296 ships comprising of tanker ships, chemical ships, bulk carriers and container ships. The Turkish merchant shipping fleet comprises of 1156 ships when taken in consideration ships which have 1000 grt and over capacity. Sample size was found adequate and the sample has been appropriately chosen to represent the target population of interest. Thus different ship management companies were chosen to achieve greater validity to research conclusions. This study represents 25% percent of Turkish merchant shipping fleet.

The participants to this study were 1 general manager and 24 human resources managers who have worked mostly as a/an master, officer or engineer in on board ships. 80% of the participants had human resources education through courses or had a bachelor degree in human resources field.

3.4 Data analysis and findings

The data gathered from questionnaire was analyzed with the SPSS 20.0 program.

The variables affecting the compensation system of seafarer are shown in the order in Table 1. The most effective factor on determining compensation of seafarer has been found as "senior managers". The variables of "economic situation of ship management companies", "situation of the freight market" and "supply-demand of seafarers" have been found as in the top four.

On the other hand, the variables of "labor unions" and "non-governmental organizations" have been found as the lowest effect on the compensation system of seafarers.

Table 1. Effects of Variables on Compensation of Seafarers

Variables	Mean*[1]	Std.Dev.
1. Senior Managers	4.12	0.781
2. Economic situation of ship management companies	3.96	0.889
3. Supply-demand balance of seafarer	3.92	0.997
4. Rate of freight market	3.88	0.881
5. Wage policy of other ship management companies	3.76	0.831
6. Human resource manager	3.76	0.879
7. Administrative board	3.64	0.952
8. Owner	3.56	1.003
9. Department manager	3.56	0.820
10. Labor unions (ITF)	2.48	1.357
11. Non-governmental organizations	1.68	0.802

*[1]: strongly disagree – 5: strongly agree

The most popular pay strategies in the labor market are "lead", "meet" or "lag" the markets. The strategy is determined with the relative position of the proposed compensation package according to the compensation package proposed in similar jobs. If the proposed compensation package position is above the level of in similar jobs, strategy called as a "lead the market". According to Koss (2008); "meet" the market is the most common pay strategy. Table 2 reveals that which pay strategy is used by Turkish ship management companies. As seen in the Table 2, "meet the market considering ship type" is stated 11 times and "lag whole market" is not stated by ship management companies. This result shows that, "meet the market" strategy is the most commonly used strategy in the Turkish Seafarers Market.

Table 2. Pay Strategy of Turkish Ship Management Companies

Variables	Frequency
1. Meet the market considering the ship type	11
2. Meet the whole market	6
3. Lead the market considering the ship type	5
4. Lead the whole market	2
5. Lag the market considering the ship type	1
6. Lag the whole market	0

Table 3 reveals that how ship management companies carry out their compensation survey. The most preferred method of survey has been found as "communicate with the other ship management companies" (4,24). At least the preferred method of survey has been found as "with a professional survey company" (1.40).

131

Table 3. How to Carry out Compensation Survey

Variables	Mean*[1]	Std. Dev.
1. Communicate with the other ship management companies	4.24	0.723
2. Feedback of employed seafarers	3.80	0.913
3. Prepared researches	2.92	1.256
4. Professional survey company	1.40	0.645

*[1]: never – 5: always

The reasons to do a survey carried out by the companies are shown in the order in Table 4. Ship management companies are mostly doing compensation survey to compare the compensation packages with other ship management companies, to retain qualified seafarers, to achieve fairness in the company and to acquire qualified officer. On the other hand, the variable which has the lowest mean with 2.86 has been found as "to avoid paying over market".

Table 4. The Reasons to do a Compensation Survey

Variables	Mean*[1]	Std. Dev.
1. To compare compensation packages	4.24	0.597
2. To retain qualified seafarers	4.20	1.118
3. To achieve fairness	4.00	1.080
4. To acquire qualified seafarers	4.00	1.155
5. To motivate seafarers	3.92	0.862
6. To determine wage policy	3.84	0.943
7. To reduce turnover	3.80	1.042
8. To compete with other with other ship management companies	3.44	1.044
9. To avoid paying over market	2.68	1.282

*[1]: strongly disagree – 5: strongly agree

In Table 5, the benefits of effective compensation management were presented. As seen from the Table 5, effective compensation policy attracts qualified seafarers, provides long term employment, enhances motivation, enhances seafarers' loyalty, and reduces turnover rate. Mean of these five items are over 4.00 and it can be said that a effective compensation policy is indispensable management tool for ship management companies.

Table 5. The Benefits of the Effective Compensation Policy

Variables	Mean*[1]
1. Attracting qualified seafarers	4.52
2. Providing long term employment	4.48
3. Enhancing motivation of seafarers	4.40
4. Enhancing loyalty	4.32
5. Reducing turnover rate	4.32
6. Enhancing working period on the sea	3.12

*[1]: strongly disagree – 5: strongly agree

In Table 6, the effects of seafarer's qualifications on determining compensation were examined. The "service times of seafarers in the company" and "competency" have the biggest effect on the compensation in the Turkish ship management companies. It has found that, eighty eight percent of ship management companies are making additional

payment called as a "seniority pay" considering to the service times of seafarers. Other factors in the survey have nearly the same effect on the compensation which are measured and examined during the recruitment process of seafarers.

Table 6. The Effects of Seafarer's Qualifications on Determining Compensation

Variables	Mean*[1]	Std. Dev.
1. Service times of seafarers in the company	4.16	0.943
2. Competency of seafarers	3.96	1.098
3. Profession knowledge	3.68	1.406
4. Personality traits	3.44	1.356
5. Attitude and behavior	3.44	1.258
6. Compliance with colleagues	3.44	1.227
7. Performance assessment conducted on board	3.36	1.221
8. Superior's opinions about seafarers	3.36	1.254
9. Language skill	3.32	1.249
10. References	3.28	1.339
11. Educational status	3.12	1.236
12. Graduation school	2.92	1.256

*[1]: ineffective – 5: very effective

The question of "which factors are more important than the level of wage?" was asked to the respondent taking into account the view of seafarers. It is determined that any of respondents are considering that the level of wage is the most important factor for the seafarers. % 84 percent of respondents indicated that "reputation of company", "confidence in company" and "paid in time" variables are more important than the level of wage. The other factors are listed in the order in Table 7.

Table 7. The Factors more Important than the Level of the Wage for Seafarers

Variables	Frequency
1. Reputation of ship management company	21
2. Confidence in ship management company	21
3. To be paid in time	21
4. Working conditions	18
5. Provision	14
6. Duration of contract	14
7. Social environment on board	13
8. Passage area	11
9. Condition of labor market	8
10. Fleet age	7
11. Do not have an opportunity to work for an another company	3
12. Obligation to work	1
13. There is not an important factor than wage	0

4 CONCLUSIONS

The most effective factor on determining compensation of seafarer has been found as "senior managers". Also, economic situation of ship management companies and situation of supply-demand balance are the other important factors which effect mostly seafarers' compensation package. Companies have to choose a pay strategy to compete with their rivals in the market. In this study, it is found that the most of ship management companies use "meet the market pay strategy" to survive in shipping market. Before choosing a pay strategy, ship management companies conduct a compensation survey to compare the compensation packages with other ship management companies.

In the study, it has found that eighty eight percent of ship management companies are making additional payment called as a "seniority pay" considering to the service times of seafarers. Other qualifications of seafarer such as "competency", "profession knowledge" and "language skill" have nearly the same effect on determining compensation of a seafarer.

The most of respondents stated that "reputation of company", "confidence in company" and "paid in time" are more important factor than the wage level for seafarer.

To sum up, this study revealed that compensation management is one the most important strategic management tool for ship management companies. Most of ship management companies use effective compensation management to attract and retain qualified seafarers, to provide long term employment, to enhance the motivation of seafarers and to reduce turnover rate. By this way, they maximize their profit, minimize expenses which arise from the recruitment process of a seafarer and obtain competitive advantage against their rivals.

REFERENCES

Bhatia, K. 2010. *Compensation Management*. Mumbai: Himalaya Publishing House.
Gonuldas, H. E. 2006. *Compensation Management and Application in Finance Sector*. Unpublished Master Thesis. İstanbul: Marmara University, Institute of Social Sciences.
Koss, S.K. 2008. *Solving the Compensation Puzzle: Putting Together a Complete Pay and Performance System* .USA: Society for Human Resource Management.
Sherman, A.W.& Bohlander, G.W.1992. *Managing Human Resource*. Eight Edition. Cincinati, Ohio: South Western Publishing Company.
Turker, F. & Er, I. D. 2007. *Investigation the Root Causes of Seafarers' Turnover and its Impact on the Safe Operation of the Ship*. TransNav - International Journal on Marine Navigation and Safety of Sea Transportation, Vol. 1, No. 4, pp. 435-440.
Werther, W. B. Jr & DAVIS, K. 1994. *Human Resources and Personnel Management*. 4th edition . Singapore: McGraw Hill International Editions.

Human Resources and Crew Manning
STCW, Maritime Education and Training (MET), Human Resources and Crew Manning, Maritime Policy,
Logistics and Economic Matters – Marine Navigation and Safety of Sea Transportation – Weintrit & Neumann (Eds)

Analysis of Parameters and Processes of Latvian Seafarers' Pool

R. Gailitis
Latvian Maritime Administration, Riga, Latvia

ABSTRACT: Maritime industry plays a key role for economy of the European Union. During recent years increasing attention is paid to the education and training of seafarers as the seafarers and their knowledge are essential to sustainable development of the maritime cluster, as indicated in the maritime strategy of European Union. However, for successful implementation of the strategy understanding of current trends in the pool of seafarers is prerequisite. Therefore the aim of this article is to analyse parameters and processes of Latvian seafarer's pool, based on information from database of Maritime Administration of Latvia Seamen Registry. The structure of database and data collected there gives possibility to analyse the processes taking into account the global changes in shipping and their impact to the structure of the pool. Such analysis in combination with the calculations of economic value of seafarer's pool creates the framework on which the decisions about the implementation of maritime strategy in Latvia can be taken.

1 INTRODUCTION

Nowadays wide attention within maritime society is paid to the shortage of seafarers. It is subject for lot of discussions going on in maritime industry. The worldwide supply of seafarers in 2010 was estimated to be 624,000 officers and 747,000 ratings, while the current estimate of worldwide demand for seafarers (in 2010) is 637,000 officers and 747,000 ratings (BIMCO/ ISF 2010). As one of the response to shortage of seafarers the International Maritime Organization in 2008 launched campaign "Go to sea" to attract entrants to the shipping industry (IMO 2008).

However it takes time for such a campaign to give results. And it is difficult to assess the results if there is no available information about changes in number of entrants. The Task Force on Maritime Employment and Competitiveness in their report underlined the need for reliable data to assess the scope and scale of the problems regarding shortage of seafarers (EC, 2011a).

European Union (EU) policy papers also underlines the importance of shortage of the seafarers as the seafarers and their knowledge, skills and competences are perceived as important resource for sustainable European maritime cluster development. European Maritime Safety agency is going to create unified database collecting the information regarding the issued certificates of competency by EU member states. While this data system is not launched it is difficult to predict what will be the possibilities of such a system. The problem with the data is clearly stated in Study on EU seafarers' employment - It is clear that detailed data on maritime employment is scarce, sometimes outdated and often not reliable. Moreover, the great differences from a country to another in data collect and presentation of results prevent all serious analysis on employment structure and evolution (EC, 2011b).

Therefore this article will show the experience of Seamen Registry of Latvian Maritime administration (LMA) in analysis of parameters (number, age, qualification) and processes (inflow, outflow) of Latvian Seafarers pool.

2 DATABASE OF LATVIAN SEAMEN REGISTRY

The basis for any analysis is information. To analyse processes and parameters of pool of seafarers Seamen Registry of Latvian Maritime Administration uses information from database of Latvian Seamen Registry. The main aim for database of Latvian Seamen Registry is to serve for seafarers' certification purposes according national and STCW requirements. The database contains information not only about the issued certificates but also about the employment, education, training and seamen discharge book. See table 1.

Table 1. Data areas, sources and main information fields of database of Latvian Seamen Registry (LMA 2013)

Data area	Main data provider	Main fields
Personal data	Seafarer	Birth date, place of living, nationality
Certificates of competence and other qualification documents	Issuing organization, seafarer	STCW reference, validity of endorsement
Education	Education institution	Name of programme, qualification, STCW reference, graduation year
Training	Training organization	Training course, STCW reference, validity of course, course date
Employment	National crewing agencies	Ship name, flag, employment, period, capacity on board

The database is based on the seafarer as data unit and related records of individual person. Each person in database has his own information blocks, where records regarding his education, training, employment certificates can be found. Main data providers are the Seamen registry, seafarers, companies and training centres.

3 NUMBER OF SEAFARERS IN LATVIA AND THEIR AGE

The number of seafarers in Latvia is calculated every year with reference date 1st January. The base for the number of seafarers is taken the number of valid documents of competence or qualifications such as certificates of competency or certificates of qualification. The validity period of endorsement or qualification document is five years, therefore it is assumed if the endorsement is not revalidated seafarers has been left the pool of active seafarers. The similar approach is used in UK where the active number of officers is based on data on those seafarers holding Certificates of Competency (CoC) (Department of Transport 2013). The difference with the UK approach is in determination of number of ratings as in UK these data comes from a membership survey conducted by the Chamber of Shipping but in Latvia the validity of qualification document is used.

The data about employment cannot provide fully complete picture of active number of Latvian seafarers as approximately 10% of seafarers are employed directly or through foreign crewing companies, which are not obliged to provide data about employment to Latvian Seamen registry. Also part of seafarers doesn't sail regularly, therefore it is difficult to assess employment data accuracy level and they are used as secondary indicator. In calculations the last valid certificates is taken as indicator showing the qualification and its level as person can have two or more valid certificates on hand at the same time.

The active or employed seafarer is person who holds a valid certificate and therefore can be employed on board. Therefore the published figures about the number of seafarers reflect the number of persons with valid certificates even they are not employed on board in particular year. The size on structure of active seafarers' pool in Latvian on is given in table 2.

Table 2. Size and structure of seafarers' pool (LMA 2013)

Total number of seafarers	12,970
1) Merchant fleet seafarers:	11,960
1.1) Deck department	5650
Officers	2500
Ratings	3150
1.2) Engine department:	4890
Officers	3040
Ratings	1850
1.3) Catering department (cooks, stewards)	1400
2) Inland fleet seafarers & personell of fishing vessels	1010

92% of seafarers are merchant fleet seafarers. From them 47% are classified as deck department seafarers, 41% are classified as engine department seafarers, but 12% are classified as catering departments' seafarers. As the shortage of seafarers is referenced to shortage of officers detailed structure of officers is shown in table 3.

Table 3. Structure of officers according the department (LMA 2013)

Deck officers from officers		45%
Officer in charge of navigational watch (from deck officers)	A-II/1	29%
Master on ships of 3000 GT or more (from deck officers)	A-II/2	39%
Chief officer on ships of 3000 GT or more (from deck officers)	A-II/2	24%
Engine officers from officers		55%
Officers in charge of an engineering watch on ships with 750kW propulsion power or more (from engine officers)	A-III/1	21%
Chief Engineer Officers on ships with 3000 kW propulsion power or more (from engine officers)	A-III/2	33%
Second Engineer Officers on ships with 3000kw propulsion power or more(from engine officers)	A-III/2	21%
Non STCW officers (from engine officers)		19%

Share of deck officers is 45% from all persons with valid officers' certificates. Main part of all officers both in engine (54%) and deck department (63%) has management level certificates which allow them to sail on ships with 3000 GT/3000 kW propulsion power or more. The number of seafarers doesn't provide information about how much seafarers are close to retirement age or how much seafarer started their career recently. To have overall picture the figure 1 shows the average age structure of merchant fleet seafarers.

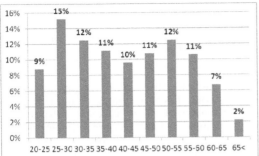

Figure 1. Age structure of merchant fleet seafarers (LMA 2013)

Analysing the average age structure it can be seen that proportion of merchant fleet increases till age of 30 years and then decreases. Active number of seafarers decreases starting from age 60. It can be assumed that active seafarers are in age range from 20 -65 years (approximately 96% of seafarers are within this age range). Also it is clear that seafarers over 65 years old will retire in closest years. Parallel the seafarers with different qualifications from different departments can be analysed. The table 4 shows how much seafarer from all seafarers with particular qualification can be found in age group.

Table 4. Age groups of seafarers in deck and engine department (LMA 2013)

Age group	20-30	30-40	40-50	50-60	60<
Masters on ships > 3000 GT	1%	21%	21%	42%	15%
Chief officers on ships > 3000 GT	18%	45%	17%	16%	4%
Watch officers on ships > 500 GT	53%	26%	12%	7%	2%
Deck ratings	35%	22%	18%	18%	6%
Chief engineers on ships > 3000 kW	0%	19%	29%	35%	17%
Second engineers on ships > 3000 kW	9%	36%	29%	20%	6%
Watch engineers on Ships > 750 kW	40%	35%	17%	9%	1%
Engine ratings	27%	18%	20%	25%	10%

Approximately half of the ratings both in engine and deck department are in age till 40 years. Largest group of watch officers and engine officers is in age till 30 years. The largest groups of chief engineers and captains are in age from 50 till 60 years. Relative number of ratings is quite similar in age groups from 30 - 60 years which can be related to the limited possibilities of the growth in comparison with engine and deck officers where amplitudes of relative numbers are higher. The relative number

changes differences in different capacities show the career path on board from watch deck/ engine officer to the master or chief engineer.

4 INFLOW OF SEAFARERS

However these parameters reflect only how much active seafarer are in particular age group. To understand the inflow process the overall analyses of persons who have received document of competence issued by Latvian Seamen Registry is applied in table 5. The time period is chosen 2005 – 2012 to have view on inflow changes in last year's.

The number of persons reflects how much persons received their first document of competence at given year. As it can be seen approximately 82% of all persons were in age till 30. Also half of the persons received the document related to deck department (rating or watch officer). It can be also concluded that inflow has been decreased in last year's. To understand what will be the impact on smaller inflow to the total number the whole pool both active and ex seafarers should be considered and outflow of the pool due to the age and other reasons.

Table 5. Data of inflow of merchant fleet seafarers (2005-2013) (LMA 2013)

Year	Number of persons	In age till 30	Department		
			Deck department	Engine department	Catering department
2005	540	83%	56%	35%	9%
2006	420	83%	43%	34%	23%
2007	340	86%	55%	29%	15%
2008	320	81%	46%	25%	30%
2009	520	79%	43%	26%	30%
2010	460	77%	49%	28%	23%
2011	400	85%	50%	31%	18%
2012	390	85%	52%	20%	28%

5 OUTFLOW OF SEAFARERS

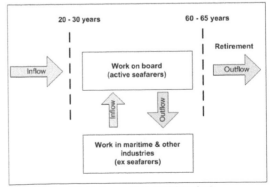

Figure 2. Simplified inflow outflow model (author)

Outflow of the pool is linked with two factors (see picture 2). One factor is retirement when seafarers reach the age of 65 as it can be concluded from age structure of active seafarers. (Only two percent of active seafarers are in age over 65.) Other factor is demand from maritime companies ashore which are willing to employ persons with seagoing knowledge and experience and employment possibilities in other industries which can utilise the skills of seafarers in the same time providing competitive or better employment conditions.

Determination of persons who will retire in next five years is quite simple. From active seafarers 1050 seafarers are in age over 60 years. From them 15% are persons with certificate "Masters on ships > 3000 GT", 16% are persons with certificate "Chief engineers on ships > 3000 kW" but 36% are persons with deck and engine rating certificates. Therefore it can be assumed that outflow rate due to the age at this size and composition of pool is approximately 200 persons per year.

It is more difficult to determine the demand from maritime companies ashore or to quantify pulling factors from other industries. Therefore pool of all persons which have ever received document of competence and are in age till 65 should be considered. According the data from database there are 19 150 persons who have received a document of competence to serve on merchant vessels and now are in age range from 20 till 65 years. Therefore as ex seafarers in total can be considered 7700 persons as number of active seafarers is 11 450 within age range from 20 - 65 years. The composition of ex seafarers is given in table 6.

Main part of ex seafarers is deck and engine ratings. Approximately 28% or 2120 persons from ex seafarer had competence document as officer. Therefore it can be assumed that approximately such demand exist ashore for persons with officers qualification, educational background, experience and skills. However database doesn't allow following to their employment ashore to see if they are still working within maritime industry or they are working in other industries. Other factor which can be derived from database is activity level of particular qualification (see table 7). This parameter shows how much persons from all persons with such a qualification are still active.

Table 6. Composition of ex seafarers (LMA 2013)

1. Deck department	53%
1.1. Deck officers	12%
1.2. Deck ratings	42%
2. Engine department	38%
2.1. Engine officers	16%
2.2. Engine ratings	22%
3. Catering department	9%

Table 7: Activity level of particular qualification seafarers (LMA 2013)

Masters on ships > 3000 GT	80%
Chief officers on ships > 3000 GT	79%
Watch officers on ships > 500 GT	70%
Deck ratings	49%
Chief engineers on ships > 3000 kW	84%
Second engineers on ships > 3000 kW	74%
Watch engineers on Ships > 750 kW	57%
Engine ratings	52%

The highest activity is for management level officers while lowest for ratings both in deck and engine departments. These data shows that demand from maritime companies ashore to employ persons with officers' maritime knowledge and experience is fully covered and companies have possibility to find right person for employment ashore. It can be also seen that activity level goes down with the relative position on board which underlines factor that ratings are more mobile in obtaining the qualification than officers and for them it is much easier to take decision for employment ashore. Hypothetically it can be linked with relatively shorter time period (several months) to obtain ratings qualification instead of several years which is case for officers. Other factor which influences the number is employment possibilities ashore in combination with wage level which in some industries are comparable to wage level on board for rating. Of course additional data analysis should be carried to verify these assumptions.

Table 8. Outflow at particular age group (2005 - 2010) (LMA 2013)

Age group	Deck officers	Engine officers	All officers	Ratings
20-25	26%	27%	27%	35%
25-30	23%	20%	21%	37%
30-35	20%	15%	17%	33%
35-40	19%	14%	16%	30%
40-45	21%	22%	22%	26%
45-50	19%	22%	21%	28%
50-55	26%	21%	23%	33%
55-60	45%	48%	47%	61%
60-65	62%	65%	65%	84%
65-70	88%	84%	85%	100%
Average 35% till 60	25%	24%	24%	

From previous analysis it can be determined that main inflow is till age of 30 years and the outflow due to the retirement is after 65 years. Also the size of demand for ex officers is determined. However these data is not sufficient to determine what is outflow at particular age group. Therefore the data about employment of Latvian seafarers was compared in 2005 and 2010. The sample of seafarers employed on different types of ships in 2005 is 10 980 seafarers. Comparing with data about employment in 2010 the size of sample is 10 050. It gives possibility by comparing the data about the seafarers who are employed in both years calculate how much of them have probably left the pool (see table 8). As it is said earlier valid certificate of competence is not always linked to employment on board therefore the data of employment is used in this case.

The data shows that persons till age of 30 have higher probability to leave the pool than persons in age range 30-55 years. Starting from 55 years persons have high probability to leave the pool, which can be linked to the coming retirement age and motivation to go to work ashore. Those data again shows the tendency that ratings are more elastic to leave the employment at sea than officers as outflow rates for ratings are higher than for officers. Also it can be assumed that higher rates in age group from 20 - 30 means that for other industries or maritime companies is much easier to attract persons at that age. For all seafarers including the catering department (which is no showed on table 8) the average outflow rate is 32% which means that additionally to the retirement on average one third of active seafarers leaved the pool due to the demand factors ashore. However those figures in table 8 should be considered with care as the data in table reflects maximum outflow values.

6 CONCLUSIONS

From showed analysis it can be concluded that for present composition of Latvian seafarers' pool the inflow is sufficient to cover the retirement outflow. The other important factor which influences the number of available or active seafarers is outflow due to the demand from maritime companies ashore or other industries which can offer the competitive employment conditions. The present analysis doesn't offer possibility to determine average value for outflow ashore which could give possibility to determine if inflow is sufficient to cover the demand from ashore.

Data about the qualification documents offers the possibility to determine the available resource of seafarers both at sea and ashore (ex seafarers). In combination with demographical profiles it gives possibility to determine outflow due to retirement age, however the data doesn't provide possibility to see the outflow process to the maritime industry therefore additional data should be used to have clear view on such processes.

However it is clear that in determination of shortage of seafarers demand from ashore also should be considered and included.

7 FURTHER RESEARCH

The shortage of seafarers is complex issue as trade and shipping is global therefore it is linked with different factors like local economic development, attraction of other industries which can influence the inflow of youngsters to seafarers' profession. For example this is the case in Philiphines where part of youngsters' prefer other courses and programmes instead of maritime programmes as these programmes are considered to be equally financially rewarding. (Magramo, Bernas, Calambuhay, Eler, 2010). This factor is related to the fact that salaries offered by the same shipping company can have different attractiveness for seafarers coming from different countries. It is predicted that in future the majority of marine officers worldwide will be employed from the less developed countries (Glen, 2008; Sencila, Bartuseviciene, Rupšiene , Kalvaitiene, 2010). Therefore to have clear picture on the supply and demand for seafarers the changes in internal factors of seafarers pool like number of active seafarers, ex- seafarers, inflow level, outflow level should be analysed in context with environmental factors like attractiveness of salaries offered by shipping companies, employment in region, emigration possibilities, attractiveness of maritime education. For such research data extracted from database of Seamen Registry of Latvian Maritime Administration have considerable value as they represent the primary source of bulk data. Even the data reflects local processes of Latvian seafarers' pool the outcome of research will give possibility to apply gained knowledge for global processes influencing the shortage of seafarers.

REFERENCES

BIMCO/ISF 2010. Highlights from the Manpower 2010 Update.
https://www.bimco.org/en/News/2010/11/~/media/About/Press/2010/Manpower_Study_handout_2010.ashx. [Retrieved 05 02 2013].

European Commission 2011 a. Report of the Task Force on Maritime Employment and Competitiveness and Policy. http://ec.europa.eu/transport/modes/maritime/seafarers/doc/2011-06-09-tfmec.pdf. [Retrieved 05 02 2013].

European Commission. Directorate-General for mobility and transport 2011 b. STUDY ON EU SEAFARERS EMPLOYMENT. FINAL REPORT. http://ec.europa.eu/transport/modes/maritime/studies/doc/2011-05-20-seafarers-employment.pdf. [Retrieved 05 02 2013].

Glen., D. 2008. What do we know about the labour market for seafarers? A view from the UK. *Marine Policy* 32 (6): 845-855.

Magramo M., Bernas L., Calambuhay J., Eler G 2010. The Role of the Maritime Institutions on the Shortage of Officers. *TransNav - International Journal on Marine Navigation and Safety of Sea Transportation*, Vol. 4, No. 4, pp. 449-451.

Sencila V., Bartuseviciene I., Rupšiene L., Kalvaitiene G 2010. The Economical Emigration Aspect of East and Central European Seafarers: Motivation for Employment in Foreign Fleet. *TransNav - International Journal on Marine Navigation and Safety of Sea Transportation*, Vol. 4, No. 3, pp. 337-342.

UK Department for Transport. 2013. Statistical release Seafarer Statistics 2012. https://www.gov.uk/government/uploads/system/uploads/attachment_data/file/70026/seafarer-statistics-2012.pdf. [Retrieved 05 02 2013].

IMO Go to sea! Campaign document. http://www.imo.org/OurWork/HumanElement/GoToSea/Documents/Gotosea!campaigndocument.pdf. [Retrieved 05 02 2013].

Human Resources and Crew Manning
STCW, Maritime Education and Training (MET), Human Resources and Crew Manning, Maritime Policy,
Logistics and Economic Matters – Marine Navigation and Safety of Sea Transportation – Weintrit & Neumann (Eds)

The Meaning and Making of the First Filipino Female Master Mariner: The Story of Capt. Ramilie Ortega

M.L. Arcelo, L.D. Gellada, G.M. Eler & M.M. Magramo
John B. Lacson Foundation Maritime University-Arevalo, the Philippines

ABSTRACT: Throughout history, women have always aimed for a recognized place in society. Although society has prescribed that a woman's place is usually at home, taking care of the children, many feminist movements force the issue of women's rights to come into people's awareness. These frontrunners which have helped redefine and consolidate the nature of women's contributions to society show that progress has been made. Today, women have access to education and can promote themselves much more easily than in the previous years. Women's changing role happens because women nowadays are more educated than in the past. This research is a qualitative type of research utilizing the narrative as an approach in data gathering. It highlighted the story of the first female master mariner in the Philippines.

1 INTRODUCTION

The United Nations in its desire of promoting women's employment and the integration of women into all levels of political, economic, and social development worked cooperatively with the IMO in producing a strategy for integrating women into the maritime sector. Hence, in 1988 , it began to implement its Women in Development (WID) programme concentrating on equal access to maritime training through both mainstream programmes and gender specific projects. One of the immediate impacts of the programme has been the rise in the percentage of women students taking part in the highest level of maritime training.

The SIRC/ILO survey in 1995 revealed that women constituted less than 8 percent in the total number of students at the World Maritime University. In the overall, the participation rate of women in seafaring remains low. It had about 1 to 2 percent of the world's 1.25 million seafarers. Moreover, most of these women are form developed countries. The study by Belcher, et al.(2003) found out that women continue to constitute a very small part of labor force of seafarers.

With women reaching out and grabbing the most coveted jobs in the world, I'm sure everybody will agree that this truly is the era of women's liberation. From being a super homemaker to maneuvering airplanes, from commanding respect as a tough boss to making way into the outer space, there isn't a job in the world where women look out of place, except the profession of Merchant Navy.

While women are breaking the traditional barriers in every walk of life, merchant navy jobs are surprisingly conspicuous by women's absence. And though there might be a handful of women earning their bread and butter working on board, merchant navy still remains a male dominated industry.

What exactly is the reason behind lack of women crew in merchant navy is a tricky question with more than one answer. Life onboard is tough, something that breaks even the most stoic ones. Women are often felt out of place on ships primarily because of the nature of the job which might require quite a lot of manual labor. And since women are often looked as the "weaker-sex", people are skeptical of whether a woman has what it takes to endure such harsh conditions of the house. Though we have come a long way from those pre-historic times, the idea of women leaving their homes for months together, is something not many can relate to. Many men find it intimidating and frustrating when a women leaves home for so many months to go away on a job.

There may be times when the choice, of not venturing into the sea, might be a woman's own. Even if a woman continues to sail after marriage, chances are, she will give it all up when she starts a family. Staying away from family might be something a woman may take into account but leaving children behind can be very tough for her.

The situation is different for women who work the typical 9 to 5 jobs as they see their children at the end of the day, but with a job like merchant navy, seeing their children is not possible for months at end. And for women no job in the world is worthy of this kind of sacrifice.

According to Singh(2011) it will be a while before we see women venturing out at sea without any hesitation and fear. Though times are changing and many companies are hiring women crew, to see the whole industry change, can take a lot of time and effort. The technological advances on the ships ensure that the hard-work and manual labor is significantly reduced, hence making it easier for women to handle things onboard. Also with changing outlook towards working women, many husbands may actually appreciate if their wives sail and travel around the world. At the end of the day, it doesn't matter what brings more and more women to this profession as it'll surely be a very welcoming change.

2 METHOD

This study utilized the descriptive-qualitative type of research. A single participant was the main focus of this research. It made use of the narrative approach in gathering the data for this research by conducting an in-depth interview of the participant. Transcriptions of the video-taped interview proceeding was done and a print-out was made. From the print-out of the interview proceedings the manuscript was analyzed and themes were organized.

2.1 *Choice of a career*

The choice of Bachelor of Science in Marine Transportation (BSMT), popularly known as nautical was the first choice of Capt. Ramilie Ortega, a native of Cabatuan, Iloilo. This seems contrary to the notions of those who know since her father is a mariner too, and a Chief Engineer. Yet, it was her own decision to take the nautical course after having heard a group of faculty conducting career guidance when she was in the fourth year. It was imprinted in her mind that she could make her dream a reality, that it was her dream to work on a ship, and that she considers the job to be challenging. She looked at a woman working on a ship unique as well as special and since she wanted to be special, to be different and unique from other girls she wanted to be a seafarer. Probably she is one in a million girls, she decided to take the nautical or BSMT course. Taking this course for her would be fulfilling a dream, a dream to be an officer on board an ocean-going vessel.

For Ramilie then, there is no looking back. She did not hesitate to take the BSMTcourse despite the fact that her father, a Chief engineer was the first person to object her plan of taking up the course. He knows how difficult life on board ship is; for he exactly knows what life is all about at sea being a seafarer himself. He was the first person who told her that this profession is not for a girl like her; that a girl's place is at home. This he would always tell his daughter, Ramilie. But they were not aware that she had already enrolled in the BSMT course although she had already passed the entrance examinations given by University of the Philippines and Central Philippine University.

She found herself alone in a line full of men for the ROTC section. She had to find a way how to enrol among these men; she knew that this was part of the trials she would be facing now and in the future. On the line were big boys and she didn't care because she knew that it was just the beginning, that she was likely to encounter more of this situation in the future. She was fully aware that this was just a part of the many challenges she would be encountering in her profession as she was already beginning to conquer a man's world. She believed in herself and she was pretty aware that she would be able to surpass this challenge.

Ramilie's choice of this profession dates back in high school. Her mother is from Cebu and everytime they would take a boat or a ship to Cebu during vacations she would wonder how a ship reached its port of destination. At that time, she knew that sea travel was different, that it was not like travelling a road or a land trip and taking the highway. She would always wonder how a vessel from Iloilo City reached its destination in Cebu considering it was in the middle of the sea. It came to be a big question for Ramilie from then on; that she would like to find out for herself. This was one big motivation for her to take the nautical course despite the fact that her parents especially her father, were not aware until after a year when her father discovered for himself that she enrolled in the nautical course. Her father thought she was enrolled in nursing for she was also wearing an all-white uniform for the nautical course. When her father was around during his vacation from sea, she would be wearing white rubber shoes instead of the black shoes for the nautical course. She wore summer white with white t-shirt and the white rubber shoes. She would take her black shoes off and left it at the locker room whenever her father was around. Actually her father dreamed of her to take any of the medical courses, for he would be willing to do everything to support her to a good medical school.

2.2 Her father's frustration

Everytime Ramilie went home from school with white pants, white t-shirt and white rubber shoes her father thought she was taking up a medical course. At one time when her father was on vacation late in the second semester of her first year taking up the nautical course, her father noticed that most of her friends were boys. Whenever these friends were at home doing group study or doing projects, she was always with friends who were either male, if not gays. Although she had girl classmates, they were coming to their house very rarely. Thus, her father started to think if her beloved daughter was indeed taking up a medical course. On one occasion, her father followed her in going to school. And then he found out for himself that his loving daughter was enrolled not in a medical course but in a nautical course.

After her father had discovered that Ramilie was enrolled in a nautical course, he was so disappointed with his daughter. Since it was her daughter's choice and not his, he told her that she had to finish the course but never expect help from him, who is a seafarer himself. "Let us see if you can board a vessel" quipped her father. Male graduates find it difficult to land a job at sea nowadays, how much more for somebody like you who is a girl. Let us see, but do not expect help from me", he repeatedly stated. He further told her to look for employment herself after graduation. Ramilie took this as a challenge and since it was a statement coming from her father, perhaps it was the biggest challenge in her life. Incidenta lly, it was also at this year when one female cadet taking up marine engineering course got pregnant. It added up to the challenge she had at hand.

2.3 Dealing with boys in school

Ramilie experienced pressure from her classmates when she was in First Year. Since the first day in class, boys would come close to her and try to win her attention. Passing through the classrooms of the senior cadets seemed like eternity. Everybody would be looking at her whenever she would transfer rooms in between classes. Students she met on the hallway would be asking for her name, her number and asked many other things personal. But in all these she had never encountered fellow male students who made nasty or dirty comments against her. It never occurred to her that they had become rude and ill-mannered. Boys were just too good and behaved trying to impress her and win her attention. She could not deny that as a normal growing up girl, she too had crushes towards the opposite sex. Among the many boys who tried to win her heart she fell in love for the first time when she was already in the second year of her schooling. He was

an NROTC Officer, a well-respected cadet among the students. With him she felt secured because many of the students had high regard and respect for an NROTC Officer like him. He was Capt. Ramilie's first boyfriend. But it was a relationship that had to be kept a secret; otherwise her parents would be scolding her. Ramilie was the typical growing up girl who at her young age had a boyfriend.

2.4 Life on board for the first time

After graduation, Ramilie joined a tanker vessel. She was able to successfully hurdle the qualifying examination given by Philippine Transmarine Carriers (PTC) before her graduation. Upon learning that Ramilie would be working on a tanker, her father was not happy and again did not talk to her. He joined a tanker vessel before and had since transferred to a non-tanker vessel. If he could not work for a longer time on board a tanker though a veteran at sea, how much more for a neophyte like his daughter. He knew how difficult life would be for her especially on a tanker. Her father wanted her to work on a cruise ship, for it would be much easier in a passenger ship than in a tanker.

Her first flight was to Italy. She was scheduled to join her vessel in this country. She was thinking then if she could be able to handle these matters, being away from her mother and siblings and the kind of life that awaited her. Upon arrival in Italy, she had to take a service boat that took her to where the vessel was anchored. When she saw the vessel she was asking herself if she could handle the task ahead of her. And she told herself there was no turning back, that she had to face the kind of life she had dreamed of, the life that she ever wanted to have. It was a product tanker, about 65,000 GRT. Everybody onboard had been waiting for her arrival; they were quite intrigued as to the entry of a Filipina girl joining the vessel.

Her father started to notice her dedication to her work and somehow she was able to get his approval when she took the board examination for Officer in charge of a Navigational Watch or the Operational level. He slowly realized that this is the kind of life that his daughter wanted to have, be a dedicated and motivated seafarer like him, a Chief Engineer. It was at this point when her father started to appreciate the work of his beloved daughter.

As a first class cadet, Ramilie indeed experienced a little difficulty of being away from her parents. Having grown up in Cabatuan, a town about 30 kilometers from the city, she was not used to living away from her parents. Joining her ship for the first time was a little awkward for her. Being close to her mother she would be easily taken cared of whenever she felt ill or was not feeling well. They had to move to the city during her college days, because they had to be near their mother.

On her first ship, Ramilie was clueless. She had no idea what awaited her life on board ship for the first time. It was really difficult to deal with men on board, especially on her first ship. Would she talk to them or not? How should every situation be dealt with? As to her work on board, she did not know where to start. Then Capt. Pecaoco came to her rescue, assisted her in her work on board. He had been very supportive to her. On her first ship Ramilie found out that when she talked to crew alone, others would make a story out of it. When there was just the two of them talking they would make malicious talks about them. This first officer who also came from JLFMU-Arevalo (formerly Iloilo Maritime Academy) had been a big help for her and was even very protective of her, being the only woman on board. For this, she was so thankful. When she frequently talked to a guy some crew would think that she already liked the guy she was talking to. But she was talking to them out of courtesy and respect to everybody on board ship. This she experienced for the first five to six months. She would spent some nights crying after hearing some talks she liked this guy or that guy because they spent some time talking. She did not know how to handle the situation; all she did was to cry. Until such time, she was able to adjust and handle each situation. She stopped talking to guys alone in the dayroom or in the galley. She never talked to them alone in her cabin. Whenever she talked to these guys, she saw to it that it was in a group especially during parties.

At last she knew how to handle the presence of guys on board ship. She was able to finish fifteen months on board, although her contract was good for nine (9) months only. She had asked for an extension from the office because she wanted to take the board examination for OIC-Navigational Watch upon her return to the Philippines. She remembered how short her hair was when she joined the vessel for the first time and how long it had grown after fifteen months on board. She was also happy that she found consolation and comfort from the Master who had been supportive of her; a captain who was like a father to her. He had been a captain for a long time now, a seasoned captain.

2.5 Sexual advancement/harassment?

When asked as to whether she experienced sexual harassment, Ramilie related her story when she was already an officer, a second mate then. She was not sure if it was a form of sexual harassment or what. But one thing she was sure, somebody did some nasty thing with her personal belongings. It was around 1500 H in the afternoon just before approaching port when she went down to her cabin from her watch on the bridge to prepare for 'standby operations' when she discovered that her cabin was

locked. She noticed a pair of shoes outside her cabin and remembered that it was usually the time when the mess boy cleaned her cabin. She was alarmed because the usual practice when a mess man cleaned the cabin of the officers, it was kept open since it was a private place. She thought for a while that maybe he was doing some vacuum cleaning near the door that was why it was closed. It was not normal for a mess man to lock the door in cleaning an officer's cabin; so blood was rushing in Ramilie's veins. She could not explain how she felt but she thought it was really bad, she felt something strange was going on. She tried to open the door but it is locked and she does not have her key with her. She then knocked and it took the mess man another three minutes to finally open the door. She was fuming mad when the door was opened and all she could utter was "get out of my cabin." Her personal things were not locked so she was trembling and was really mad and she then focused her attention on the sofa, and on her bed trying to see something. But because time was of the essence to return to the navigating bridge since it was already standby for docking operations, she just closed her door and locked it. She could not keep her mind off with what had just happened.

After the docking operations and when the ship was already moored securely in port, she told the Chief Officer of the incident. She still felt bad of the incident but before telling the incident to the Chief Officer she went back to her room and checked her personal belongings. She made a quick inventory and found out that nothing was lost in her underwear and other intimate apparel. Ramilie finally got her composure back and was able to make a joke out of the incident while retelling it.

An incident of a similar nature happened three weeks before that in another ship. The pajama of the deck cadette was wet with something sticky and watery. This was what Ramilie was entertaining in her mind when that incident happened to her.

As a result of the incident the mess man was called to report to the Chief Officer and was made to explain. He said that it was how he cleaned the room of the officers and it was for Ramilie a strange practice under normal circumstances. While everything was in Ramilie's hand to send the mess man home because of what he did, she also pitied him and gave him a second chance to stay on board and reform. Considering that the mess man was also from Iloilo, she was compelled to forgive him and he could continue to finish his contract.

2.6 Being a young officer and a woman

Ramilie was a third officer when she was 21 years old, and she found herself working with an able-bodied seaman (AB) who was between 45 to 48 years old. Some of them were too weak to move on the deck and she had to do what she ordered in some instances. She would help them pull ropes and even the fire wire and she had to lend her a hand whenever the work was being carried by an older seaman. At this stage in her career sometimes she was ashamed to ask these AB to do the work, for some of them had been on board vessels for 20, 25, or even 30 years. Work on deck for Ramilie was not a strange thing. She had done all types of job on deck when she was a cadet and when she was also an AB. She had done exactly what was being done by a common deck hand on board. For her it was important that an officer knew the job that he wanted done by the AB and other seafarers. If an officer did not know how to do the job himself then he had no business giving the orders to do a particular job. If an order was given back to her, she was confident that she could handle the job for she had experienced it herself. When she was Chief Officer, she had 3rd officer and 2nd officer who were foreigners who were just giving orders but they themselves did not know the job. One reason she was really proud of Filipino officers was that they know how to do the job that they wanted done by the ratings. They had gone through these jobs before and this made things easier for Filipino officers. And she proudly claimed she could do any job on deck even if she is a female. If ever there were things or job she could not handle, she always consulted the master of the vessel. She recalled a particular incident when she had this pump man who was new to the company and he had no idea that she had been in this company for eight years already. He kept on complaining about the job order that she would give him, and usually the job was not completed at the end of each day. He would complain that she was giving her jobs which were difficult, when actually they were not. She had previously done the job and she knew pretty well it can be done at the end of the day. She talked to the concerned pump man and told him that it was his job and he must accomplish it. Then finally she brought the matter to the master when it seemed she could not compel him to do the assigned job. The captain called for the pump man and they had a talk in the presence of the master. Finally, they were able to settle things and it was cleared for the pump man that he had to do the assigned task to him.

Ramilie thought that the pump man begrudged her for the incident; but she was so happy to know that everything was smoothly handled in that particular dialogue. She reminded all the crew that they had to follow her orders as Chief Officer. They did not have to like her just because she was beautiful and sexy (she said jokingly) but because she was the Chief Officer. They just have to follow her orders. Some crew may not like her because she was a lady or whatever the reason maybe, they just had to respect her for her position as Chief Officer

on board. For her, respect was everything and it was to be everything especially that she was a woman, the only woman on board. And she made sure that she gained the respect of everybody in any of the vessels she joined.

Today, Ramilie is a holder of a master mariner license. She can take command of a vessel anytime soon. She has gained the respect of everybody on board since the first day she joined the vessel until now as a Chief officer on a chemical tanker.

2.7 On falling in love

Ramilie could be a master mariner licensed officer now and acting as Chief Officer on board a tanker, and many of the crew around her know she is a woman. But nevertheless, we asked her if she had fallen in-love as an officer not the casual puppy love when she was in high school or in her early years in college. These authors asked her whether she is a lady or a lesbian and the quick response was: "I am a lady." She felt in love the second time when she was a second mate. She met this guy from Finland, the nephew of one of the partners of the company she was working for, a designated person ashore. As guest on board the vessel where Ramilie worked for as a second mate, they had been together in this trip for three weeks. He had asked his uncle to board this vessel for he wanted to find out how life was at sea. He started to find her attractive and sweet for he would frequently visit the navigating bridge every time Ramilie was on duty. He would ask about life and family, and even personal life. And then came the time for him to disembark and he told her he was interested in her. But being a, she did not mind it. She did not consider it as a big deal for never did it cross her mind that she would fall in love with a foreigner. She never ever dreamed spending her life with a foreigner. She knew it was impossible to be with this guy for they were a world apart. And so it was time to say goodbye for this Finnish guy. Ramilie still had three weeks to complete her contract.

As soon as she completed her contract on board and she was already on her second week on her vacation in the Philippines Ramilie, received a phone call from the guy in Finland to pick him up at the airport. She immediately told him she still had to arrange for a ticket in order to go to Manila. And she asked him, why? He was so interested to see her again and he was so serious to be together with her, that he had to come all the way to Iloilo to show her how much he cared for her. Making a long story short, they became lovers. She fell in love with this guy, and the relationship lasted more than three years. Actually, they were engaged to be married.

2.8 The wedding proposal

A very romantic proposal happened on board. Everyone on board knew that they had been together for he was allowed to sail on board as Ramilie's visitor or guest. He could sail with her anywhere in Europe. There was a party and I could not recall anymore what the event was, but there he was with a chocolate cake.... she said. She loved chocolate cakes and it was a Finnish cake. And what was inside the cake was the engagement ring and he kneeled and asked for her hand in marriage in front of the Captain, in front of all the crew. "It was the most romantic experience I had ever felt with the cake and the kneeling during the proposal" Ramilie quipped. And so they were engaged, but it seemed that Ramilie was not yet ready to tie the knot. There were still many things she wanted to accomplish; she thought that she had to finish what she had started in her career. She was still not satisfied with what she had achieved so far. While she had been a Chief Officer, she still wanted to be in command of a vessel. Maybe by then, she would be ready to marry the guy she loved. And so the guy started to doubt Ramilie's love for him. Maybe she was not really in love with him why she was not marrying him yet. And when he called her up in Finland after his vacation in the Philippines, he finally asked her if she would marry him or not. If she was not ready to marry him, fine with him for there was this other girl who was ready to marry him anytime. For her, it was a very shallow reason. All that she was asking for him was a little understanding. She had come this far in her career and she could not see any reason why he would not understand that this was her dream. This was her dream to be a full pledged master mariner, to be in command of a vessel. No doubt that she loved this guy but for her he was immature enough not to understand. Ramilie started to doubt her feelings for him, too

2.9 The break up

It was a painful break-up considering that the relationship lasted for almost four years. The first two years were really very good but slowly the real character of the guy had surfaced. She tried very hard to adjust to the cultural differences between them until it was no longer bearable on her part. His upbringing was totally different considering that he was with the Finnish army before. His true colours came out, being strict especially on her ways. For him they could not talk while eating. So because she really would like to talk she had to hurry up in eating her food. She did not like eating and at the same time not communicating, just looking at each other. But still they could not talk because he was slow in eating and she had to wait for him to finish eating in order to talk to him. For him, eating should be done in a formal way, just like in the military. And one more thing, he did not like eating in malls. He would prefer fine dining in exclusive hotels and restaurants while she was happy eating in fastfoods. Sometimes she felt like a robot and operated by a remote control and she could not be her real self anymore.

And the worst was concerning family matters. They would argue a lot, especially money matters. For him, there was no need to extend financial assistance to her parents. It is common in the Filipino culture that children have to extend financial help to their parents and other relatives when they have the capacity. Ramilie would usually give financial assistance every time she would come home after her contract. He would get mad whenever he knew that Ramilie gave a certain amount of money for her mother. Her parents are not receiving monthly allotments from her because she is self-allottee. The orientation of a Finnish family and that of a Filipino is a major cultural difference that is a common source of disagreements between them.

She also noticed that her boyfriend had a bad temper especially when he got drunk. He could not be controlled when he was mad especially when he was drunk. He did this even in the presence of Ramilie's mother. Her mother was a witness how he smashed a bottle after they had argued over something. And so her mother was afraid for her safety. Never did her mother think of something like this for her daughter. But it was something that her mother never pushed her to end their relationship. It was left for her to decide, they had hoped that she would see this as an eye opener. The real attitude and personality of her boyfriend surfaced after two years of relationship. And so she decided to let go of him. She told him to go ahead with his plans of marrying the other woman who had been waiting for him and was always ready to marry him. Ramilie must admit the break up was painful but she was able to handle it.

2.10 A freak accident on board

For reasons until now she could not remember, she met an accident on board. She fell on the stairs around six steps and from the manifold to another six steps on deck. And she also lost her hard hat as she fell on the stairs with her hair slightly hitting the steps. She was thankful she had not suffered from any head injury. She was subjected to an MRI test and several tests were also undertaken. She stayed in Finland for two weeks and after that she was asked if she wanted to undergo therapy in Finland but she opted to go home to the Philippines despite the trouble of having to travel alone in a wheelchair. Indeed it was not easy for her to travel alone not only because she would travel in a wheelchair but that it would be a long flight and she would also be transferring from one flight to another. Despite all the difficulties of her travelling alone in a wheelchair, she decided to be brought back to the Philippines because she knew that the moment she was already back in the country she could be taken cared of by her family. She was confined at a hospital in Manila after which she was transferred to Iloilo for further therapy. Her therapy in the hospital in Iloilo was the beginning of her second love story.

2.11 *The therapy and falling in love again*

When asked how she would describe herself as a woman, she simply said she is a normal woman who wears eyeliners, blush-on, make up, etc, just like what any ordinary girl or lady would usually wear. But she only wears the regular attire of a woman when she is on vacation.... she never wears shorts on board even board shorts. She always wears long pants or jeans and t-shirts. She really never thought that she would fell in love again after the accident.... this time with her therapist.

2.12 *On carrying herself as an officer and earning her subordinates' respect*

Earning the respect of her colleagues and subordinates was something that Ramilie had always wanted to maintain. By not wearing shorts on board ship was one way of bringing across the message to the opposite sex she meant business as an officer. She said that men are men and women should not give any hint to the guys if they want to be respected on board ships. She never drank alcoholic beverages on board. She only took light beer when on vacation in the Philippines. Whenever somebody offers her something hard to drink during parties on board, she never accepted it. She would tell them, "we can drink when on vacation but never while on board."

2.13 *On values and discipline acquired in school*

When asked as to what values she had acquired while she was in school, she immediately replied discipline, patience, respect and loyalty. For her these are the core values that officers must possess in order to be successful in the profession. Respect for your colleagues, superiors and even subordinates must be imbibed by all individuals aspiring to become a master mariner. These she said she owes a lot to the school where she came from, the John B. Lacson Colleges Foundation, now a maritime university. She owes this to her professors in the university who unselfishly shared their knowledge to their students. She takes pride in being a graduate of the university (JBLFMU) that she never experienced being looked down by graduates of other maritime schools in the country and even by foreigners whom she had worked with on board ships. Loyalty pays.... the company that you have worked for will always remember the contribution of an officer and therefore will not forget also the career development of its seafarers. Be loyal to your company and your company will take care of you and your profession.

2.14 *On young girls aspiring to become officers*

She advises young ladies today who aspire to become officers someday to go for it. However, they should consider this profession seriously. She is not asking them to follow what she did while she was on board but to see to it that they will be able to gain the respect, because the moment that you have earned it from them you will forever have it and see to it that you will not lose the respect you have gained. Some crew may not like you at first especially because you are a woman on board, but that crew must respect not only the person but the authority that goes with the position as an officer on board. The work on board for her is not difficult, what is difficult is dealing with these people on board. Sometimes you have subordinates who are twice your age working as able bodied seaman or maybe ordinary seaman. Sometimes it is also difficult to deal with other nationality and an aspiring officer must be patient in dealing with them.

3 FINDINGS OF THE STUDY

1 Women can face discrimination even getting into seafaring work. In some countries, for example, maritime education and training institutions are not allowed to recruit women to nautical courses. Women tend to enrol on navigation rather than engineering courses. Even once trained, they may have to face prejudice from ship owners who won't employ women.

2 Once employed, women seafarers may also face lower pay even though they are doing work equivalent to that of male colleagues. Women may also be denied the facilities or equipment available to male workers, which is a form of discrimination.

3 Women seafarers may also have to deal with sexual harassment or even abuse while at sea. Many maritime unions now have policies covering sexual harassment.

4 Motivation and determination are the qualities of the first Filipino female master mariner to become successful in her chosen career. Conquering a man's world is not easy but when an individual has the proper motivation and a strong determination, he/she can make his/her dream a reality.

REFERENCES

"A Woman Chief Engineer from Brazil Describes Her Interesting Life"; Marine Insight News Network: March 14, 2011 Belcher P. Sampson H., Thomas M. Viega J. Zhao M., 2004. Women Seafarers: Global employment policies, ILO, ISBN 92-2-113491-1, UK

BIMCO/ISF, 2005, Manpower Update: The Worldwide Demand of and Supply for Seafarers Institute fro Employment Research, University of Warwick.

Magramo, M and G. Eler(2012) "Women Seafarers: Solution to Shortage of Competent Officers?:Transnav- International Journal on Marine Navigation and Safety of Sea Transportation, Gdynia Maritime University: Gdynia, Poland.(Vol.6, No.3,September 2012)

The Role of Women Seafarers in Turkish Maritime Industry, Elif BAL and Ozcan ARSLAN(2012). Maritime Transportation and Management Engineering Department,Istanbul Technical University, Turkey.

Seafarer's International Research Centre (SIRC 1999). Seafarers. Cardiff University, Wales UK

Skei, O. M. (The growing shortage on qualified officers), a paper delivered at the 8th Asia-pacific Manning and Training conference, November 14-15, 2007. Manila Philippines.

"Women in the Maritime Industry", SPC Women in Fisheries Information Bulletin # 14: September 2004.

Women Seafarers - Global Employment policies and practices, International Labour Office, 2003, ISBN 92-2-113491-1.

Human Resources and Crew Manning
STCW, Maritime Education and Training (MET), Human Resources and Crew Manning, Maritime Policy,
Logistics and Economic Matters – Marine Navigation and Safety of Sea Transportation – Weintrit & Neumann (Eds)

Hiring Practices of Shipping Companies and Manning Agencies in the Philippines

M.M. Magramo, W.P. Ramos & L.D. Gellada
John B. Lacson Foundation Maritime University-Arevalo, the Philippines

ABSTRACT: This study aims to determine the hiring practices; methods of recruitment, process and criteria for selection that are being adopted by different ship-manning agencies in Iloilo City for the employment of seafarers. The respondents were chosen by purposive type of sampling. Generally, the shipping companies and manning agencies in the Philippines regardless of category adopt almost the same hiring practices. This may be due to the nature of business and profession. This finding implies that the employment process involved the hiring of the best-qualified people for the right job. The finding that shipping companies and manning agencies get their recruits from the same sources reflects the common practice of the agencies to wait for the applicants to come to their offices to apply.

1 INTRODUCTION

The Philippine maritime industry has been called as one of the country's sunshine industry earning millions of dollars in seafarer remittances. In fact, the country has been known globally as the top supplier of seafarers in global labor market, both officers and ratings.

However, Filipinos have mastered the scientific and technological advances in navigation, engineering and communication so as to give them the edge of employment and thwart effort to render them redundant or dispensable. Domestic and world trade have made seafarers the world most globalized force to date, subject to terms and situation than span the world's maritime spaces.

Seafarers on the other hand, are in the lead position to address global environment and human security issues that threaten world peace and development.

In like manner, Filipino seafarers are widely recognized for industry, dedication, adaptability, and reliability. Presently, the enforcement of the Standards for Training, Certification and Watchkeeping (STCW) 1978 as amended in 1995 has caused a change in shipping practices.

Consequently, the most crucial factor in their competitiveness is the quality underlying their education; training, certification and requirements of efficiency and new technologies have likewise the focus in high quality, skills and competences.

Moreover, a high level of quality in these areas does not only ensure economic competitiveness but also contributes to the goals of safety and environmental protection in maritime sector. With these, seafarers play a vital role in maintaining stability and promoting sustainable growth in maritime sector. Shipping companies and manning agencies find difficulties in hiring competent and productive seafarers. Despite these difficulties, the shipping companies and manning agencies have to employ more seafarers to sustain the great demand for officers on board.

Shipping companies and manning agencies not only screen and select competent and highly trained crew, but also ensure that the processing and administration of crew comply with the owners' quality systems, which range from ensuring the maintenance of personnel records to evaluating and monitoring mandated training courses. Given the great task of employing seafarers, the shipping companies and manning agencies are loaded with the task of recruiting qualified and competent seafarers.

It is on this premise, that the researchers felt a need to conduct present investigations to determine the hiring practices of the shipping companies and manning agencies so that the seafarers and future seafarers would be aware to such matters in order for them to comply such qualifications.

The primary objective of this study is to identify the hiring practices adopted by the shipping companies and manning agencies based in Iloilo City in terms of recruitment, selection and the actual employment of seafarers.

1.1 Objectives of the study

This descriptive investigation is aimed to determine the hiring practices adopted by the shipping companies and manning agencies in the Philippines in terms of (a)recruitment taking into consideration the sources of recruits and the methods used in the recruitment of the crew; (b) selection with respect to the selection criteria adopted, the process involved, the references considered for the selection and the factors considered during the interview; and (c) the areas discussed between the employer and the newly hired crew during the actual employment process.

1.2 Significance of the Study

The knowledge of the existing hiring practices in terms of recruitment, selection and the actual employment of seamen and the graduates of maritime schools and universities by the shipping companies and manning agencies will be beneficial to the administrators, faculty and students in the maritime programs. This will help them find ways to identify the improvements needed in the production of quality graduates who will make them globally competitive seafarers.

The findings of this research could provide insights to the faculty and administrators of maritime schools and universities of the different factors that influence the employment of their graduates so that necessary interventions could be done.

The result of this research could create awareness on the part of the students taking maritime courses to strive hard and acquire knowledge and skills needed for employment and to become globally competent seafarers.

To the managers of the shipping companies and manning agencies, the result of this research would be of valuable guide for them in streamlining their hiring policies so that their expectations can easily be met in consonance to the international standards.

Factors Affecting the Employment of Filipino Seafarers

"There is not a Filipino who has not a remarkable inclination for the sea, nor is there at present in all the world a people more agile in maneuvers onboard, or who learn so quickly nautical terms, and whatever a good mariner ought to know" (de Viana, 16th century on his logbook). This quote is referring to the natural ability of the Filipino seafarers in navigation and seamanship.

Technical developments present an opportunity to examine the manning requirements of vessels, evaluate the precise shipboard needs, and develop working practices more aligned to the vessel trades. Further, the changing profile of the fleet creates different skill demands with an accompanying change in recruitment requirements (Ebsworth, 1999).

1.3 Changes in Shipping Industry

The enforcement of International Safety Management(ISM) Code of the IMO brought major changes in shipping industry. The main purpose of this code is to provide an international standard for safe management and operation of ships and for oil pollution prevention. This was also a sign for the manning agencies to improve their standard because as stated in ISM code chapter 6.2, "the company should ensure that each ship is manned with qualified, certified and medically fit seafarers in accordance with national and international requirements".

The Standard of Training, Certification and Watchkeeping (STCW) 1978 Convention as Amended in 1995 had provided standards that all member states have to comply and demonstrate that their training and certification systems are subject to independent evaluation incorporating quality standards. Wilfredo Ramos, 2000 states that this amendment caused changes in shipping scenario, it requires a greater demand for quality crew and that demand puts pressure on manning agents.

Ramos (2000) mentioned that the traditional means of recruitment were still practiced. He found out that there were not much differences in the hiring practices. His work comes into the conclusion that the manning agencies need to explore other methods in the recruitment of the qualified crew.

2 METHOD

This study aimed to determine the hiring practices; methods for recruitment, process and criteria for selection are being adopted by ship-manning agencies in the employment of seafarers.

The descriptive method of research was employed in this study which involves giving out of questionnaires.

2.1 The Respondents

The respondents of this study involved the 17 branch managers or officer in-charge of different shipping companies and manning agencies based in Maanila which are registered by Philippine Overseas Employment Administration (POEA) whose contract will not expire prior to the completion of this research or before March 2012.

3 FINDINGS

The findings of the present investigation included the following:

1 The shipping companies and manning agencies in the Philippines when categorized according to years in existence, type of vessels, nationality, company ownership and scope of operation do not vary in their adopted hiring practices. They generally employ announcement of vacancy and recruitment of applicants; selection and screening of applicants; and hiring, placement and orientation of new crew.

2 The shipping companies and manning agencies in the Philippines regardless of category considered the top three sources of recruits namely: walk-ins, call-ins, write-ins and personal contacts and advertisement. Only few considered job posting, job fairs, bonding by scholarships, government agencies, open house and school job placement bureaus as their sources of recruits.

3 The recruitment methods employed by the shipping companies and manning agencies in the Philippines when categorized according to years in existence, type of vessels, nationality, company ownership and scope of operation were similar. The most common recruitment methods used were by blood relationship with ex-crew, personal contacts, walk-ins, call-ins, write-ins and computerized system. The least methods used were newspaper advertisement, bonding by scholarships, direct mail recruitment, job fairs, job posting, open house, government agencies, school placement bureaus and radio and television.

4 The selection criteria adopted by the shipping companies and manning agencies in the Philippines regardless of category were almost the same. These were the competences, experience, interview results, license, technical knowledge and skills and written/oral examination. The other criteria used were the academic records, physical appearance and age. The least criteria adopted were the aptitude tests, physical fitness, school graduated from, sex and safety awareness.

5 The shipping companies and manning agencies in the Philippines regardless of category considered the same selection criteria very important. These were competences, experience, interview results, license, technical knowledge and skills and written/oral examination. The least considered criteria were the school graduated from and the physical appearance.

6 The shipping companies and manning agencies in the Philippines regardless of category adopted the same selection processes important in hiring of their crew. These are filling up of application form, initial screening, preliminary interview, final interview and medical examination. They do not considered tests as important selection process in hiring of seafarers.

7 The shipping companies and manning agencies in the Philippines regardless of category considered employment references and character references important basis in the selection of their crew. Civic references were not considered important in the selection of their crew.

8 The shipping companies and manning agencies in the Philippines regardless of category looked for certain qualities or factors from the applicants during the interview. These factors were assessment of safety awareness, assessment of attitudes and physical appearance, assessment of competence and skills, assessment of work or technical knowledge and assessment of the applicant's fitness for the job. They least considered the level of knowledge of the English language from the applicant.

9 The topics discussed between the shipping companies and manning agencies managers or presidents and the new crew when categorized according to years in existence, type of vessels, nationality, company ownership and scope of operation were the same. They discussed about the job position, nature of work, salary, allowances, fringe benefits, working condition, safety familiarization and pre-departure briefing and orientation. They have least concern about the company expectations of the new crew.

4 CONCLUSIONS

Based on the findings, the following conclusions were drawn:

1 Generally, the shipping companies and manning agencies in the Philippines regardless of category adopt almost the same hiring practices. This may be due to the nature of business and profession. This finding implies that the employment process involved the hiring of the best-qualified people for the right job.

2 The finding that shipping companies and manning agencies get their recruits from the same sources reflects the common practice of the agencies to wait for the applicants to come to their offices to apply. Walk-ins, call-ins, write-ins, personal contacts and advertisement were considered the famous source of recruitment of the competent seafarers.

3 The recruitment methods employed by the shipping companies and manning agencies in Iloilo city do not involved much travel and expenses.

4 The selection criteria adopted and preferred by the shipping companies and manning agencies in the Philippines in the selection of their crew are reflective of the quality of the seafarers they would like to deploy. Majority of the agencies preferred applicants who are competent, well

experienced, smart, licensed, knowledgeable and skillful and can communicate well. They also considered applicants who are young, academically prepared, and with pleasing personality.

5 The shipping companies and manning agencies in the Philippines regardless of category considered the same selection criteria very important. These were competences, experience, interview results, license, technical knowledge and skills and written/oral examination. The least considered criteria were the school graduated from and the physical appearance.

6 The shipping companies and manning agencies in the Philippines regardless of category adopted the same selection processes in the selection of their crew. These are the filling up of application form, initial screening, preliminary interview, final interview and medical examination.

7 The shipping companies and manning agencies in the Philippines regardless of category considered employment references and character references important basis in the selection of their crew. These references were considered to check the employment background and the personality of the seafarers. They also considered educational references to check the academic background of the applicant and the professional references to check the social aspect of the prospective seafarers.

8 The shipping companies and manning agencies in the Philippines assessed their applicants very thoroughly during the selection process. They choose applicants who are most competent, skillful, and knowledgeable for the job. Also they considered safety conscious, with no attitude problem and with pleasing personality.

9 The managers/owners of shipping companies and manning agencies discussed important areas of concern with the newly hired crew. This is important for the seafarers to understand the different conditions of their employment to prevent any problem that may arise later and misunderstanding between the company and the seafarers.

5 RECOMMENDATIONS

Based on the findings and conclusions, the researchers presented the following recommendations:

There is a need for the shipping companies and manning agencies to make known their hiring practices especially to the graduates of maritime schools and universities.

Ship owners and manning agents or managers should reach out the sources of the seafarers which is the maritime schools or universities that trained the cadets for the maritime profession.

The personnel officer or manning agents need to explore other methods in the recruitment of the qualified and competent seafarers.

There is a need for the company or manning agency to expand farther their personal contact by exploring idea of bonding with the probable seamen through scholarships. Giving support to the cadet while still in school develops an identity, a bond of belonging to the benefactor-company.

There is a need for the shipping companies and manning agencies to support the cadetship program in the development of future officers. If the companies would only employ experienced seafarers, who will give a chance to the newly graduates to be trained and prove themselves worthy seamen?

The government agencies involved in the education and training of the Filipino seafarers need to work closely and harmoniously in order to fully satisfy the requirements of the international convention to make the Filipino seafarers competitive in the world maritime market.

The shipping companies and manning agencies should evaluate their hiring practices and give opportunities for the apprentice cadets to develop their skills.

Maritime schools should inculcate among their cadets the values of discipline, loyalty, safety awareness and consciousness, cleanliness, industry and belongingness as the key to a successful maritime career.

Maritime schools and universities should explore other possible means of providing their cadets the necessary education and training that would prepare them for shipboard practice.

Maritime institutions should coordinate with the shipping companies and manning agencies in the development of the new programs and the advancement of maritime education and training in order to produce seafarers who are competitive in the international or world market.

REFERENCES

Hiring Practices, Problems, and Concerns of Ship-Manning Agencies in the Philippines
Mish, 1990
Downing, 1982
Webster, 1988
Ramos, 2000
Social Security (Seafarers) Convention, 1987
Branch 1980
Personnel Management, Organization and Training Manual
Random House, 1997
Encarta, 2009
http://www.clarku.edu
WWW.Google.Com
WWW.Yahoo.com

Human Resources and Crew Manning
STCW, Maritime Education and Training (MET), Human Resources and Crew Manning, Maritime Policy,
Logistics and Economic Matters – Marine Navigation and Safety of Sea Transportation – Weintrit & Neumann (Eds)

Development of Maritime Students' Professional Career Planning Skills: Needs, Issues and Perspectives

G. Kalvaitiene & V. Senčila
Lithuanian Maritime Academy, Klaipeda, Lithuania

ABSTRACT: The particular dynamics of vacancies and movement of labour force, rapid changes of activity content are characteristic to maritime sector and its labour market. It urges to activate the research of future seafarers' professional career planning and the increase of professional career planning skills development. Educational and training institutions have not only to ensure the preparation of qualified specialists but also systematic and streamlined education for career. Therefore the aim of the article is to define needs, issues and perspectives of maritime students' professional career planning skills development. Maritime students' professional career planning skills development is analyzed using a systematic approach. In the empirical part, having conducted the surveys of Lithuanian and foreign and Lithuanian experts, quantitative research, the relevance of maritime students' professional career planning skills development.

1 INTRODUCTION

Strategic documents and recommendations of the European Union Maritime Transport Policy until 2018 (2009) emphasize that in order to ensure safe and competitive navigation it is very important to maintain the high level of seafarers' education and professional competence. Therefore the EU states have to envisage the appropriate system of seafarers' education organization and apply various means to achieve this goal. "Community actions should aim, in particular, at adopting measures facilitating lifelong career prospects in the maritime clusters, giving special consideration to developing advanced skills and qualifications of EU seafarers to enhance their employment prospects and ensuring that seafarers have good career paths..." (p.5). Professional seafarers' education and training shall provide the future seafarers the most advanced skills that would open plenty of employment opportunities (Green Paper, 2006). Strategic documents mention the promotion and support of research related to the human factor (Strategic Goals and Recommendations for the EU Maritime Transport Policy until 2018, 2009; Green Paper, 2006).

Services of career design are important to the functioning and efficiency of education and labour market systems. Services of career design are named as a means competitiveness, employment, the growth of economy and social stability (Guidelines

for the Employment Policies of the Member States, 2007, Draft Joint Employment Report, 2009/2010, etc.).

Educational and vocational training institutions have not only to ensure the qualified specialists' education, but also provide the opportunities to gain knowledge, skills and attitudes system necessary for the development of their career – *systematic and streamlined development for career* (Reardon, etc., 2000; Sampson, etc., 2004; Pukelis, 2007, Gottfredson, Duffy, 2008, Savickas, etc., 2011). European Commission Memorandum emphasizes that the objective of the vocational guidance, career guidance and counseling as well as services of career planning development is to ensure that every person could easily access high quality information and advice related to lifelong learning and professional activity opportunities.

The particular dynamics of vacancies and movement of labour force, rapid changes of activity content are characteristic to maritime sector and its labour market. It urges to activate the research of future seafarers' professional career planning and the increase of professional career planning skills development. MET institutions have not only to ensure the preparation of qualified specialists but also systematic and streamlined education for career.

The aim of the research is to define needs, problems and perspectives of maritime students' professional career planning skills development.

The objectives of the research:
1 To define the conception and mission of professional career planning skills development.
2 To disclose the expression of maritime students' professional career planning skills.
3 To determine the methods and means of maritime students' professional career planning skills development.

The research methods: the analysis of scientific literature, expert interview method and focus group method.

The methods of 15 expert interviews (8 Lithuanian and 7 foreign experts from Poland, Finland, Latvia, Estonia, the Netherlands and the United Kingdom) and focus groups (7 Maritime students from Lithuanian Maritime academy) were chosen for quantitative research. The expert interviews were carried out applying the methods of semi - structured interview and questionnaire survey in September 2012; the focus group discussion took place in October, 2012.

15 experts can be divided into two categories:
– the rectors (directors) of maritime educational institutions (10 experts);
– the representatives of shipping companies and organizations employing seafarers (5 experts).

The aim of focus group discussion was to evaluate the maritime students' experience related to professional career planning, to investigate and determine the practical aspects of students' professional career planning skills development.

2 PROFESSIONAL CAREER PLANNING SKILLS AND THEIR DEVELOPMENT

The aim of professional career planning skills development is to motivate a personality to improve the professional activity and to develop the personal professional self-expression. The perception of the development of professional career planning skills as a continuous process is important (Reardon, 2000; Sampson, 2004; Pukelis, 2007, Gottfredson, 2005, Gottfredson, Duffy, 2008, Savickas, etc., 2011).

Pukelis (2007) distinguish the following main professional career planning skills: *self-cognition, self-development skills, cognition of labour world, social skills and lifelong learning skills.*

This research refers to the provisions of systems theory, because the interrelation between maritime students' professional career planning skills development process partners cannot be static, the process is dependent on the factors of the varying environment: *labour market, its specifics, the legal regulation of the profession and its peculiarities, the training process participants' characteristics.*

The relevance of maritime students' professional career planning skills development is based on the needs of labour market, its specifics and the legal regulation of seafarers' education. The identification of labour market needs helps to answer the following questions: What is the need for seafarers in labour markets? What characteristics of maritime labour market and seafarer's profession can be distinguished? What are the peculiarities of scafarcr's professional career? What knowledge, skills and provisions that form the content of maritime students' professional career planning skills development are necessary?

Different international, European Union and national legal acts envisage the seafarers' education and development ensuring the quality of them. These acts not only define the content of the seafarer's qualification, opportunities to acquire the qualification, point out the institutions responsible for training and its supervision. Legal regulation of seafarer's profession also reveals the peculiarities of seafarer's profession and skills that are important in the process of maritime students' professional career planning skills development (Kalvaitienė, 2012).

In order to ensure professional career planning skills development, human material and technical resources as infrastructure and methodical means are necessary for a successful educational process. Lecturers, supervisors of apprenticeships, their professional maturity are very important in the process of maritime students' professional career planning skills development (Figure 1).

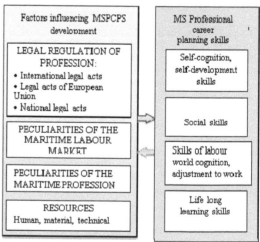

Figure 1. Maritime students professional career planning skills (MSPCPS) and factors influencing MSPCPS development

3 RESULTS OF INVESTIGATION

3.1 *Problems arising for maritime students during the employment and start of the maritime career*

While carrying out the content analysis of experts' responses, the following problems related to maritime graduates; employability and the skills of professional career planning.

– The problems of the maritime labour market peculiarities

"Students lack information about the threats related to the job search through crewing agencies. Some students have to be directed while searching for the first job. "(E10).

„Some crews are not good for a career start. "(E13).

„Problems depend on the type of a vessel, duration of a contract, are related to the marital status... What the crew is, what its national peculiarities are "(E9).

– The problems of the maritime profession peculiarities

„Experience of skills and knowledge acquired at sea are not easily acknowledged in case of the seafarers' move to the shore "(E14).

– Difficulties of adaptation (social skills)

„....a student is frequently not ready to stay alone, to communicate in a foreign language "(E1).

„ a new place, a new environment, a ship, her crew, interpersonal communication, subordination, rigidity, discipline - everything is new for them "(E3).

– A lack of self-representation skills (*self-cognition, self-development skills*)

„ Representation skills such as an ability to represent oneself, to show the own professional knowledge are necessary when interacting with a company "(E3).

„They do not have skills „to sell" themselves "(E4).

– Communication problems in the multinational environment (*social skills*)

„Communication problems in the multinational market "(E14).

– Groundlessly high expectations (*self-cognition, self-development skills, Skills of labour world cognition*)

„ Young specialists want „here and now". They emphasize salary. The attitude is short-term. There should be an attitude to strive for the professional competence first and only then to solve the issues of the amount of salary. A balance between experience and high salary is necessary. "(E4)

– The dependability of career success on personal qualities (*self-development, adjustment to work and lifelong learning skills*),

„As in many cases, cadets' personal qualities, their interest in their speciality, their wish to pursue further career determine a lot. Those cadets, who willingly accept responsibility, are proactive, have specific goals during their seagoing training, and are usually successful employees in their professional career "(E7).

"All lacks are seen during the first seagoing training, and if a person is motivated to work at sea, he is interested to learn and to acquire knowledge and skills "(E2).*

Some experts maintained that they can not point out any difficulties of getting employment and starting seafarer's career:

„ I do not see any big problems in employing and starting maritime career "(E2).

„No problems. There is a high demand for seafarers and all graduates get employed quickly "(E11).

3.2 Methods and means of maritime students' professional career planning skills development

The carried out research revealed the experts' opinion towards the methods and means of maritime students' professional career planning skills development.

The following subcategories were distinguished by using the method of content analysis:

– Involvement of social partners, employers

„Communication with crewing companies, ship owners, conclusion of contracts... Direct and good relations with employers are important. Then the graduates can adapt easier" (E9).

"Visits and seminars in maritime institutions" (E10).

– Preparation for a seagoing training

"A student should be prepared in advance in the institution training seafarers. It is critical to acquaint him what is waiting ahead of him... He should be acquainted well what environment he is going to get into, what he will have to do; the tasks, diploma project tasks, what he has to take with him, starting from the personal hygiene means, etc. have to be explained to him " (E3).

"...Information and a list of necessities should be provided, they have to be instructed on the aims of the training, other niceties related to the elements of culture (forbidden dress code when entering navigation bridge, etc.), meeting the rules of statutory relationships" (E1).

"The appropriate program and supervision on board during the training" (E10).

"Training on board, regular change of ships" (E13).

"Students' seagoing training has to be carried out in training ships or the training has to take place in merchant fleet" (E6).

– The development of professional career planning skills through the study subjects

„The formation of some skills should be integrated into study subjects, e.g. CV writing in English lessons. Psychological aspects should be revealed in the subject of Management psychology " (E1).

„Preparation for seafarer's profession can be integrated into subjects by involving certain topic into curriculum..., because the specificity of

seafarer's work greatly differs from the specificity of the work onshore " (E2; E9).

„Lecturers should devote at least 5 – 7 minutes for the development of a personality, id est discuss what to pay attention to, when at sea, acquaint the future specialist with his future challenges and environment " (E3).

„Multicultural skills can be developed during lectures of English... English speaking guests having practical experience (seafarers, lawyers, employees of port companies) can be invited..."(E4; E10).

– The utilization of students' experience

„Transfer of later year students' experience to other students. I always recommend the first year students girls to associate with the student girls from the latter years who have already returned from training" (E1).

„Students communication with their peers is important ... " (E9).

– The use of lecturers' experience

„While communicating with students I talk about what is awaiting of them, how to prepare for that, how to behave in one or another situation. Uppermost, it is a transfer of lecturers' personal experience " (E2).

"Lecturers have to form students' responsibility by means of their own example, their experience and teaching methods ... " (E1).

„...they have experience and, of course, they acquaint with such episodes which are not described in the books both from the speciality subjects and personal features... Lecturer's authority is very important " (E3;E9).

– The utilization of graduates' experience

„...more conversations with the graduates, the young ones, who could share their experience of climbing career ladder, are needed... (E1; E3).

– Informational methodological publications for students

„Informational publication for students who are preparing for seagoing training is necessary " (E1).

„There is a lack of methodological publications on nuts-and-bolts how to prepare oneself for maritime career "(E2).

– The use of active learning/teaching methods

„Group work, presentation of the work to the group. It develops abilities of representation, communication" (E4; E3).

"... there are a lot of methods, courses, couching systems, professional and character development opportunities " (E11).

"Diploma projects oriented towards the problems of marine industry. The analysis of case studies, the use of simulators" (E10).

– International exchange

„To carry out the programs of students' international exchange " (E5).

3.3 Students' group discussion opinion on the need for the methods and means of professional career planning skills development

The members of group discussion distinguished methods and means, which, in their opinion, would help to prepare for the professional seafarers' career and to acquire the skills of professional career development:

– Preparation for professional seagoing training

„Preparation is necessary. Before leaving for training, I did not know many things...", "Lectures are necessary to discuss questions just before training".

– The use of of later year students' experience

„It is a very good thing: the students, having completed their seagoing training, deliver presentations".

– The variety of information sources

„...the information about the companies, the links to the companies, etc. are necessary online"; "There are seafarers' chat rooms online, where you can find a lot of information. This can be shared"; "There were a lot of videos on different topics, computer tests onboard. It is very interesting and necessary. Such is needed".

The students' group discussion subcategories distinguished by the means of content analysis correspond to the experts' opinion on the methods and means of maritime students' professional career planning skills development.

4 CONCLUSIONS

The development of professional career planning skills: self-cognition, self-management (self-education), labour world cognition, adjustment to work, lifelong learning, and social skills is defined as a process enabling persons to plan their professional career as a consistent and managed process. Therefore the mission of professional career planning skills development is to encourage a personality to be responsible for the planning and improvement of the own professional activity.

The external factors: legal regulation of seafarer's profession, marine labour market, the peculiarities of seafarer's profession and the resources (human, material, technical) indispensable for the development have a direct influence on maritime students' professional career skills development.

The following problems related to the graduates' employability and their professional career planning: the problems of the maritime labour market peculiarities and the maritime profession; difficulties of adaptation; communication problems in the multinational environment (a lack of social skills); a deficiency of self-representation skills (insufficient self-cognition, self-development skills);

groundlessly high expectations (a deficiency of labour world cognition skills); the dependability of career success on personal qualities (adjustment to work and life-long learning skills).

Maritime education and training institution have to anticipate human, material and technical resources necessary for the development of maritime students' professional career skills.

The carried out research allowed the establishment of lacking professional career planning skills and the possible methods and means of maritime students' professional career planning skills development: the integration of professional career planning skills development into the subjects of studies; the preparation for vocational seagoing apprenticeship; the resort to the higher years students' and graduates' experience; the use of lecturers' experience; the engagement of social partners and employers; the use of active teaching/learning methods; international exchange; informational methodical publications for students.

REFERENCES

Green Paper: Towards a future Maritime Policy for the Union; A European vision for the oceans and Seas. 2006. Brussels: Commission of the European communities. http://ec.europa.eu/maritimeaffairs/pdf/com_2006_0275_en _part2.pdf.

Guidelines for the Employment Policies of the Member States. 2007. Council of the European Union, 2007, July 7. http://www.consilium.europa.eu/documents/access-to-council-documents-public-register.aspx?lang=EN.

Gottfrendson, L. S. 2005. *Carrer development and counseling. Putting theory and research to work.* Ed. Steven D. Brown Robert W.Lent.

Gottfredson, L. S., & Duffy, R. D. 2008. Using a theory of vocational personalities and work environments to explore subjective well-being. *Journal of Career Assessment,* 16 (1), p. 44-59.

Pukelis, K. 2007. *Profesinės karjeros planavimo gebėjimų ugdymo C metodika.* Vilnius.

Reardon, R. C., Lenz, J. G., Sampson, J. P., Peterson, G. W. 2000. *Career development and planning: A comprehensive approach.* Stamford: Brooks/Cole and Thomson Learning.

Memorandum on Lifelong Learning. 2000.

Kalvaitienė, G. 2012. Model of maritime students professional career planning skills development. Doctoral dissertation. Vytautas Magnus University: Kaunas.

Sampson, J. P., Reardon, Jr. R., Peterson, G. W., Lenz, J. G. 2004. *Career counseling and services.* Belmont: Thomson Brooks/Cole.

Savickas, M. L. Nota, L., Rossier, J., Dauwalder, J-P., Eduarda Duarte, M., Guichard, J., Soresi, S., Van Esbroeck, R. & van Vianen, A.E.M. 2011. Life designing: A paradigm for career construction in the 21st century. *Journal of Vocational Behavior,* 75 (3), pp. 239-250.

Strategic Goals and Recommendations for the EU Maritime Transport Policy until 2018. 2009. Briusel. http://europa.eu/legislation_summaries/transport/waterborn e_transport/tr0015_en.htm.

Human Resources and Crew Manning
STCW, Maritime Education and Training (MET), Human Resources and Crew Manning, Maritime Policy,
Logistics and Economic Matters – Marine Navigation and Safety of Sea Transportation – Weintrit & Neumann (Eds)

The Role of Human Organizational Factors on Occupational Safety; A Scale Development Through Tuzla Region Dockyards

U. Mörek, L. Tavacioglu & P. Bolat
Istanbul Technical University, Maritime Faculty, Turkey

ABSTRACT: Occupational safety gains much importance in the world as well as in Turkey with the growing industry. As number of injuries some of which are resulted by death occur in the recent years, mainly in so called shipyards, governments and labor unions focused on the work safety incurred by occupational accidents which caused great concern.

Normally the risk concerned is defined and eliminated by applications such as risk management and risk transfer in order to keep under control the practices after results of risk analysis are evaluated.

Despite the laws and control of local government on the occupational safety, a safety culture shall be established within the organization. Safety applications shall be described in the within created organizational culture of the firm. Organizational values are to be operated effectively and relationship between organizational culture and values maintained.

In this study, a deep research is aimed into effect of organizational culture structuring on occupational safety especially in Turkish Maritime Industry. In order to measure the level of organizational culture in shipyards and awareness related to thereof, a questionnaire was applied which is developed as the first scale in the industry. By achieving above, a base is aimed to be created for the policy for Maritime Industry, thus improving awareness on organizational culture related to occupational safety in addition to review and application of theoretical and academic studies.

1 INTRODUCTION

An organization is considered as the smallest society and continues to operate with subsidiary groups and member of cultures. Shipyards are the core firms of this kind, where slipways, dry and floating docks, high capacity cranes, workshops with machinery fittings, containing chemical industry as well, are combined (Tezdogan and Taylan, 2009: 10).

In addition, nature of the works performed in the shipyards is two times more dangerous than works in other industries (Bell, 2005).

Safety plays an important role in shipping and occupations at sea are high-risk compared with most other occupations.

Considering the dangers, occupational accidents cause important production loss and this situation increases costs. After occur the occupational accidents and/or illnesses, the amount of money spent for diagnosis and treatment may increase as well. However, protective approaches mostly can be provided with considerably less costs. (Bilir, 2005: 9).

Sharon Clarke (1999) in his research, stated that despite positive attitudes regarding safety of organization members, the common interest is not at the same level in all hierarchical degree.

According to Schwartz and Davis, organizational culture is the sum of beliefs and expectations shared by the organization members and forming the rules that shapes the behaviours of groups and individuals in the organization. However, according to K. Szymanski, different companies are using different methods and variety of tools and techniques. Most members within the organization share organizational culture, in another attempt, such as beliefs and values about people, work, the organization and the community.

Of course, a reliable and valid measure of safety orientated scale offers great advantages for shipping oriented firms, classification societies and insurers and can help overcome limitations associated with traditional safety measures, such as rates of time lost as a result of accidents and accident investigation reports. As accidents are rare events compared to work hazards and near misses, they do not allow for an evaluation of employee risk exposure, and they are invariably retrospective (Glendon and McKenna, 1995). An important aim of a positive safety culture in an organization is to influence safety behavior, which in turn may result in fewer incidents and accidents.

It is considered that shipbuilding and ship repair industry can be developed only in a way where safety measures are taken through full commitment to organizational culture. In this respect, the purpose of this study is to create a scale applicable to determine role of organizational factors and culture structuring occupational safety (Iri, 2007: 1).

Occupational safety is the technical and systematic studies to search and prevent mechanical, facility-wise and material related defects during performing work in the yards (Koc, 2004: 6)

Safety orientated questionnaire consists of the cultural and contextual factors creating attitudes and behavior influencing occupational health and safety. It relates closely to ISO 9000 and ISO 14000 standards as well as the International Safety Management (ISM) code and organizational learning.

However, measurement issues in occupational health and safety are receiving more attention from academics and practitioners and many business executives demand simple, low cost measures for bench-marking purposes or for use as measures in a balanced scorecard (Mearns and Håvold, 2003).

2 DEVELOPING A SCALE

As is known, the connection between the feature to be measured and scale items is related with the validity of scale tool. Pre-studies are required for scale items to measure the necessary features (validity of content) or in order to determine the power of relevant items (McGartland et al., 2003).

Other factors effecting the validity of the scale tool are the points to be taken into account for scale reliability, such as the intelligibility of scale items, suitability of target group, etc. The necessary opinions received from experts during pre-study process play an important role for the determination and review of the scale developed.

Some researchers show that lengthier questionnaires and interviews can reduce response rates and lower the data quality, while others claim there is no effect from questionnaire length (Herberllein and Baumgartner, 1978; Kanuk and Berenson,1975; Yammarino et al., 1991). Most of the research claiming that survey length does not have any effect on response rates is older and it seems now, as simplicity and use of time are more important for a higher response rate than a few years ago. According to Newell et al. (2004), lower response rates seem to be an effecting survey. One

of the main reasons given for not completing the relevant forms was survey length. Because of the potential for reduced data quality, as well as lower response rates and higher administrative costs associated with larger questionnaires, it makes sense to try to reduce questionnaire length as far as possible without compromising quality.

For above mentioned reasons, two of the most important decisions to be made in scale construction are the number of factors and the number of items on each factor. (Hair et al., 2006). In addition, the parallel analysis criterion (PC) (Montanelli and Humphreys, 1976) has been highly recommended by several researchers (Hayton et al., 2004). Item analysis makes it possible to increase the overall quality of a scale while shortening it, either by eliminating unsatisfactory items or by removing redundant ones. Spector (1992) recommended eliminating items that have a low Item/Total correlation as well as items that increase the Cronbach's alpha as a first step.

DeVellis (1991) suggests that either an odd or even number of choices can be used for the response scale, depending on the phenomenon being investigated and the goals of the investigator. According to Harzing (2005), response style in a questionnaire based cross national research and cultural country level characteristics also influenced acquiescence and extreme response styles. English language questionnaires seemed to reach a higher level of middle responses, however questionnaires in respondent's native language are resulted in more extreme response styles.

When constructing the questionnaire using existing scales and items, it has been assumed that an underlying structure is already present in the data. Validity of the questionnaire should be established to ensure that the scales in fact measure what they purport to measure (Hair et al., 2006). An important part of scale validation is the testing and interpretation of the structural models.

In this part of study, the effect of organizational culture over occupational safety and application of this culture in the firms are analyzed. Normally, scales are developed either by experimental process or by hypothetical process (Tezbasaran, 1989 and Torgerson, 1958).

The scale can be used in benchmarking as a key performance indicator, or as an indicator in a balanced scorecard type of management tool.

In order to test hypothesis, one of the most effective research methodology, general research model, is selected. General research model consists of many units in the surrounding universe, some of which are selected as a model group or sample, or totally used, in order to reach a judgment regarding selected universe. The universe of this research is the workers in Tuzla region shipyards.

The KMO and Barlett values are shown in the Table-1 here below. The values given prove that it is convenient to perform factor analysis.

Table 1. KMO and Bartlett's Test

Kaiser-Meyer-Olkin Measure of Sampling Adequacy.		,847
Bartlett's Test of Sphericity	Approx. Chi-Square	2532,580
	df	903
	Sig.	,000

3 FACTOR ANALYSIS

Table 2. Total Variance Explained

C.	Initial Eigenvalues			Extraction Sums of Squared Loadings			Rotation Sums of Squared Loadings		
	T	%V	C%	T	%V	C%	T	%V	C%
1	9,727	46,318	46,318	9,727	46,318	46,318	9,419	44,852	44,852
2	1,803	8,586	54,904	1,803	8,586	54,904	1,934	9,210	54,062
3	1,623	7,728	62,633	1,623	7,728	62,633	1,800	8,571	62,633
4	,958	4,560	67,192						
5	,818	3,895	71,087						
6	,755	3,594	74,681						
7	,721	3,432	78,113						
8	,596	2,836	80,949						
9	,567	2,700	83,650						
10	,489	2,330	85,980						
11	,431	2,054	88,034						
12	,400	1,905	89,939						
13	,355	1,692	91,631						
14	,351	1,670	93,301						
15	,316	1,503	94,805						
16	,245	1,165	95,970						
17	,237	1,130	97,100						
18	,206	,981	98,081						
19	,154	,732	98,813						
20	,130	,620	99,433						
21	,119	,567	100,000						

C. – Component
T – Total
%V – % of Variance
C% – Cumulative %

As per Table-2, after calculations and correlation in the SPSS program, total variance is defined as % 62,633 which is considered as an agreeable value for the researches in the social science.

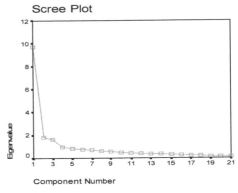

Diagram – 3

4 DATA COLLECTION AND SCALE TOOLS

As per the description made by ILO (International Labor Convention), occupational safety and health is to improve physical, mental and social moods and maintain this level as well as carry out ergonomic studies to prevent any harm to the lives of workers. (Yasan and Kucuk, 2005: 4).

Occupational safety deals with necessary technical regulations in order to remove dangers related to worker life and physical integrity (Demircioglu and Centel, 1995: 286).

For collecting necessary useful data, a questionnaire was applied to workers in Tuzla region shipyards. The questionnaire developed for this purpose contains 43 questions. (see Annex-I). In the questionnaire, questions are with multiple choices and a 5-point Likert scale is used, [Always, Often, Sometimes, Seldom, Never]. The questions mentioned in the questionnaire hereby are related to occupational safety and organizational culture.

The questions used in the questionnaire mentioned in Annex of this article, are inspired by and in conjunction with the questionnaire form used in the article "From Safety Culture to Safety Orientation" by Jon Ivar Havold and Erik Nesset.

Permission was obtained from the managing director or the yard owner at the participating shipyards. Each questionnaire had an introduction to the project on the first page, with information about the use of the data and the importance of answering the questionnaire as honestly as possible.

5 METHODOLOGY

Even occupational accidents do not end up with death, it happens not only to inexperienced workers, but also to technicians and engineers, meaning that it cuts down the hypothesis that occupational accidents occur due to professional inability and lack of education (Ozdemir, 2009: 55).

The main purpose of occupational safety and culture is to protect its employees and provide a safe working environment by making risk assessment for possible dangers that may arise due to work performed (Gokpinar, 2004: 20)

The universe of the study is Tuzla shipyard workers, where 124 of them in 6 different shipyards were sent the questionnaire. 100 of these workers responded with their answer while 74 of them are unskilled/skilled workers, 26 of them are engineers.

The reason why the questionnaire is aimed to be applied in yards with numbers of workers more than 50 is that such yards are considered to have procedures for providing occupational safety with their improved organizational departments solely in charge.

The questionnaire is applied to workers in Tuzla region shipyards in September and November 2012.

As per information received and data collected from workers through questionnaire, results are analyzed by means of cross correlation. Having the aim of testing validity of the statistical results obtained, an α coefficient "structural validity coefficient" defined by Baykal (1994) shall be used for easy calculation. In order to verify the reliability of the components (scales), calculations of alpha values for each scale are an important factor.

Table 3 - Rotated Component Matrix

	Component		
	1	2	3
S11	,828		
S29	,808		
S28	,803		
S37	,768		
S38	,766		
S34	,743		
S35	,739		
S14	,730		
S10	,726		
S27	,722		
S7	,722		
S26	,713		
S30	,704		
S33	,700		
S4	,692		
S18	,658		
S43		,792	
S42		,719	
S41		,666	
S39			,788
S40			,779

The relevant scale consists of 3 factors and 21 items in total. The factor load in each factor is shown in Table-3 with yellow color. It can also be observed which item is located in the corresponding factor.

6 RESULTS AND FINDINGS

There is consensus according to Jerez-Gomez (2005) that while constant learning is of importance and organizational learning as the essential instruments for the firm operations and performance (Senge, 1990; Garvin, 2000), there is an ongoing debate about how a learning capability can be efficiently developed.

The practices in the human resources departments of firms are a fundamental tool for organization's learning capability development such as employee education and training, rewards, and motivations. However, according to reviewed literature, firm competencies cannot be observed directly thus providing unique competitive advantages for the firms in their product development activities (Fowler et al., 2000; Wang et al., 2004; Iansiti and Clark, 1994).

Another firm competency is the emotional capability of the firm (Huy, 1999). Emotional capability is inherently an individual-level concept and attributable to organizations as well as to individuals (see, Fineman, 1993; Rafaeli and Worline, 2001; Ashforth and Humphrey, 1995; Domagalski, 1999).

According to Bion (1961), people share common assumptions about an organization and their feelings and emotions are determined by internal individual characteristics, organizational structure, routines, and culture (Giddens, 1984). Therefore, in an organizational structure, culture is created or shaped by the collective actions and interactions (Fineman, 1993; Giddens, 1984).

Firm innovativeness is a multidimensional construct involving several aspects such as product, process, and market, technological and strategic planning (Wang and Ahmed, 2004; Hurley and Hult, 1998).

If a firm is capable of managing its individuals and their collective emotions, it can develop a shared vision and collective understanding on operations and tasks from skills and abilities generated in the culture. In addition, providing freedom for workers to give opportunities to express their ideas without guilt and fear, increases the firm's capabilities in all aspects (Huy, 2005) thus increasing personal commitment and loyalty (Mathieu and Zajac, 1990). Emotional capability helps organizations to monitor, evaluate and use their employees' emotions to contribute to the product innovation process as well.

In this respect, the questions for the developed questionnaire shall be in compliance with managerial commitment, trying to determine openness of employees, revealing systems perspective and questions to address knowledge transfer and integration.

As per our calculations Alpha = 0, 9443 shows us the general reliability of the whole scale which is a high value for the reliability analyses.

Table 4. Coefficients

Factor 1 :	
Alpha coefficient	0,9361
Factor 2 :	
Alpha coefficient	0,7475
Factor 3 :	
Alpha coefficient	0,7917

For factor 1, the reliability coefficient is Alpha = 0, 9361, factor 2 the reliability coefficient Alpha = 0, 7475 and for factor 3 the reliability coefficient Alpha = 0, 7917. We have observed that all relevant coefficients found are sufficient values for the research performed.

The dynamic of display freedom, identification, and experiencing are significant predictors for a questionnaire, when openness and experimentation as a dependent variable is selected. When the knowledge transfer and integration are selected as dependent variables, the dynamic of display freedom is the significant predictor. Regarding the control variables, the results show that larger the firm size, the higher the management commitment, and knowledge transfer and integration. It appears that when the number of employees increases in the firm, the need for managerial support and leadership commitment and internal spreading of knowledge through verbal and non-verbal communications increases.

In addition, people in the firm can work together in a coordinated fashion, have generalized knowledge regarding the firm's objectives, and are well aware of how they contribute to achieving the overall objectives.

People in the firm show loyalty to the organization, express their deep attachment to salient organizational characteristics such as values and beliefs, and stay together due to the mutual benefits and emotional bonds, which are an integral part of the questionnaire, is looking for.

The organization can develop a climate with a culture established for accepting new ideas and points of view, both internal and external, and favor experimentation, promote joint actions, and foster relationships based on the exchange of information and shared mental models and people can constantly renew, widen, and improve their knowledge.

So that questionnaire hereby attached to this study is generated from the portfolio in the existing industry by multiple cross checks with the emotional, cultural and organizational structures in the Tuzla Dockyards Region and the first questionnaire developed thereof.

REFERENCES

Baykal, A. (1994). "Davranış ölçümünde yapısal geçerlik göstergesi." Türk Psikoloji Dergisi, 33, 45-50.
Veneziano L., Hooper J. (1997). "A method for quantifying content validity of health-related questionnaires". American Journal of Health Behavior, 21(1):67-70.
McGartland, R. D., Berg-Weger, M., Tebb, S., Lee, E. S., Rauch, S. (2003). "Objectifying content validity: Conducting a content validity study in social work research". Social Work Research, 27(2), 94 - 104.
Tezdoğan, T., Taylan M. (2009). "Tersanelerdeki İş Kazalarının İstatistiki Olarak İncelenmesi", Gemi ve Deniz Teknolojisi, TMMOB Gemi Mühendisleri Odası yayınları, Sayı 180, Nisan, s. 10–16.
Bell, V. (2005). Shipyard Work Safety, The Fabricator, http://www.thefabricator.com/article/safety/shipyard-work-safety--, 22.02.2011
Bilir, N., Yıldız, N. (2004). "İş Sağlığı ve Güvenliği: Temel Bilgiler, İş Sağlığı ve Güvenliği İçinde", Hacettepe Üniversitesi Yayınları, Ankara.
İri, A. (2007). "OHSAS 18001 İş Sağlığı Ve Güvenliği Yönetim Sistemleri Ve Bir İnşaat Firmasında Uygulanması", Basılmamış Yüksek Lisans Tezi, İstanbul Teknik Üniversitesi Fen Bilimleri Enstitüsü.
Beyer, J., Trice, H.M. (1993). "Corporate Culture: The Culture of Work Organizations", Prentice-Hall.
Koç, E. (2004). "Orman Ürünleri Endüstrisinde Çevre Sorunları, İş Sağlığı ve İş Güvenliği", Çalışma ve Sosyal Güvenlik Bakanlığı İş Sağlığı ve Güvenliği Dergisi, Sayı:22, Yıl:4.
Yasan, G., Küçük S.(2005). "İş sağlığı ve Güvenliği – Risk Değerlendirme, Son Gelişmeler Işığında, İş Sağlığı ve Güvenliğinde Teknik ve Hukuki Boyut Eğitimi Notları", İstanbul Sanayi Odası Eğitimleri.
Demircioglu, M. ,Centel T. (1995), "İş hukuku", Beta Basım, İstanbul.
Özdemir, N. (2009), "Gemi Sanayinde iş Güvenliği Yönetimi ve OHSAS 18001 Uygulaması", Basılmamış Yüksek Lisans Tezi, Yıldız Teknik Üniversitesi Fen Bilimleri Enstitüsü
Gökpınar, S. (2004), "İşçi Sağlığı iş Güvenliğinin Temel ilkeleri", iŞ Sağlığı Ve Güvenliği Dergisi, Sayı:19, Mayıs-Haziran.
Schein, E.H. (1988).Organizational Culture, unpublished artical
Akgün A. E., Keskin, H., Byrne, J. and Aren, S.(2007]. "Emotional and learing capability and their impact on product innovativeness and firm performance", Technovation, 27 (9), 501-513,
Håvold,J. I. and Nesset, E. (2009), "From safety culture to safety orientation: Validation and simplification of a safety orientation scale using a sample of seafarers working for Norwegian ship owners", Safety Science, 47, 305-326.
Guldenmund, F. W.(2000). "The nature of safety culture: a review of theory and research", Safety Science Group, Delft University of Technology, Kanaalweg 2b, NL-2628 EB Delft, The Netherlands
C. Chiera-Ungureanu and P.-E. Rosenhave (2012), "A Door Opener: Teaching Cross Cultural Competence to SeaFarers", TRANSVAV, International journal on Marine Navigation and Safety of Sea Transportation, Volume 6, Number 4, December 2012
K. Szymanski (2007), "Risk Management – Do We Really Need it in Shipping Industry" TRANSVAV, International journal on Marine Navigation and Safety of Sea Transportation, Volume 1, Number 2, June 2007

ANNEX I

This questionnaire is prepared in order to measure
the effect of organizational culture on occupational
safety in shipping industry.

Table 5. Role of Human Organizational Factors on Occupational Safety

	Always	Often	Sometimes	Seldom	Never
1 Personnel employed by yard is subject to occupational health and safety training					
2 The yard I'm working now has much safer conditions than my previous workplace					
3 Our yard has occupational safety instructions					
4 Our yard is often controlled for its occupational safety					
5 Our yard has a time schedule related to commission plan					
6 Time schedule related to commission plan can be easily applied					
7 The equipments related to occupational safety is provided regularly within the yard					
8 There is emergency planning in the yard					
9 All employees in the yard are trained for first aid during emergency cases					
10 The employees in the yard are aware of their assigned positions					
11 The rules and regulations related to occupational safety are definite and clear					
12 Our yard has forms available which employees can report the relevant dangerous situations related to occupational safety					
13 Employees within the yard re in cooperation with each other for sharing information related to occupational safety					
14 Employees within the yard can recognize the difference of behaviors if complies with occupational safety procedures or not					
15 Employees within the yard feel responsibility to obey occupational safety rules					
16 Employees within the yard do not hesitate to contact top management in order to inform non - observants to occupational safety rules					
17 Workload prevents me from performing the work safely					
18 Employees within the yard can express their distress easily					
19 Employees within the yard have common team spirit					
20 Employees who works in accordance with occupational safety are awarded in the yard					
21 Employees who do not work in accordance with occupational safety are fired in a short time					
22 I take initiative in case needed related to occupational safety					
23 I believe the necessity that rules related to occupational safety within the yard shall be defined in advance					
24 Procedures related to my duties enable to perform my work in a professional way					
25 Recording accidents provides safe working environment					
26 Yard management holds its own employees responsible for occupational safety					
27 Yard management makes effort in order to be kept informed regarding problematic situations related to occupational safety					
28 Yard management is willing to keep records of any accident					
29 Yard management tries to receive feedback in every respect regarding occupational safety					
30 Yard management keeps all kind of occupational safety equipment in stock					
31 Accidents can be prevented by good luck					
32 If any accident is meant to happen, such accident cannot be prevented					
33 Yard management searches the reasons of accidents objectively					
34 Yard management continuously issues the instructions related to occupational safety					
35 Yard management continuously observes our working environment in order to control compliance with occupational safety procedures in the workplace environment					
36 Yard management often provides its employees with occupational safety guide					
37 Yard management always tracks if its employees comply with occupational safety procedures or not					
38 Yard management regularly carries out inspection regarding occupational safety					
39 I am open to any project that will improve occupational safety quality in my working environment					
40 I am interested in occupational safety					
41 All employees are familiar with reporting procedure for injury cases					
42 Employees contribute to the decisions taken with their ideas in order to improve occupational safety quality					
43 Employees can contribute during forming occupational safety procedures with their ideas					

Human Resources and Crew Manning
STCW, Maritime Education and Training (MET), Human Resources and Crew Manning, Maritime Policy,
Logistics and Economic Matters – Marine Navigation and Safety of Sea Transportation – Weintrit & Neumann (Eds)

Ilonggo Seafarers' Edge Among Other Nationalities of the World

R.A. Alimen, R.L. Pador & M.G. Gayo, Jr.
John B. Lacson Foundation Maritime University-Molo, Iloilo City, Philippines

ABSTRACT: The study aimed to determine if the Ilonggo seafarers have an edge over other nationalities. Furthermore, the study determined the Ilonggo seafarers' communication and work ethic, Ilonggo seafarers' attitude towards work, Ilonggo seafarers' relationship with other seafarers. The descriptive qualitative method of research was employed in this investigation. According to Good and Scates (2000), descriptive research involves collecting of data in order to test hypothesis or answer questions concerning the edge of Filipino seafarers against other nationalities. Descriptive research according to Evans (2009) is concerned with the description of the existing distribution of variables, as opposed to theory building. Or, in plain language, descriptive studies focus on answering the basic W questions: Who, what, when, where. The fifth W, "why" falls outside of the scope of descriptive research, that by definition must not concern itself with the effect that one variable has on another. Qualitative data is extremely varied in nature. It includes virtually any information that can be captured that is not numerical in nature. In this investigation, in-depth interviews were utilized.

1 BACKGROUND OF THE STUDY

According to the 2007 national census, the population of province excluding Iloilo City is 1,691,878. If Iloilo City is included, the population is 2,110,588 (NSO, 2011). People from Iloilo are called Ilonggos. There are two local languages spoken in the province: Hiligaynon sometimes called Ilonggo, and Kinaray-a. Hiligaynon and variants of it are spoken in Iloilo city and a few towns of the province. Spanish is strictly a local language, at least in a historical way, but the numbers of natural Spanish speakers have declined strongly after WWII, and due to this, there are today many Ilonggos who do not consider it a local language.

By the end of the war, Iloilo's economy, life and infrastructure, was damaged. However, the continuing conflict between the labor unions in the port area, declining sugar economy and the deteriorating peace and order situation in the countryside and the exodus of Ilonggos to other cities and islands that offered better opportunities and businessmen moving to other cities such as Bacolod and Cebu led to Iloilo's demise in economic importance in southern Philippines.

By the 1960s towards 1990s, Iloilo's economy progressed although slowly but surely. The construction of the fish port, the international seaport and commercial firms that invested in Iloilo marked the movement making the city as the regional center of Western Visayas.

The completion of the new Iloilo Airport of International Standard in 2007 has enhanced better business opportunities that have affected local, national and international markets in agriculture, finance, tourism and other vibrant sectors of the Philippine economy.

2 STATEMENT OF THE PROBLEM

This study was thus conceived to determine the edge of the Ilonggo seafarers among other nationalities on board. This study sought to answer the following questions:
1 Do Ilonggo seafarers have an edge over other nationalities?
2 What makes Ilonggo seafarers better than the other nationalities in terms of communication and work ethic?
3 What is the Ilonggo seafarers' attitude towards work?

4 What is the Ilonggo seafarers' relationship with other seafarers on board?

5 What are the views/other observations they have in terms of other Filipino seafarers or seafaring as their profession?

3 CONCEPTUAL FRAMEWORK

The conceptual framework of the present study was illustrated below.

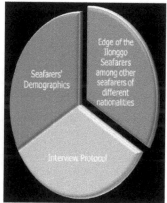

Figure 1. Conceptual Framework

4 RESEARCH DESIGN AND METHODOLOGY

The descriptive qualitative method of research was employed in this investigation. According to Good and Scates (2000), descriptive research involves collecting of data in order to test hypothesis or answer questions concerning the edge of Filipino seafarers against other nationalities.

Descriptive research according to Evans (2009) is concerned with the description of the existing distribution of variables, as opposed to theory building. Or, in plain language, descriptive studies focus on answering the basic W questions: Who, what, when, where. The fifth W, "why" falls outside of the scope of descriptive research, that by definition must not concern itself with the effect that one variable has on another.

Qualitative data is extremely varied in nature. It includes virtually any information that can be captured that is not numerical in nature. In this investigation, in-depth interviews were utilized. This included individual interviews (e.g., one-on-one). The data were recorded utilizing audio recording and written notes. In this interview, the researchers requested some individuals to interview the seafarers who are currently on vacation. The purpose of the interview is to probe the ideas of the interviewees about their edge as Ilonggo seafarers as compared to other seafarers of other nationalities.

5 THE RESEARCH RESPONDENTS

The respondents of this study were the ten (10) seafarers who were purposively chosen for this study. The seafarers were currently on vacation and the researchers had a chance of interviewing them.

Table 1. Distribution of the Respondents

Respondent	Civil Status	Number of Children	Number of Years on Board	Vessel
1	M	1	5	International
2	M	2	7	International
3	M	1	2	International
4	S	0	2	International
5	S	0	2	International
6	M	3	30	International
7	M	1	10	International
8	M	2	25	International
9	M	2	8	International
10	S	0	8	International

6 THE RESEARCH INSTRUMENT

The researchers prepared an interview schedule and started interviewing the seafarers.

Interview - Interviews are among the most challenging and rewarding forms of measurement. They require a personal sensitivity and adaptability as well as the ability to stay within the bounds of the designed protocol set by researchers. In this study, the researchers utilized the following questions during the interview: Do you think that Ilonggo seafarers have an edge against international seafarers, in terms of communication? Work ethic? Attitude towards work? Relationship with other crew on board?

What are your views/other observations do you have in terms of other Filipino seafarers or seafaring as your profession?

7 STATISTICAL TOOLS

Frequency count and percentage were used to describe and determine the purpose of the present study.

8 RESULTS OF THE STUDY

The results revealed that:

When the respondents were asked about their edge in terms of communication, fifty percent (50%) indicated that their edge is their fluency in the English language; 30 percent said that their being good communicators is their edge; and 20 percent said that their spontaneity in communication has become their edge in terms of communication compared to other seafarers of other nationalities.

When the respondents were asked about their edge in terms of work ethic, fifty five percent (55%) indicated that their edge in their work quality, "clean and concise in work"; 20 percent said that their love for work is evident; 15 percent said that they are diligent and hardworking, and 10 percent said that their creativity and intelligence has become their edge in terms of work ethic compared to other seafarers of other nationalities. Senator Legarda (2010) supports this saying that "to ensure they comply with the international requirements and maintain their edge over their foreign counterparts, the Filipino seafarers will be given access to affordable and quality educational advancements and training courses to better improve them in their work and maintain such edge."

When the respondents were asked about their edge in terms of attitude toward work, fifty five percent (45%) indicated that their attitude towards work is edge in their work quality, "clean and concise in work"; 35 percent said that their love for work is evident; 15 percent said that they are diligent and hardworking, 10 and 10 percent said that their creativity and intelligence has become their edge in terms of work ethic compared to other seafarers of other nationalities.

The Filipino seafarers' attitude towards work is evident in the words of Captain Idemoto (2006) in, "New Oasis in Manila for World's Seafarers" stating that at least 35,000 Filipino marine officers are needed within the next five to 10 years in the Japanese based seafaring industries based on the report of the Department of Labor and Employment (DOLE). In addition, it was stated that the deployment of Filipino seafarers is growing at an annual rate of 10 percent, according to the DOLE. The department said foreign employers usually prefer Filipino seafarers because of their ability to speak the English language well. Maritime Industry Authority (Marina) regional director Glenn Cabañez said international shipping firms employ Filipinos because of their competence, skills as well as their attitude toward work. Japanese ship owners employ more than 20,000 Filipino seafarers onboard their vessels, seafarers who have made a significant contribution to the development of Japan's maritime industry.

Young (2010) says that there is now high regard for Filipino seamen. And with the strong supervision and support from (your) government, they will continue to demonstrate competence, professionalism and even bravery as risks and hazards are so inherent in their jobs," Young told the Bulletin.

Relationship with other crew on board was perceived by the seafarers as one of their edges compared to other nationalities. When the respondents were asked about their relationship with other crew on board, fifty five percent (45%) indicated that they are "a complete package," 35 percent said that "they can get easily along with others;" and 20 percent said that they are "loyal in friendship."

One of the seafarers interviewed that their profession is a unique profession that has contributed not only to world trade but has also become a process of making new friends in their interactions with their crew members. Almost 90 percent of the world trade is sea-borne and that staying on board ties them at work for weeks and even months. Thus, developing better relationship with crew mates can offer a better way to eradicate boredom and homesickness.

When the seafarers were asked about their views and observations about the seafaring profession, two themes emerged: their love and passion for their work and their belief that seafaring can better their socio-economic situation.

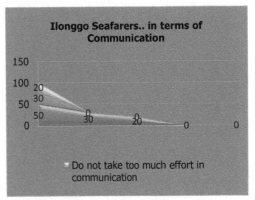

Figure 2. Ilonggo Seafarers' Edge ... in terms of Communication

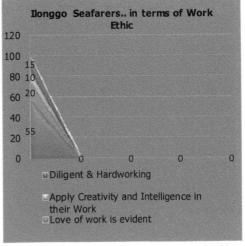

Figure 3. Ilonggo Seafarers' Edge in terms of Work Ethic

167

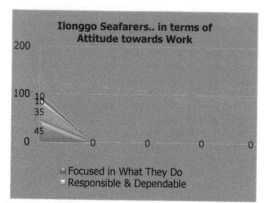

Figure 4. Ilonggo Seafarers' Edge in terms of Attitude towards Work

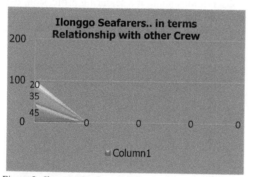

Figure 5. Ilonggo Seafarers' Edge in terms of Relationship with other Crew

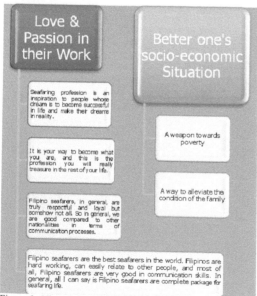

Figure 6. Views/other Observations of Filipino Seafarers of Seafaring as a Profession

9 CONCLUSIONS

The Ilonggo seafarers' edge in terms of communication in terms of being communicators and spontaneity has been a remarkable mark of their better communicators as seafarers.

In terms of work ethic, their edge in their work quality were their cleanliness and conciseness in work, love for work, diligence and hardwork, and their creativity and intelligence have become their edge in terms of work ethic compared to other seafarers of other nationalities.

In terms of their relationship with other crew on board, their being "a complete package," "ability to get along well with others," and "loyalty in friendship" were their edge compared to other seafarers.

Two themes emerged about Ilonggo seafarers' views of their profession. These two were their love and passion for their work and their belief that seafaring can better their socio-economic situation.

10 RECOMMENDATIONS

Based on the findings advanced in this study, the following are recommended.

In view of the positive responses derived for the seafarers with regard to their attitude toward work, the result of this study will be made available or will be presented to faculty members at JBLFMU to enhance more the students and their choice of their profession.

Maritime English will be enhanced especially in the teaching inside the classroom to help students develop a certain level of communicative proficiency to better equip them in their future work.

More studies on other factors and areas will be considered to further validate the results of this investigation.

REFERENCES

Dickson, A.F. (1996). Seafaring: A Chosen Profession. U.S.A.: Amazon.

JSU Maritime Journal (March, 2006). Good News for Filipino Seafarers. Vol 3 (2).

Philippine Information Agency. 13 April 2010. "Old Iloilo capitol now a national historical site". Retrieved 16 April 2010.

Population and Annual Growth Rates for Region, Provinces and Highly Urbanized Cities Based on Censuses 1995, 2000 and 2007. 2007 Census. National Statistics Office. Retrieved 16 July 2011.

Young, G. (June 24, 2010). Pinoy Seafarers Among World's Best Group. The Bulletin. International Transport of Workers Federation (ITF).

Human Resources and Crew Manning
STCW, Maritime Education and Training (MET), Human Resources and Crew Manning, Maritime Policy,
Logistics and Economic Matters – Marine Navigation and Safety of Sea Transportation – Weintrit & Neumann (Eds)

Influence of Emotional Intelligence on the Work Performance of Seafarers

K. Kilic, L. Tavacioglu & P. Bolat
Istanbul Technical University, Maritime Faculty, Turkey

ABSTRACT: Although there is a growing interest on the effect of Emotional Intelligence (EI) in organizations, there is very little research conducted to support its influence in maritime transportation science. The seafarers on a ship can be directly influenced from communications, behaviors and psychological situations and should continually control their emotions in a closed system of a ship. Thus in this study; it is aimed to understand the effect of emotional intelligence on the work performance of the seafarers. The cohesiveness and performance were investigated through seafarer's relationship with Emotional Intelligence which was assessed by the Schutte Emotional Intelligence Scale (SEIS)

1 INTRODUCTION

The acceleration of marine traffic due to the globalization forces the maritime regulatory agencies and policymaking institutions to systemize maritime safety and security. This systemizing is generally based on the maritime accidents or incidents which had negative impacts on environment or/and which were resulted with loss of life such as in accidents of Torrey Canyon, Exxon Valdez and Titanic. When the investigation reports of maritime accidents are analyzed, it is obviously understood that defects in human factor through the operational tasks during maritime transportation process come out an important cause for marine accidents.

Researches related with human factors show that human factor-based error could root in the loss of information and capabilities during the information processing cycle of an operational task, which can refer to giving a response to environment after acquiring and transforming data to information. Thus, performing a given task or solving an operational problem is a function of capacity of information processing which can also be defined as cognitive ability. Cognitive ability is associated with four main capacities (Mc Master, 1996);
– Capacity of learning
– Capacity of acquiring knowledge
– Capacity of adaptation to unfamiliar conditions

Capacity of configuration of knowledge for future events

Apparently, in the complex environment of a ship as a workplace, environmental, psychological and physiological alterations that occur during execution of these four main abilities would probably affect the cognitive process of the individual who is performing the task during a voyage. However, academic studies show that some individuals have more vulnerability than the other individual's vulnerability to the challenges within the same operational environment (Suls, 2001). So when the reasons beyond this difference in vulnerability of individuals are inquired, Emotional Intelligence (EI) can arise as a reason as it can be related to the ability of adaptability to the changing conditions by managing the emotions of himself/herself and the others (Bar-on, 1997; Matthews & Falconer, 2002). Emotional intelligence is directly about personality traits and is open to development through experience in the emotional brain (LeDoux, 2003).

It is a reality that working onboard has very tough and stressful environment. It is also a closed system and in this closed system, seafarers continually should understand each other, should be careful in **hierarchical** ranking and should manage their thought and emotions. It needs controlled behaviors as well as focusing and judgment skills of a person. Accordingly, it is a fact that EI becomes one of the topics that should be researched among the seafarers for an effective, efficient and safe mode of transportation. Through this fact, in this study, it aimed to discuss EI within the maritime transportation process by conducting a preliminary survey among the seafarers. In the paper, firstly literature review is given. Secondly the survey is explained and the results are shared.

2 LITERATURE OVERVIEW

2.1 *Emotional Intelligence:*

At the beginning of the 20th century, a new concept was developed from Intelligence Quotient (IQ). Firstly, Edward Lee Thorndike, an American psychologist, developed different concept as "Social intelligence" to explain the dimensions of success in the 1920s. Social intelligence was described as "the ability to understand and manage men and women, boys and girls, to act wisely in human relations" by Thorndike (Thorndike E.L,1920). The popularity of this concept approach had been decreased until 1980s. Then, as per studies of Gardner, interpersonal and intrapersonal intelligence had become popular again at 1980s.

In 1990's, EI became popular and researches on EI has grown dramatically in academia, psychologists, education, media and business circle. That popularity had inspired academicians to research this new approach.

Firstly, Salovey and Mayer (1990) mapped the new concept, EI. Salovey and Mayer (1990) stated that EI was the ability that was used for perceiving and integrating the emotions for processing the thoughts, understanding the emotions and regulating them to achieve personal growth.

The ability to use emotions is the ability to harness emotions to facilitate cognitive activities such as information processing and decision-making. The ability to understand emotions represents the ability to comprehend emotion language, the distinctions among discrete emotions, and the causes and consequences of emotions. Finally, the ability to manage emotions is the ability to change emotions in one self and others.

Salovey and Mayer emotional intelligence model is based on four factors:

1 Appraisal expression of emotions both verbally and nonverbally,
2 Management of intrinsic emotions and emotions of others,
3 Emotions as a form of knowledge to cultivate intellectual and emotional growth,
4 Ability to generate emotions to assist problem solving.

However, EI has become popular outside the academia by Daniel Goleman who was a science journalist, author and psychologist. He had written for The New York Times, specializing in psychology and brain sciences for twelve years.

Goleman's book "Why EQ is more important than IQ?" became best-seller in 1995. So "Emotional Intelligence" became popular all around the world with the development of communication technology. Goleman explained that individual success in terms of managerial performance depends more on EQ than IQ.

Goleman describes EI in order to describe individuals' ability to monitor their own and others' feelings and emotions, to discriminate among emotions, and to use this information to guide thinking and action. Reuven Bar-On has also contributed to emotional intelligence researches with explaining it with the sets of abilities. As a result, EI has been defined as ability (Salovey P. & Mayer J.D., 1990), a combination of skills and personal competencies (Goleman D.,1995) or a set of traits and abilities (Bar-On R., 1997).

Bar-On described Intelligence as a mix not only Emotional but also Social Intelligence (ESI). Bar-on (1997) identified that the common points of interrelated emotional and social competencies were used to understand ourselves and others, to express ourselves and to overcome the daily demands by describing ESI with the five abilities which were:
– interpersonal skills
– intrapersonal skills
– adaptability
– stress management
– general mood

After 1997, a lot of academic researches for EI has been conducted such as; (Cooper & Sawaf, 1997), (Dulewicz & Higgs, 1998), (Huy, 1999), (Wong & Law, 2002), (Gardner,2003), (Reus & Liu, 2004), (Cote & Miners, 2006), (Kafetsios & Zampetakis, 2008).

2.2 *Work Performance:*

Performance can be defined as an achievement or an accomplishment. It can also refer to "the amount of useful work accomplished by a thing compared to the time and resources used"(Fletcher J.L., 1993). This definition shows that high performance can produce results much better than expected both in individuals and in organizations.

Performance is the result presented by not only a person bur also a group or an organization. Mathis and Jackson (2005) defined performance, as it was what had been done or not done by employees. Mangkunegara (2007) also claimed that performance was the result of the quality and quantity of an accomplished work of an employee. Due to these definitions that an employee in performing a task could be evaluated in accordance with by a given performance level, e.g. the employee's performance could be determined from the result of the achievement of targets in the specific period within the organization.

Most researchers are still debating on the definition on the performance. However, there are different general models of work performance and the determinants of work performance.

Campbell et al. (1993) stated that performance was a multi-dimensional phenomenon that was comprised of various latent factors, which were declarative knowledge, procedural knowledge, skill and motivation. Landy and Conte (2007) also introduced that performance was a behavioral concept that could be described as something that people can do and observe. From a working environment perspective it can only include the specific actions and behaviors which are relevant and applicable to the organization's goals and which can be measured in terms of the individual employee's proficiency.

Landy and Conte (2007) also brought into consideration the term of effectiveness and productivity as performance indicators. They explained the productivity is "the ratio of effectiveness (output) to the cost of achieving that level of effectiveness (input)".

Thus work performance can be considered as a contribution to the goals and objectives of any system (Schermerhorn, Hunt & Osborn, 2005). Accordingly work performance can be related with individual or group of that system.

Organization's success or failure depends on work performance of the individuals in that organization. Muchinsky (2003) presented that work performance was the set of worker's behaviors that could be monitored, measured, and assessed achievement in individual level. It can be said that these behaviors are also important for the organizational goals. It should be emphasized that individual work performance is important factor to push forward to make organization excellent.

Work performance is maybe the most important dependent variable in industrial and organizational psychology. Companies has focused on improving worker productivity, which is one of the work performance measures (Borman, 2004). In 1996, Greguras described work performance as the extent to which an organizational member contributes to achieving the objectives of the organization (Greguras et.al, 1996). Keller (2006) indicates that, if organization expects and wants the best from employees, they will give their best.

On the other hand motivation is one of the most effective factor to enhance effective work performance of workers in organizations. Motivation is a basic psychological process. According to Luthans (1998); motivating is the management process of influencing behavior based on the knowledge of what make people tick. Luthans (1998) asserted that motivation was the process that arouses, energizes, directs, and sustains behavior and performance. So, it is the process of stimulating workers to action and to achieve a goal. Motivation is the one way of stimulating people is to employ effective, which makes workers more satisfied with and committed to their works. Of course reward is another important factor but it's not ev. For example money is not the only motivator in some circumstances. There are other parameters which can also serve as motivators.

Job satisfaction is an important motivation way for the employees. Employees are the workers of the organization and organization will be strong correlatively if the employees have enough motivation to achieve organization's targets. Spector defines Job satisfaction as "how people feel about their jobs and different aspects of their jobs. It is the extent to which people like or dislike their jobs" (Spector P,1997). As we said before motivation is positively correlated with job satisfaction and work performance is correlated with job satisfaction too. And emotional intelligence is related with the job satisfaction and so does the work performance.

3 STUDY

In this study; it is aimed to understand the effect of emotional intelligence on the work performance of the seafarers'. The cohesiveness and performance are investigated through their relationship with the Emotional Intelligence, which is assessed by the Schutte Emotional Intelligence Scale (SEIS).

3.1 *Method*

Schutte Emotional Intelligence Scale is used in this study. Schutte constructed the questionnaire in 1998 to measure the EI. This scale was 33-item scale. Then eight items had been added into this scale and made this questionnaire 41-item scale which is called as three-factor structure as Emotional Clarity, Emotional Attention and Emotional Repair.

Seafarers' emotional intelligence was assessed using a 41-item questionnaire (Schutte, 1998) whose items were generated based on the trait model of emotional intelligence. The respondents were asked to rate each item using a 5-point Likert scale that ranged from "strongly disagree" (1) to "strongly agree" (5). This questionnaire is translated into Turkish by Tatar et.al, 2011.

Job performance data is supplied by the questionnaire which made by researchers in 2013 including many factors like intra-group corporation, using others' experience, transferring experience, trying to introduce new ideas, participating in personnel training and using the proper way and style at work. The reliability of this scale was 0.82.

The first section of the questionnaire requested demographic information about participants including age, marriage and education. The basic variables include the emotional intelligence and job performance and their sub-variables.
- Emotional intelligence includes self-awareness, self-motivation, empathy and social skills.
- Job performance includes intra-group corporation, conveying their experience, using others' experience, trying to introduce new ideas, participating in personnel training and using the proper methodology.

3.2 Participants

The cross sectional study which was done from August 2012 to January 2013. We had participants who were working in ships. Questionnaires were distributed among the participants. From the 150 participants, 105 participants were enrolled (response rate 86%).

The range of participants' age was from 22 to 38. About 35% of them were married and 20% of participants graduated from university. In order to analyze the data, we applied SPSS software .

3.3 Results

The data analysis was carried out with the IBM SPSS Statistics 20.0 programme. Figure 1 and Table 1-6 gives the results of statistics analysis related with this survey.

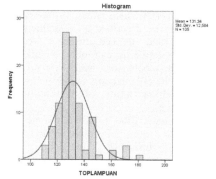

Figure 1: Histogram of Total points

Table 1. Descriptive Statistics

	N Statistic	Minimum Statistic	Maximum Statistic	Mean Statistic	Std. Deviation Statistic	Kurtosis Statistic	Std. Error
Total Points	105	111	183	131,34	12,584	3,740	,467
Valid N (listwise)	105						

Min: 111, max:183 points has been observed.

Factor analysis was conducted to the data that handled through the mentioned questionere. As a result, three factors were obtained.

Table 2. KMO and Bartlett's Test results

Kaiser-Meyer-Olkin Measure of Sampling Adequacy.	,710
Bartlett's Test of Sphericity Approx. Chi-Square	240,293
Df	45
Sig.	,000

At the initial stages of factor analysis, some pre-tests such as Bartlett Test and Kaiser-Meyer-Olkin tests were applied to understand the eligibility of factor analysis for the data.

Bartlett test (Bartlett Test of Sphericity) that tests the hypothesis "a correlation matrix is equal to unit matrix". Rejection of hypothesis means that there is a correlation between the variables and factor analysis can be applied to the variables. For this study, it was found that the correlation matrix of main cluster was not equal to the unit matrix and the sphericity criteria was validated (p<0.05). Kaiser-Meyer-Olkin (KMO) Value was also applied for understanding the reliability of factor analysis to this study. According to KMO criterion, low KMO

values show that factor analysis would not give true results for given clusters. For this study, KMO value criteria were also validated through sample size, observed correlation and partial correlation coefficients.

Table 3: Sorting of Conceptual Significance of Factors

Optimism/Coordinate the Mood	Factor 1
Use of Emotions	Factor 2
Evaluation of Emotions	Factor 3

Then Kolmogoroc-Smirmov test was used to test the normality of factors to determine an appropriate method for finding the correlations between these three factors.

Table 4. Kolmogorov-Smirnov Test results

		Factor 1	Factor 2	Factor 3
N		105	105	105
Most Extreme	Absolute	,102	,130	,112
Differences	Positive	,102	,130	,112
	Negative	-,070	-,078	-,064
Kolmogorov-Smirnov Z		1,044	1,336	1,145
Asymp. Sig. (2-tailed)		,225	,056	,145

As the results of Kolmogorov-Smirnov test were p>0.05 for three factors, normal distribution was assumed and Pearson Correlation was used to obtain the correlation between the factors.

Table 5: Pearson Correlation Results

		Factor1	Factor2	Factor3
Factor1	Pearson Correlation	1	,441**	-,085
	Sig. (2-tailed)		,000	,391
	N	105	105	105
Factor2	Pearson Correlation	,441**	1	,334**
	Sig. (2-tailed)	,000		,024
	N	105	105	105
Factor3	Pearson Correlation	-,085	,334**	1
	Sig. (2-tailed)	,391	,024	
	N	105	105	105

**. Correlation is significant at the 0.01 and 0.05 level (2-tailed).

As it could be seen from Table 5,

Factor 1 and Factor 2 is in % 44 positive correlation between each other.

Factor 1 and Factor 3 has no significant correlation.

Factor 2 and Factor 3 have %33 positive and significant correlation.

Consequently we can interpret these results as optimistic approach or coordinating their moods in the seafarers can help them in using their emotions in their jobs. Most of the seafarers that can use their emotions during their work performance, can also success in evaluating the emotions of others and himself/herself while performing a task.

After factor analysis, some important questions were analyzed with Kendall's tau-b correlation coefficient that can give effective results for two variables. These questions and their correlations are given below in Table 6.

4 DISCUSSION

Emotional Intelligence is the first coined term by Salovey and Mayer in 1990. Salovey and Mayer defined emotional intelligence as "the ability to perceive emotion, integrate emotion to facilitate thought, understand emotions and to regulate emotions to promote personal growth" and then Goleman described EI as to" describe individuals' ability to monitor their own and others' feelings and emotions, to discriminate among emotions, and to use this information to guide thinking and action."

When we focus on maritime transportation, emotional intelligence can emerge as an important issue among the seafarers on a board. As mentioned previously, ship is a closed and complex system that involves much interaction and communication between the crew.

This study is conducted to get some emotional intelligence data from seafarers.

The mean of EI score is 131,34, meaning that according to the defined scale of EI questionnaire is greater than moderate. This result can indicate that maritime sector firms are successful in selecting work force with high EI. But this indication should also be supported with the future studies.

Due to the general gender who works as seafarer is male, we didn't show gender in demographic information analyses. Only 15 of the 105 participants were women, so we could not have any result according to the gender. However, it could be omitted if we take into consideration seafarers' gender are male generally.

In general, EI contains the sort of skills, which are educable, flexible and variable in the course of time. This course of action, therefore, will increase employees' ability to adopt with work place and facilitate proper work relationship, which leads to improving efficiency and job performance.

Table 6. Some important questions correlations

Question 1	Question 2	Kendalls Statistics p values	Correlation
"I am not affected from my mood while I am solving a problem."	My emotions don't facilitate my business while I am trying to generate new ideas."	0.023	No
"I am aware of the emotions that I feel"	"I have difficulties to understand the nonverbal messages such as body language, facial impressions of other people"	0.01	No
"I generally misunderstand what happens in social life"	"I have difficulties to understand the nonverbal messages such as body language, facial impressions of other people"	0.2	Yes
"I generally misunderstand what happens in social life"	"I can't understand sometimes when somebody is serious or kidding while speaking"	0.25	Yes
"I can understand what the other people are feeling by just only looking them"	"People are generally rely upon me."	0.145	Yes

The results shows us that the seafarers who can coordinate the his/her emotions and mood during a work performance, can also understand the other's emotions. These seafarers who can use their emotions during a work performance can also evaluate his or her emotions effectively. In additions, from the question correlations we can also understand that seafarers can also gain the confidence of the other people which is very important skill for a ship environment. The other important result of the question correlation is that the lack of emotional intelligence can be a reason for misunderstanding between the individuals. Thus we can say that emotional intelligence can also a driving factor for a good communication skills and peace in the ship environment.

We suggest further studies for understanding the emotional intelligence issues on ships as sea environment has its own theories for the individuals.

REFERENCES

Bar-On R, (1997). Bar-on emotional quotient inventory, a measure of emotional intelligence, technical manual. Toronto: Multi-Health Systems.

Borman W.C., (2004). The concept of organizational citizenship. Current Directions in Psychological Science 13 (6), 238–241

Campbell, J. P., McCloy R. A., Opper S. H, Sager C. E. (1993). A theory of performance. In N. Schmitt, W. C. Borman, & Associates (Eds.), Personnel selection in organizations: 35–70.

Daus, C. S., & Ashkanasy, N. M. (2005). The case for an ability-based model of emotional intelligence in organizational behavior. Journal of Organizational Behavior, 26, 453–466

Fletcher J.L. 1993. Patterns of High Performance: Discovering the Ways People Work Best. San Francisco, USA: Berrett-Koehler Publishers.

Goleman D. (1998). Working with emotional intelligence. New York: Bantam.

Greguras, G. J., Ployhart R. E., Balzer W. K. (1996). Performance appraisal training program for Wood County Council on Alcohol and Drug Abuse. Bowling Green, OH: Bowling Green State University, Institute for Psychological Research and Application.

Johnson, P.R. & Indvik, J. (1999). Organizational benefits of having emotionally intelligence managers and employees. Journal of Workplace Learning 11, pp.84-88

Keller R.T. (2006). Transformational leadership, initiating structure & substitutes for leadership: A longitudinal study of research & development project team performance. Journal of Applied Psychology, 91(1): 202-210.

Landy, F.J., Conte J.M. (2007). Work in the 21st Century: An Introduction to Industrial and Organizational Psychology. 2nd ed. Oxford, UK: Blackwell Publishing.

LeDoux,J.(2003) The Emotional Brain. Phoniex, Arizona.

Locke, E. A., Latham G. P. (1984). A Theory of Goal Setting and Task Performance. Englewood Cliffs, NJ: Prentice Hall.

Luthans F. (1998). Organisational Behaviour. 8th ed. Boston: Irwin McGraw-Hill.Manzoni, J.

Mangkunegara A. (2007). Manajemen Sumber Daya Manusia Perusahaan. Cetakan Ketujuh. Bandung : Remaja Rosdakarya

Mathis R.L., Jackson J.J. (2005). Human Resorce Management. Essential Perspectives. Third Edition. South Western:Thomson Corporation

Matthews, G., & Falconer, S. (2002). Personality, coping, and task-induced stress in customer service personnel. In Proceedings of the Human Factors and Ergonomics Society 46th Annual Meeting (pp. 963–967).Santa Monica, CA: Human Factors and Ergonomics Society.

McCLoy R.A., Campbell J.P., Cudeck R. (1994). A Confirmatory test of a model of performance determinants. Journal of Applied Psychology, 79, 493-505

McMaster M. D. (1996). The intelligence advantage: organizing for complexity, Newton. MA: Butterworth-Heinemann

Muchinsky, P. M. (2003). Psychology Applied to Work (7th ed.). Belmont, CA: Wadsworth.

Rosete, D. & Ciarrochi, J. (2005). EI and its relationship to workplace performance outcomes of leadership effectiveness. Leadership Organizational Development, 26, 388-399

Salovey P, Mayer J.D. (1990). "Cognition and Personality" Emotional Intelligence, Imagination, Cognition and Personality, 9:185-211.

Salovey P., Mayer J. D. (1997). Educational development and Emotional intelligence. New York, NY: Basic Books.

Schermerhorn J. R., Hunt, J. G., Osborn R. N. (2005). Organizational Behavior. New York:Wiley.

Spector P. (1997). Job Satisfaction: Application, Assessment, Causes and Consequences. Thousand Oaks. CA: Sage.).

Suls, J. (2001). Affect, stress, and personality. In J. P. Forgas (Ed.),

Handbook of affect and social cognition

(pp. 392–409). Mahwah, NJ:Erlbaum.

Tatar A., Tok S., Saltukoğlu G. (2011). Adaptation of the revised schutte emotional intelligence scale into Turkish and examination of its psychometric properties. Bulletin of Clinical Psychopharmacology, Vol: 21, N.: 4, 2011 - www.psikofarmakoloji.org.

Thorndike E.L. (1920). Intelligence and its use-Harper's Magazine, 140, 227-235.

Waldman D.A , Spangler W.D. (1999). Determinants of Work performance, Human Performance, Vo. 2, No.1,pp.13-32.

Wong C. S., Law K. S. (2002). The effects of leader and follower emotional intelligence on performance and attitude: An explanatory study. The Leadership Quarterly, 13(3), 243-274.

Zeng, X. & Miller, C. (2002), Emotional Intelligence and Personality. IOOB, 2002.

Chapter 4

Terrorism and Piracy

Terrorism and Piracy
STCW, Maritime Education and Training (MET), Human Resources and Crew Manning, Maritime Policy,
Logistics and Economic Matters – Marine Navigation and Safety of Sea Transportation – Weintrit & Neumann (Eds)

Characteristics of Piracy in the Gulf of Guinea and its Influence on International Maritime Transport in the Region

K. Wardin & D. Duda
Polish Naval Academy, Gdynia, Poland

ABSTRACT: Piracy is one of the major problems of maritime transport in the XXI century. It occurs in several parts of the world but it has become a real difficulty since 2008 in the Horn of Africa region (Somalia waters) and in the Gulf of Aden as well as the whole Indian Ocean. Unfortunately the successes of Somali piracy has shown other nations that it can be an easy way of living, and for this reason international society can observe an enormous outbreak of unlawful and pirates' activities in other regions. One region of a big importance for international commerce, lines of maritime transportation and energy security is a western part of Africa and especially the Gulf of Guinea. Countries located in this part of the world, Nigeria, Benin, Cameroon, are reach in natural resources exclusively oil and gas, which makes this region very important for the USA and Europe, particularly that the situation in the Far East is so unstable and unpredictable. The article describes illegal actions in the waters of Niger Delta and the whole Gulf of Guinea showing at the same time the influence of these acts on maritime transport in this part of the world.

1 INTRODUCTION

Piracy and armed robbery against ships is still one of the main concerns of the international society around the African continent. The reason for this unstable situation is the fact of a very poor economic situation in this part of the world as well as the existence of most of world's failed or very weak states on this continent. West Africa is one of the poorest and least stable political and economic parts of Africa. The Gulf of Guinea is one of the most trouble regions in term of maritime crime and piracy. There is no universally agreed geographical definition of the Gulf of Guinea. The region is defined as the part of the Atlantic Ocean southwest of Africa. Almost too obtuse to be a gulf, the region encompasses over a dozen countries from West and Central Africa.[1]

[1] There are: Angola, Benin, Cameroon, Central African Republic, Ivory Coast, Democratic Republic of Congo, Equatorial Guinea, Gabon, Gambia, Ghana, Guinea, Guinea-Bissau, Liberia, Nigeria, Republic of Congo, São Tomé and Príncipe, Senegal, Sierra Leone and Togo., C. Onuoha, *Piracy and maritime security in the Gulf of Guinea: Nigeria as a microcosm,*
http://studies.aljazeera.net/ResourceGallery/media/
Docments/2012/6/12/201261294647291734Piracy%20and%20Marit me%20Security%20in%20the%20Gulf%20of%20 Guinea.pdf, 20.11.2012.

Because the situation in the region, in the assessment of the political and economic conditions of experts from many countries in the world, is constantly worsening, a number of organizations and international institutions are involved in monitoring the states in the region. The organizations include the United Nations (UN), the World Bank, and the British Department for International Development as well as the CIA (Central Intelligence Agency). Using selected indicators they identify a list of countries that are in the worst situation in terms of politics and economy. The list written by *Found for peace* organization and published by *Foreign Policy* magazine is described as failed, falling or weak states' list and consists of states which are located in the region of the Gulf of Guinea starting from Guinea and Sierra Leone through Liberia, Ivory Coast, Ghana, Togo, Benin, Nigeria, and Cameroon.

Growing insecurity in the Gulf of Guinea has attracted the attention of policy makers and the academia on a global scale as a result of the spate of piracy and smuggling that takes place in the region. Under international law, piracy constitutes crimes against the security of commerce on the high seas for private ends. This includes attacking, robbing and pillaging of ships on the high seas. Although cases of the Gulf of Guinea are not exactly the cases of maritime piracy from the law point of view, but maritime security is critical for countries within this region to achieve optimum benefits and unfortunately it has been declining constantly for the past few years.

While media headlines abound and the mass ongoing conventional naval campaign against Somalia's maritime piracy has dominated the minds of the world's maritime security experts, security conditions in the Gulf of Guinea continue to decline rapidly.[2] The region's potent criminal cocktail which has developed over time encompasses the full spectrum of nefarious activities starting from:

- Illegal weapons and human trafficking;
- Sea-based & Narco-terrorism;
- Violent maritime piracy.

The mixture of failing states in the region, flourishing of organised crime and the lack of effective action of appropriate services leads to believe that the problem in the near future will exceed its scope, complexity and strength even coup Somali pirates. It cannot be forgotten that the region itself is an incubator of drugs trafficking as well as the illicit smuggling of arms and people. Organised, maritime criminal activity in the Gulf of Guinea is diverse and seemingly all pervasive. Piracy and robbery are now so well established as to represent a considerable issue for Nigeria, Cameroon and Benin. In 2009, the International Maritime Organization (IMO) substantiated 28 attacks in their waters but it is believed the real figure is to be twice that. In 2010 the Nigerian Navy alone received over 100 reports of pirate attacks on ships in their waters.[3] Table 1 shows the scale of the problem in certain countries according to IMB piracy and armed robbery against ships reports.

Table 1, Number of attacks in the Gulf of Guinea region in 2006-2012, Source: *Piracy and armed robbery against ships, annual reports* January – December 2008, 2009, 2010, 2011 and 2012.

	2006	2007	2008	2009	2010	2011	2012
Ghana	3	1	7	3	-	2	2
Cameroon	1	-	2	3	5	-	1
Ivory Coast	1	-	3	2	4	1	3
Guinea	4	2	-	5	6	5	2
Nigeria	12	42	40	29	19	10	24
Benin	-	-	-	1	-	20	4

As the table shows the problem is the most warring in Nigeria as the number of the attacks stays on the same level or even raises for the past years making the region a perfect incubator for piracy, providing both resources and safe haven.

Surrounded by some of Africa's most proficient oil producers, including Nigeria, Angola, Gabon, Ghana, and Guinea, the Gulf is a major transit route for oil tankers on their way to international markets, especially the US market. Nigeria seems to be the most troublesome country in the region and on the other hand most important among oil producers.

2 POLITICAL SITUATION IN NIGERIA AND THE REGION

Since 1999 after the elections, ending almost 33 years of military rule, Nigeria had a civilian leadership of Oluseguna Obasajno, which unfortunately caused production of independent militia led by former governors of the Nigerian provinces. In turn, the arising of the political spirit turned that many Nigerians started to manifest the highest aspirations to the governmental offices. Both of these groups very quickly came to the agreement and began to work together. This led to the political fragmentation of the country, corruption, lack of legitimacy and stable governments.

When in May 2007, Yar'Adua, the leader of People's Democratic Party (PDP), became the President of Nigeria everybody predicted new, positive deal in this country. However, the new leader was not able to cope with internal stalemate, which was the reason for even bigger chaos in the country. Dismal economy also began to reflect negatively on neighboring countries of Nigeria. In the effect, the absence of a coherent and reasoned strategy of the state increased the scale of violence and crime in the Delta of Niger, that was the reason to (for the entire 2008 year) many encounters between urban squads and branches of government.[4]

As the former British colony Nigeria is the 12th largest producer of petroleum in the world and the 8th largest exporter, and has the 10th largest proven reserves. (The country joined Organization of the Petroleum Exporting Countries - OPEC in 1971). Petroleum plays a large role in the Nigerian economy, accounting for 40% of GDP and 80% of Government earnings.[5] Because Nigeria is a major exporter of oil, not surprising that the number of attacks by "militants" on the oil rigs increased in 2008 by 1/3 in comparison to 2007, when there were 68 attacks (2008 amounted to 92). This was reflected on the mining of oil and oil production in Nigeria which has dropped by 20% both, on- and off-shore, since 2006 as a result of the rise of piracy in the region, costing the Nigerian economy approximately 202 million US$ between 2005 and 2008. Illegal

[2] D. Mugridge, *Piracy storm brews in West Africa: Gulf of Guinea under maritime siege*, http://www.defenceiq.com/naval-and-maritime-defence/articles/piracy-storm-brews-in-west-africa-gulf-of-guinea-u/, 10.11.2011.
[3] C. I. Obi, *Nigeria's Niger Delta. Understanding the Complex Drivers of Conflict*, http://www.nai.uu.se/research/nai-foi%20lectures/calendar2009/obi.pdf, 20.12.2011.

[4] F. Kąkol, *Metody zwalczania bezpieczeństwa statku i platform wydobywczych w kontekście wzrostu piractwa w obszarze Zatoki Gwinejskiej*, Naval University of Gdynia, Gdynia 2012, p. 9-10.
[5] *The World Factbook, Nigeria* https://www.cia.gov/library/publications/ the-world-factbook/geos/ni.html, 15.12.2012.

bunkering (oil theft) costs Nigeria's oil output 10,000 barrels per day (bpd), equivalent to 10 % of daily output and 1.5 million US$. This money could supply a fighting force of 1,500 individuals with weapons and ammunition for two months, and can contribute to sustaining pirates and other attackers, who can afford to become more sophisticated in their activities. It is estimated that oil to the value of between 300 and 400 billion US$ has been stolen in Nigeria in the last half decade.[6]

Apart from being Africa's largest oil producer and exporter, Nigeria is also a producer of natural gas, accounting for an estimated output of 31 billion cubic meters per year. Natural gas exports account for about 4 billion US$ worth of earnings annually. Most of the natural gas is produced from the Niger Delta or its coastal waters. However, this oil and gas rich region that generates billions of dollars worth of revenues and profits annually, is also paradoxically one of the least developed and conflict ridden parts of Nigeria.[7]

As a result of the oil boom in the 1970s, the Niger Delta has been a hub of activity for oil companies. Local groups became unsatisfied with the amount of oil wealth that was making its way back to the Delta via the government and resentment rose spurring on militant groups to take action in the form of piracy and kidnap-for-ransom. The last activity is dangerous especially for white oil rigs' workers, engineers, who are the best targets to the locals.

There are several groups responsible for such actions but the most known and important are the Nigeria Delta Peoples Volunteer Force (NDPVF), the Niger Delta Vigilante (NDV) and the best known the Movement for the Emancipation of the Niger Delta (MEND). The last group is the most active of the militant groups involved in maritime crimes. MEND "claims to be fighting for a fairer distribution of Nigeria's vast oil wealth, and as a protest to the environmental damage caused by oil production in the Delta." The group is dissatisfied with not satisfactory benefits to the region despite having it exploited for its oil reserves. Pollution and environmental degradation add insult to injury as this has had a harsh impact on local fisheries and farmland, as well as causing a number of chronic illnesses in the local population, most notably in children.[8] the magnitude and impact of MEND's attacks which, in 2007 targeted such corporations as Shell, Agip and Chevron, leading to a shut-in of 27% or 675,000 bpd out of Nigeria's estimated daily production of 2.4 million bpd, the highest levels of loss since the crisis began. The scenario worsened after Yar Adua had been elected president, and one of MEND's leaders Henry Okah was arrested the same year in Angola on charges of gun-running.

While political issues are a great motivator of piracy in Nigeria and the region, this is not the only cause. Due to high levels of poverty and unemployment, the great financial reward of pirate activity has arisen as a motivator. In fact, it is reported that the majority of attacks are now "motivated by financial and not political gain." This has drawn young people into piracy, kidnap-for-ransom and bunkering, both within the Delta and at offshore facilities resulting in a tense security situation of maritime transportation in this region. Young Nigerians take example from Somali pirates, who have gained so far satisfactory profits from piracy.

Moreover, Cameroon's piracy troubles are also linked to those of Nigeria. The rise in piracy being experienced in Cameroon is linked to the breakdown in a deal that resolved a long-running border dispute between Nigeria and Cameroon over the Bakassi Peninsula. Some of the piracy is attributed to Nigerian pirates moving southwards, but some is attributable to the militant groups, basing themselves in both Nigeria and Cameroon, wishing for secession for the Bakassi region as a result of resource location. The tactics of MEND have also spread throughout the region through organised criminal gangs and separatist movements in Angola, Benin and Equatorial Guinea[9] making the situation even more difficult than it was.

As West Africa is wealthy in energy and mineral commodities, and borders an important sea-lane; the impacts of piracy have severe economic effects. Mindful that Nigerian pirates have been known to venture farther afield in pursuit of pirate activity, it is clear that vessels passing through these waters, and oil installations based in them, are increasingly vulnerable.

The situation seemed to worsen and contributed to increasing hijackings when the Nigerian government decided to remove state oil subsidies in January 2012. That decision led to mass protests, especially in Nigeria's north, which threatened to evolve into an Arab-Spring situation. Fortunately the Nigerian President Goodluck Jonathan backtracked, still reducing the subsidy, but not reversing it entirely. It has hindered the smuggling of oil products from Nigeria across land borders, resulting in a shift to hijackings at sea.

In Nigeria's south, MEND are using the Gulf of Guinea piracy to raise the stakes for pushing Nigeria's president to meet their resource and revenue demands. In fact, they claimed responsibility for a 28 February attack on a Dutch cargo ship off Port Harcourt and the kidnapping of three crew members in 2012. MEND has not

[6] L. Otto, *Piracy in the Gulf of Guinea: Attacks on Nigeria's oil industry spill over in the region.* http://www.reuters.com/article/2009/04/15/idUSL9951481, 12,12,2011.
[7] C. I. Obi,... op. cit.
[8] L. Otto,... op. cit.

[9] Ibidem.

previously been definitively linked to piracy networks operating in the Gulf of Guinea, but this type of expansion could internationalize the Niger Delta conflict.[10] The dynamics of Gulf of Guinea piracy is complex and very fluid, making it very difficult to trace hijackings and the end user of ransom payments. Fighting the problem generally should be the case for the whole international society but it is true that some countries should be interested in this matter more.

The most important question in this part of the article is who benefits from this region most in terms of oil import. The region produces 5.4 million barrels of oil per day, and it contains 50.4 billion barrels of proven reserves. The region has the fastest rate of discovery of new reserves in the world. Nigeria now supplies 10% of US imported oil and is the world's 8th largest oil exporter. The U.S. National Intelligence Council says that the region will provide 25% of US oil by 2015. Events in Afghanistan and Somalia illustrate the dangers that come from the nexus between organised crime, terrorism and failed/failing states. While many look to Africa for an African solution to retake control of their seas, they can't achieve this without timely Western assistance.[11]

Particularly growing in nature and frequency of these afflictions is sea piracy. Thus, for a region of such geostrategic and maritime significance, ensuring that good order at sea prevails is a matter of absolute necessity for the Gulf of Guinea states and extra-regional and world powers with growing economic interests in the region. This explains why the rapid increase in incidents of piracy in the region has attracted the attention of the United Nations Security Council (UNSC). The UNSC has adopted two resolutions - 2018 (October 2011) and 2039 (February 2012) – calling for more regional coordination and logistical support to regional security initiatives to counter the growing menace of piracy in the region.[12]

Oil pollution, extreme poverty, high levels of youth unemployment, pollution, perceived discriminatory employment practices against locals by oil companies and socio-economic and political marginalization and neglect by successive administrations constitute the main grievances against the oil companies and the government. These complaints have a long history that is connected to the view of the ethnic minority groups in the Niger Delta, that they are being "cheated" out of a fair share of oil revenues because they are politically marginalized by a federal government that is dominated by bigger (non-oil producing) ethnic groups, which in partnership with foreign Oil Multinationals exploit their region, take their lands and expose it to oil pollution and environmental degradation, while the indigenes and owners of the land do not benefit from the billions of dollars generated from their region, nor do they get adequate compensation for the destruction of their livelihoods or the "loss" of their lands.

3 THE CHARACTERISTICS OF PIRATES' ATTACKS IN THE GULF OF GUINEA

The problem of piracy in the Western Africa, as mentioned above, stretches from Ghana in the North to Angola in the South, which gives all together over 2,700 km of coast but historic epicenter is Nigeria. The growing investments in the region, especially in offshore oil infrastructure, mean that coastal trading and maritime traffic are bound to increase in the region.

At the very beginning pirates used simple boats and knives. With time piracy in Nigeria has become more professional. The pirates started to operate in groups of 20 to 30 members and started to be known as really violent, as they usually deploy sophisticated arms and weapons like AK-47s. The traditional modus operandi of pirates operating in the region had largely involved the use of speedboats to attack and dispossess shipping crew of cash, cargo and valuable, when the vessel is at anchor or in harbour, but mostly close to shore. The fact that, some of the pirates were able to identify and using cost-effect method select ships to attack, has led many companies to the conclusion that gangs cooperate with masters and custom officers. Most of the attacks on the waters of Nigeria is carried out on slowly passing or anchored ships and fixed installations near the coastline or on inland waterways. In the Niger Delta and beyond Bakassi Peninsula pirates are active day and night. In Lagos virtually all attacks occur during the hours of the night. The attacks usually operate near the coastline, i.e. to 20 nautical miles are the most common, but there are also attacks significantly further away from land. Most acts of piracy occur to 30 nautical miles from the coast of Lagos and 40 nautical miles from the Niger Delta. Just a few attacks have been confirmed beyond this distance. This distance proves that the problem occurs beyond 12 nm limits of territorial waters and according to international law should be considered as maritime piracy and fought by international community.

A few years ago the difference between those incidents in Lagos and the Niger Delta was significant, particularly in regard to the level of

[10] J. Alic, *Gulf of Guinea piracy increases amid unrest and rising oil prices*, http://oilprice.com/Energy/Energy-General/Gulf-of-Guinea-Piracy-Increases-Amid-Unrest-and-Rising-Oil-Prices.html, 20.06.2012.
[11] D. Lewis, *Piracy in the Gulf of Guinea: Attacks on Nigeria's oil industry spill over in the region* http://www.reuters.com/article/2009/04/15/idUSL9951481, 12.11.2011.
[12] C. Onuoha,... op. cit.

violence. However, from the end of 2008, this difference has become less pronounced, because pirates of the Lagos became much more brutal and started to be as cruel as Nigerians. Most of the incidents in the main port of Nigeria takes place with the help of firearms, which resulted in a significant increase in the number of wounded or even killed sailors in the attacks. Acts of the piracy in this part of Africa have a unique brutality in comparison to other such events in other regions of the world, including the acts of piracy carried out by Somali pirates, which makes a distinction between those two illegal activities. The main difference between incidents in East and West Africa is that pirates in Nigeria are rather entirely confined to the limits of territorial waters.

Acts of piracy committed in Nigeria are reported to the Piracy Reporting Centre (PRC) by the masters, owners and other organizations that collect information on threats against piracy as the Risk Intelligence. The PRC is aware of the fact that some of the reported attacks are not confirmed but on the other hand under-reporting of vessels who have been victims of attacks is a big problem as well. In 2008 the majority of acts of piracy were reported in the area of Nigeria. With 40 documented attacks 27 ships were climbed in and five abducted.

As can be seen in the figure 1, piracy in the Gulf of Guinea has not been evenly distributed. The incidents of piracy in the region were rising from 2006 up to 52 attacks in 2008 and started to decrease in the next four years achieving the level of 36 incidents in 2012. The majority of the attacks recorded between 2006 and 2012 occurred in Nigerian waters (65%), even as most attacks in Nigerian waters go unreported.

Pirates are however, increasingly modifying their tactics by hijacking fishing vessels, particularly within Nigerian waters, and using same to attack other vessels operating off the coasts of neighbouring countries like Benin and Cameroun. Recent attacks have extended further out at sea and have focused largely on oil-laden vessels, to steal the petroleum product. The current focus on oil vessels is not unconnected to the hike in oil price, especially in Nigeria.[13] Pirates know very well that stilling oil or petroleum gives them opportunity to sell it on the black market and have normal life and so they keep attacking passing tankers, oil rigs and Floating Production, Storage and Offloading Unit (FPSO).

4 REASONS FOR PIRACY IN THE GULF OF GUINEA REGION

The existence of piracy in this region can be explained by several important reasons. First of all we should remember that historically the region is connected with the discoveries of Vasco da Gama and slavery period related to cotton plantations in the North America, but these go back to past times and can only be the historic background that makes the region prone to piracy. Modern piracy has its roots in several important vectors and probably the most important is the fact of existing sea lines of transportation with heavy traffic, weakness of states and luck of proper governments in many countries as well as full acceptance of this procedure by local people.

Problems of the states derive primarily from bad governance and tragic economic situation. Despite their vast oil endowments, most of the Gulf of Guinea states parade worst indices of human development such as high unemployment and poverty generated by bad governance. With declining opportunities for legitimate livelihood amidst affluence, some youths in the region are easily recruited for violent conflicts or take to criminality, especially piracy, for survival. The case of resource-conflicts in Nigeria's oil-rich Niger Delta region and Angola's Cabinda region arising from bad governance are typical examples. Luck of stability is the main reason for fighting of many counterparts which is connected with flow of arms and easy access to all kinds of weapon. It cannot be forgotten that the history of Africa is linked to many clan and territorial conflicts that make the situation even more difficult. After the period of colonialism, African countries regained their independence but fighting between clans never stopped. Logically the most important forces for governments were land forces, so when the problem with piracy occurred the countries have not had suitable forces to deal with the crisis properly and still do not have such possibilities. Table 2 shows the difference in strength of army and navy forces with comparison to the length of the coastline of particular countries.

Table 2. The strength of army and navy forces with comparison to the length of the coastline,
K. Wardin, *Współczesne piractwo morskie. Wyzwanie somalijskie oraz odpowiedź społeczności międzynarodowej*, Belstudio, Warszawa 2012, p. 81.

Country	Army	Navy	Length of coastline (km)
Ghana	6 000	1 000	539
Cameroon	11 500	1 300	402
Ivory Coast	6 500	900	515
Guinea	1 100	120	402
Nigeria	8 000	8 000	853
Democratic Republic of Congo	120 000	1 000	37

---

[13] See more: C. Onuoha,... op. cit.

The data show that countries having problems with piracy do not posses adequate maritime security forces such as navy, air force and coast guards to control territorial waters and perform international duties connected with deterring crime and piracy at sea. To a large extent, the disproportionality manifests not only in personnel strength, but also in the size of budgetary allocations to maritime forces. The result is that maritime security forces are ill-equipped and underfunded to perform interdiction operations. Some countries do not even have the possibility to control crime on land because of the luck of proper administration, bad governance or corruption amongst its government workers and it is well known that it is the root of the menace.

Nigeria is the most afflicted by maritime threats pervading the Gulf of Guinea region. Nigeria loses about 800 million US$ yearly over poaching, 9 billion US$ per annum to piracy, and 15.5 billion US$ annually due to oil theft or illegal oil bunkering. Worst still, over 70% of about 8 million illegal weapons in circulation in West Africa are in Nigeria. These security leakages are both the features of, and contributors to, the complexity of maritime insecurity in the Gulf of Guinea.[14] Moreover, in a globalised world facilitated by tremendous growth in maritime transportation, the range of maritime threats requires collaboration among national navies in defense of their coast. In the described region in particular, it is doubtful whether there is any country that has a navy powerful enough to combat piracy alone.

One of probably the most important reasons why piracy flourishes in the region is the fact that local people do not perceive this activity as illegal or crime. On the contrary they seem to be proud of this activity and consider it as compensation for injustice done by their governments. This would be probably the most difficult fact to change if international community takes up fighting piracy seriously. Human mentality is always very difficult to change, but plays in this matter the biggest role. With bad governance, violent criminality, weapons proliferation, ill-equipped and underfunded navies, and the absence of a maritime strategy coexisting with affluence, it is not surprising that piracy is brewing in the region.

5 COST OF PIRCY IN THE GULF OF GUINEA

The cost of piracy in the region comes in a variety of shades. The loss of human lives and valuable property as well as infliction of bodily injuries and trauma to innocent crews and their families are the most obvious direct impacts of pirate activities.

Many international organizations try to bring this problem into the light to make public opinion aware of this matter. On the other hand, there are also voices saying that this issue is important to only a few hundreds of people so it does not affect the whole community. It cannot be forgotten that piracy makes wanted trade and investment in the region more risky and expensive. Given the risk involved in transporting goods through the region, insurance premiums have been escalating as because shippers factor in higher risks into their operating costs. Extra equipment installed onboard vessels; to make them less prone to pirate activities is also costly. Ransoms paid to pirates to free high jacked crew members make also some costs and even though the Gulf of Guinea is not so dangerous as the Gulf of Aden and Indian Ocean, we have to admit that the costs are getting higher every year.

The International Bargaining Forum[15] noted that from April 2012, the territorial waters of Benin and Nigeria have been designated a high risk area in order to allow ship operators to make any necessary preparations.[16] The implication is that these waters will be treated the same as the high risk areas in the Gulf of Aden and near Somalia due to increased pirate attacks. It also entails that seafarers have the right to refuse to enter these waters and are entitled to double the daily basic wage and of death and disability compensation while within the areas of risk. This has impacts on the income of regional ports and the cost of goods destined for the region. As we can expect this will make a difference and costs of international maritime transport will probably rise.

For economies that depend on oil from the Gulf of Guinea, this will translate to increased pump price of petrol but not only, as oil is an important component of many modern products. There is also another side of this matter which is the fact that in globalized world and the age of outsourcing our world is linked and exists only because of cooperation, so higher prices in one part mean higher prices in other countries.

Piracy also leads to disruption of livelihood systems as well as shortages in food supply, which in case of African continent is extremely important. In February 2008, for example, the Nigerian Trawler Owners Association recalled about 200 vessels from shore due to spiralling piracy. This resulted in a temporary work stoppage for an estimated 20,000 workers, and consequently over 100% increases in the prices of seafood in the local market.[17]

---

[14] Ibidem.

[15] A global labour federation for the transportation industry representing 690 labour unions including 600,000 seafaring members
[16] *International Bargaining Forum: Piracy High Risk Area*, http://www.scoop.co.nz/stories/WO1203/S00445/international-bargaining-forum-piracy-high-risk-area.htm, 12.12.2012.
[17] See more: C. Onuoha,... op. cit.

Unchecked piracy in the region also has the potential to undermine diplomatic relations and can as well trigger the intervention of external forces, of which the ramifications may not be in the best interest of the region. African people and politicians are very sensitive in the area of their independence and any case of unwanted help can be understood as the possibility of another invasion. Anyway the intervention of foreign navies and private security companies (PSC) has not ended the threat of Somali piracy in Horn of Africa but only held it back. There is also emerging concern over human rights violations by foreign naval forces or PSC targeting Somali fishermen because it is hard for them to differentiate from pirates. The difficulty would probably be the same in case of the Gulf of Guinea.

Luck of counter piracy efforts would also give an unwanted sing for other troublesome regions where piracy has not been a big problem yet but may become in the future. This would bring even higher costs both human and economic to come and to bear, and the international community is not ready for such involvedness.

6 EFFORTS AT COUNTERING PIRACY AND IMPROVING MARITIME SECURITY IN THE REGION

The high incidents of piracy and other maritime crimes in the region have prompted several national, bilateral, regional, and extra-regional engagements to improve maritime security. Obviously some countries such as US or France are more interested in keeping the region safe from maritime crime and piracy. Generally the European Union (EU) and North Atlantic Treaty Organization (NATO) are also concerned about stability of this region.

At the national level, for instance, the Nigerian government in January 2012 transformed its Joint Task Force *Operation Restore Hope*, which was initially established to combat militancy in the Niger Delta into an expanded maritime security framework, known as *Operation Pulo Shield*. The *Operation Pulo Shield* was set up by the Federal Government to rid the maritime environment of criminal elements in the Eastern waterways eliminate, among others, pipeline vandalism, crude oil theft, illegal oil refining, piracy and all forms of sea robbery within its area of responsibility. Nigeria has strengthened its military potential in the Navy considerably over recent years. Paying particular attention to the improvement of security in the Niger Delta and along the 853 km coast where there are numerous offshore drilling. They have also bought new aircrafts (ART 42MP) to help patrolling the area of interest. On 13 May 2010, the Nigerian Navy received from the US Coast Guard *Cutter Chase* (WHEC-718), as an extra measure of defense under the Foreign Assistance Act. In addition, Nigeria also received 10 units, to support the operation in the Niger Delta. The US Navy has created two new operational bases: Nigerian Navy Ships (NNS) *Lugard* and NNS *Jubilee*. In

the next ten years the forces that are responsible for security at all territorial waters (Nigeria, Benin, Togo, Ghana) and water routes in the Gulf of Guinea will be enlarged of the 49 vessels and 42 helicopters. American HSV *Swifty* was sent to the Gulf of Guinea in the African Partnership Station. Also France sent a frigate *Germinal*, to Benin, Togo and Ghana to help to patrol the coast. Apart from that the French Navy supports, by conducting anti-piracy training in Benin and Nigerian navies. France also has introduced a program for solidarity funds for security sector reforms in the area of the Gulf of Guinea. The European Union is active in training, as well as in the exchanging information on piracy. The EU believes that action against the pirates should be based on the sovereignty and national ownership of the countries, and should be based on maritime law.

At the meeting, which was held 19 October 2011, and devoted to the problem of piracy in the Gulf of Guinea, the members of the Security Council addressed the problem of anti-piracy measures. General Mahamene Toure, Commissioner of Political, Peace and Security Matters of Economic Community Member West Africa (ECOWAS) informed about many key initiatives in the region in order to resolve the problem of piracy and stressed that the new strategy should help to combat this problem.

Some other states from the region like Republics of Benin and Ghana are also taking measures in the form of increased policing, provision of detection and surveillance systems, creation of Maritime Domain Awareness (MDA) capabilities and acquisition of requisite platforms to suppress piracy. In terms of bilateral response, the Governments of Nigeria and Benin Republic in October 2011 set up a combined maritime patrol of their waters. Code-named *Operation Prosperity*, the bilateral cooperation was the first of its kind in the region and is in tandem with the Maritime Organization of West and Central Africa (MOWCA) Coastguard Function Network Initiative. It is expected that the joint patrol will be expanded to include the navies of Ghana and Togo, in a bid to increase the span of surveillance and eliminate maritime threats.[18] The Secretary General of United Nations, Ban Ki-moon, at the meeting in 19 October 2010 said that Benin and Nigeria launched anti-piracy activities and motivated other countries which had a very positive influence on the current situation at sea.

At the regional level, measures to improve maritime security have come in the form of joint

---

[18] Ibidem.

training exercises among Navies of the Gulf of Guinea states. In February 2012, the Nigeria Navy hosted *Exercise Obangame Express*: an annual naval exercise of Africa, US and Europe. The idea was to encourage countries in the region to improve on interoperability of communications and sharing MDA information to collectively combat piracy and other maritime crimes. The US collaborates very closely and so far they have spent about 35 million US$ in the training of naval personnel in Nigeria and other countries within the region on how to combat piracy, oil bunkering and other maritime crimes that have plagued the region in recent times.

7 CONCLUSION

To sum up it is important to underline that piracy is not an easy problem to fight but it cannot be left unattended if the international community does not want to bear extra costs of pirates' activities. The factors of geographical, political, social matters together with cultural acceptance as well as the development of piracy should be a very broad issue, since the omission of any from the above factors, may prevent the proper diagnose of the problem, and thus hamper finding its suitable solution. Total elimination of the phenomenon of piracy, through constant control of the infected region using warships assisted by tracking systems, is rather impossible, but its limitation should contribute to a faster solution to piracy in the Gulf of Guinea.

At the end it has to be pointed out that a very important issue in the fight against maritime piracy in the Gulf of Guinea is the international cooperation not only in terms of coordinated patrols within the region, but also international exercises or financial support. Piracy is not the problem of the countries in which it occurs but it is the common problem of the whole community. The community has hopefully learnt its lesson in the Gulf of Aden and should lend a helpful hand to the countries in the Gulf of Guinea so it gives the proper consignment to every probable pirate in any part of the world. This crime cannot be unnoticed and must be punished. Obviously to change the situation at sea we have to change the situation inside countries, we have to bring justice and even treatment of all people as well as fair shares in natural resources. The fight against corruption must be made a top priority, as corruption leads to the loss of funds that could be used to improve human development, provide social services. Probably this would be the most difficult part of solving the problem of piracy along with changing people' mentality and acceptance of this activity, but international community cannot leave this problem unattended. Fighting piracy, in case of any region so in the Gulf of Guinea as well must start on land because this is the place where it stems from and only the elimination of its roots can give positive feedback in the future.

REFERENCES

Alic J., *Gulf of Guinea piracy increases amid unrest and rising oil prices*, http://oilprice.com/Energy/Energy-General/Gulf-of-Guinea-Piracy-Increases-Amid-Unrest-and-Rising-Oil-Prices.html, 20.06.2012.

International Bargaining Forum: Piracy High Risk Area, http://www.scoop.co.nz/stories/WO1203/S00445/internatio nal-bargaining-forum-piracy-high-risk-area.htm, 12.12.2012.

Kąkol F., *Metody zwalczania bezpieczeństwa statku i platform wydobywczych w kontekście wzrostu piractwa w obszarze Zatoki Gwinejskiej*, Naval University of Gdynia, Gdynia 2012.

Lewis D., *Piracy in the Gulf of Guinea: Attacks on Nigeria's oil industry spill over in the region* http://www.reuters.com/article/2009/04/15/idUSL9951481, 12.11.2011.

Mugridge D., *Piracy storm brews in West Africa: Gulf of Guinea under maritime siege*, http://www.defenceiq.com/naval-and-maritime-defence/articles/piracy-storm-brews-in-west-africa-gulf-of-guinea-u/, 10.11.2011.

Obi C. I., *Nigeria's Niger Delta. Understanding the Complex Drivers of Conflict*, http://www.nai.uu.se/research/nai-foi%20lectures/calendar2009/obi.pdf, 20.12.2011.

Onuoha C., *Piracy and maritime security in the Gulf of Guinea: Nigeria as a microcosm*, http://studies.aljazeera.net/ResourceGallery/media/Docume nts/2012/6/12/201261294647291734Piracy%20and%20Ma ritime%20Security%20in%20the%20Gulf%20of%20Guine a.pdf, 20.11.2012.

Otto L., *Piracy in the Gulf of Guinea: Attacks on Nigeria's oil industry spill over in the region*. http://www.reuters.com/ article/2009/04/15/idUSL9951481, 12,12,2011.

Piracy and armed robbery against ships, annual reports, January – December 2008, 2009, 2010, 2011 and 2012.

The World Factbook, Nigeria https://www.cia.gov/library/publications/the-world-factbook/geos/ni.html, 15.12.2012.

Wardin K., *Współczesne piractwo morskie. Wyzwanie somalijskie oraz odpowiedź społeczności międzynarodowej*, Belstudio, Warszawa 2012.

Neethling T., *Piracy around Africa's West and East coast: a comparative political perspective*, http://www.ajol.info/index.php/smsajms/article/viewFile/7050 5/59109.

Terrorism and Piracy
STCW, Maritime Education and Training (MET), Human Resources and Crew Manning, Maritime Policy,
Logistics and Economic Matters – Marine Navigation and Safety of Sea Transportation – Weintrit & Neumann (Eds)

Transport Infrastructure as a Potential Target of Terrorist Attacks

G. Nowacki

Military University of Technology, Cybernetics Faculty, Warsaw, Poland

ABSTRACT: The paper presents some problems referring to transport infrastructure as a potential target of terrorism attacks. The range of transport infrastructure has spread and includes railways, inland waterways, road, maritime, air, intermodal transport infrastructure and intelligent transport systems (ITS). Terrorism means acts of violence committed by groups that view themselves as victimized by some notable historical wrong. Although these groups have no formal connection with governments, they usually have the financial and moral backing of sympathetic governments. Typically, they stage unexpected attacks on civilian targets, including transport infrastructure, with the aim of sowing fear and confusion. Based on the analyses, transportation infrastructure is potentially threatened with terrorism attacks, especially road and rail infrastructure (about 23 %), and to a smaller degree the maritime and air transport infrastructure (about 2 %). Legal steps to fight terrorism have been taken on the international level; furthermore, some institutions have been established for this purpose at the UN, NATO and EU level as well as in Poland too.

1 INTRODUCTION

The infrastructure is a group of the essential equipment and institutions, necessary for a correct functioning of the economy and society (Schmid & Jongman 1988, Encyclopaedia 1982).

In view of W. Mirowski, the infrastructure is an international term, meaning a group of essential objects, equipment and institutions of a service character necessary for the correct functioning of the society and production branches of economy (Mirowski 1996).

Transport is a transfer of people, freight in space using appropriate means (Rydzkowski & Wojewódzka-Król 2008). Transport is closely linked with the remaining branches of economy. It's development is a condition for their development and vice versa – the worsening of economical or transport development is associated with worsening of the situation in respectively transport and economy.

The transport infrastructure is mainly created by the three essential groups:
- routes of all types of transport (road, railways, inland waterways, maritime and air),
- transport junctions (intersections, airports, ports, rail stations, intermodal terminals, logistics centres, etc.),
- auxiliary equipment for the direct servicing of routes and transport junctions.

The term „transport infrastructure" and it's scope have been defined in the Union legislation and have not changed for a long time, practically since 1970 till 2011. According to the Regulation (EEC) No 1108/70 of the Council of 4 June 1970 and Regulation (EEC) No 2598/70 of the Commission of 18 December 1970 and Commission Regulation (EC) No 851/2006 of 9 June 2006, the transport infrastructure means all roads and permanent equipment for the three types of transport, which are necessary for ensuring flow and safety of traffic.

The definition and scope of the transport infrastructure will be changed this year, based on the final proposal for a Regulation of the European Parliament and of the Council on Union guidelines (COM 650 final 2011/0294) for the development of the Trans-European Transport Network concerning union guidelines for the development of the trans-European transport network (planned date of acceptance - March 2012).

According to a new decree, the transport infrastructure, including that of trans-European transport network, consists of:
- Railway transport infrastructure,
- Inland waterways infrastructure,
- Road transport infrastructure,

- Maritime transport infrastructure,
- Air transport infrastructure,
- Multimodal transport infrastructure,
- Equipment and intelligent transport systems associated with the transport infrastructure.

On the 16 of December 2008 the European Commission published Announcement - Plan of the introduction of the intelligent transport systems in Europe, COM (2008)886, which found its reflection in the Directive 2010/40/EU of the European Parliament and of the Council of 7 July 2010, M/453 mandate of 6 October 2009 and Commission Implementing Decision 2011/453/EU of 13 July 2011 executive decision.

From the presented analysis of the subject literature it transpires that the scope of the transport infrastructure will widen, and additionally, in each of the infrastructure category the intelligent transport systems have been highlighted.

Transport infrastructure as a potential target of terrorist attack should be a critical input into risk management policy regarding risk allocation decisions. Terrorism attack on transport infrastructure can cause collateral damage to neighbouring environment, including massive economic and social consequences that rely upon that infrastructure.

Transport infrastructure protection might include major highways, bridges, railways, and it also typically addresses air transportation, including air traffic control operations and airport security as well as waterways.

2 CHARACTERIZATION OF TERRORISM

2.1 The term and essence of terrorism

Terrorism is not a new phenomenon, but is a variable, multi-level and dynamic one, as among the others, the forms, means and objectives of the terrorist activities, change. The terrorism phenomenon is affected by the civilisation development and scientific and technical progress, especially as far as new communication means, mass media and advanced communication technologies are concerned.

The US Defence Department defines the terrorism as unlawful use or the threat of using force or violence against people, property, to pressurize or scare the governments or societies to achieve political, religious or ideological aims (Aleksandrowicz 2008).

The term – terrorism, should be understood solely as the use of violence by the individuals or groups of people, to exert the pressure both on the government and public opinion as well as the groups of people and individual persons (Olechów 2002).

Schmidt defined the classic terrorism, as an attack of a subversive forces on the innocent individuals, aimed at causing fear, kill or injure people, and thus force the political concessions from a person or organisation not actually being a direct victim of an assault, and to which the attacked individuals do not belong. (Schmidt 1988). This criterion was used in 1988 by Alex Schmid and Albert Jongman. They made a statistical analysis of 109 terrorism definitions and concluded that the most frequently occurring elements of terrorism are (Schmidt & Jongman 1988) – Fig. 1:
- violence/force – 83,5%,
- political aspect of the phenomenon – 65%,
- fear – 51%,
- threats – 47%,
- psychological effect – 41,5%,
- existing discrepancies between the target and victim– 7,5%,
- planned, purposeful, systematic and organised actions – 32%,
- fighting, strategic and tactical methods – 30, 5%.

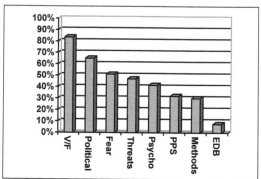

Figure 1. The most frequently occurring elements of terrorism

According to the UN, the terrorism (Latin) are various ideologically motivated, planned and organised actions of individual people or groups resulting in violating the existing legal order, undertaken for the purpose of forcing the state authorities and society into certain behaviours and benefiting actions, often violating the welfare of the outsiders. These actions are performed with an outmost ruthlessness, by various means (physical violence, use of weapons and explosions), in order to give them maximum publicity and intentionally causing fear in the society (The report of UNIC 2012).

According to the Criminal Code (The act from 16 April 2004) the crime of a terrorist character is a forbidden act punishable by an imprisonment, whose limit is at least 5 years, and which is committed in order to:
- seriously frighten many people,
- force the public authorities of the Republic of Poland, other countries or the international

organisation authorities, to undertake or abandon certain activities,
– cause serious disturbances in the political system or the economy of the Republic of Poland, other countries or the international organisation – as well as the threat of committing such an act.

The main forms of the terrorist attacks are (Pawłowski 1994):
– assault against life – is directed most often against important persons, political party leaders,
– bomber attack – exerts specific psychological pressure on the society,
– hijacking vehicle or plane,
– taking hostages, kidnapping – serves the purpose of using them as a trade over element in meeting the demands.

It needs to be noted also that 95 % terrorist attacks in the world were conducted using explosives.

According to RMS' historical catalogue of macro terrorism attacks (defined as attacks with the minimum severity of a car bomb), the terrorist violence has increased substantially since 2001 (Coburn & Paul &Vyas & Woo & Yeo 2012). More than 2 400 macro attacks have occurred worldwide since 2001, killing over 37 000 people and injuring nearly 70 000.

The Institute for Economics and Peace (IEP) has indicated in its inaugural global terrorism index of 158 countries, that only 31 have had no attacks in the ten years from 2002 to 2011 (Institute for Economics & Peace 2012). Yet although the attacks are distributed widely around the world, the majority are concentrated in just a handful of countries. Iraq ranks first weighted average of the number of incidents, deaths, injuries and estimated property damage. Other terrorist hotspots include Pakistan, Afghanistan, India and Yemen (Tab. 1).

Table 1. Global terrorism index from 2002 to 2011 (Institute for Economics & Peace. Dec 4th 2012)

Rank	Country	Score
1	Iraq	9.56
2	Pakistan	9.05
3	Afghanistan	8.67
4	India	8.15
5	Yemen	7.31
6	Somalia	7.24
7	Nigeria	7.24
8	Thailand	7.09
9	Russia	7.07
10	Philippines	6.80

The total number of terrorist attacks and terrorism-related arrests in the EU continued to decrease in 2011. The main figures for 2011 are:
– 174 terrorist attacks in EU Member States,
– Lone actors were responsible for the killing of 2 persons in Germany, and 77 persons in the non-EU country Norway.

2.2 Terrorist attacks conducted on the transport infrastructure objects

Facilities accessible to large numbers of people, such as transport infrastructure facilities and vehicles are vulnerable to attack by vandals, extremists, terrorists, and other criminals. For example, the March 11, 2004 Madrid commuter train bombings by terrorists inspired by Al-Qaeda killed nearly 200 people and wounded 1 800. And, on the morning of July 7, 2005, four suicide bombers successfully detonated bombs on the London Underground, killing 52 people and injuring over 700 (Coburn & Paul &Vyas & Woo & Yeo 2012).

In view of the Aon Global Risk Consulting experts, the most often targeted trade by the terrorists was retail trade - 24, 18% (The report of Aon 2008) - Tab. 2.

Table 2. Terrorism attacks on various trade

Various trade	%
Retail trade, gastronomy	24,18%
Land transport (road, railways)	23,36%
Mining industry	14,55%
Infrastructure (objects of the state authorities, local governments, public, religious cult)	8,2%
Construction	5,74%
Tourism	6,56%
Finances	2,05%
Air transport	2,46%
Maritime transport	2,25%
Public utility enterprises	2,25%
Other	8,4%

Affected were not only the shopping centres or supermarkets, but also small gastronomy outlets, restaurants, clubs and bars. It is possible to indicate three main reasons for the terrorist attacks in the retail sector. First of all, vast part of them is directed at the shops owned by non-Islamic people. Secondly, the terrorists often chose renowned makes as the target of their attacks, which are a symbol of the western world and capitalism. And thirdly, the retail trade is an attractive target for the attacks of the terrorist groups due to the possibility of causing significant human losses, serious difficulties in the everyday life and an effective scaring of the civilian population.

The second place on the list of the terrorist attacks takes road and railways transport infrastructure (23, 36%).

Transport infrastructure is a frequent target of terrorist attacks because of its significance in several dimensions (Zeng & Chawathe & Wang 2007). Because physical transportation networks attract large numbers of people, they're a high value targets for terrorists intending to inflict heavy casualties. Transport infrastructure is important to the modern economy, and related damages and destruction can have quick ripple effects. Operationally, transportation systems interact with and provide support for other systems, such as emergency response and public health, in complex ways. Terrorists can perceive an attack on such a link (that is, one that connects many systems) as an efficient means to create confusion, counter the countermeasures, and damage the targeted society as a whole. Furthermore, transport infrastructure can be both the means and the end of an attack, making them a critical part of almost all terrorist attacks in the physical world.

Susan Pantell has presented terrorist attacks involving transport infrastructure, especially vehicles and facilities being tallied and covered the 41-year period 1967-2007 in the USA (Fig. 2).

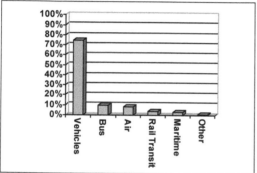

Figure 2. Terrorist incidents involving transport infrastructure in USA (Light Rail Now Project, 2008)

According to presented data, 90,3% of incidents involve land transport (74,5% – vehicles, 9,5% – buses, 6,3% - rail), 8,3% - air transport, only 0,9% maritime transport and 0,5% other.

Air transport (8,3 %, which clearly has been shown to be quite vulnerable, and, along with motor vehicles, a favoured target – and instrument – of terrorists) has been subjected to a comparatively rigorous security crackdown; on the other hand, private motor vehicles movement has been subjected to basically nothing in terms of comparable ongoing security measures.

Recent terrorist attacks on transport infrastructure have highlighted the vulnerability of road and rail networks, reinforcing the importance of good risk assessments to protect some of our biggest assets.

Throughout the European Union there are five million kilometres of road, and among them:
– 65 000 km of these being motorways,
– 212 000 km rail lines,
– as well as 42 000 km of navigable inland waterways.

According to Brian D. Taylor, director of the Institute of Transportation Studies at the University of California (UCTC 2010), acts of terrorism intersect with transport infrastructure in three ways:

– When transportation is the means by which a terrorist attack is executed,
– When transportation is the end, or target, of a terrorist attack,
– When the crowds that many transportation modes generate are the focus of a terrorist attack.

3 COMBATING TERRORISM

3.1 Tasks to combat terrorism

Transport infrastructure is inherently vulnerable to terrorist attacks, as it is an open area that gathers large numbers of people at predictable times in predictable places.

As defined earlier, transport infrastructure includes physical facilities, equipment, assets, service networks, and communication and computing hardware and software that enables information access and transactions (Zeng & Chawathe & Wang 2007).

Transport security encompasses mass passenger transport systems such as roads, bridges and tunnels, bus terminals and rail stations, trams, and air cargo supply chains and maritime ports. From the mentioned aspects, transport infrastructure security tasks should include relevant application context as follows:
– physical access management and control of employees and passengers,
– perimeter intrusion detection,
– vulnerability assessment,
– intrusion detection and access control in the cyberspace in which pertinent information systems operate and exchange data,
– related simulation and decision support tools.

Transport plays an important role in the economic growth and globalization because of connecting to other economy sectors.

Transport infrastructure security is the complex of roles, responsibilities and relationships in the different sectors (Roads, Railways, Inland waterways, Air and Maritime, Multimodal) and refers to other economy sectors. Thus transport infrastructure security management against terrorist attacks, needs multi agency and multi jurisdictional domain, regulatory, law enforcement and intelligence agencies on international and national levels.

3.2 International institutions to fight terrorism

UN has taken many efforts to fight terrorism since 1963 and accepted many conventions. Security Council created Counter-Terrorism Committee Executive Directorate (CTED) based on resolution 1535 from 2004. General Secretary appointed UN

Counter-Terrorism Implementation Task Force - CTITF in 2005.

On 8 September 2006 all UN member states accepted Global Strategy of Combating Terrorism. For the first time in history a joint stand has been agreed on combating terrorism. Accepting Global Strategy of Combating Terrorism crowns years of efforts and at the same time fulfils the obligation made by the world leaders at the World Summit in September 2005. While developing strategy, many proposals and recommendations presented by a former Secretary General Kofi Annan were taken into account.

The basis for the strategy is unequivocal, unconditional and firm condemnation of terrorism in all it's forms – used by anybody, anywhere and for whatever reasons. The strategy establishes concrete measures, that are to be taken in order to eliminate causes for spreading terrorism and to strengthen individual and joint capability of nations and United Nations in preventing and combating terrorism, protecting simultaneously human rights and legal rules.

The strategy combines many new proposals and strengthenings of the current actions taken by the member states, UN system and other international and regional institutions into a common platform of strategic co-operation.

The General Assembly reviewed the implementation of the United Nations Global Counter-Terrorism Strategy on 28-29 June 2012 at the UN headquarters in New York. The Strategy refers to activities of four pillars, which are:

1 Addressing conditions conducive to the spread of terrorism,
2 Preventing and combating terrorism,
3 Capacity-building to prevent and combat terrorism,
   Respect for human rights and the rule of law.

The terrorism creates direct danger to the safety of the NATO states and for the international stability and welfare.   The terrorist groups infiltrate and propagate in the areas of a strategic significance to the Alliance, modern technology causes increase of the danger and potential threat of the terrorist attacks, especially if the terrorists were to come into possession of the nuclear, chemical, biological or radiological capabilities.

At the NATO level operates EADRCC (Euro-Atlantic Disaster Response Coordination Centre), created in June 1998, at the NATO Head Quarters, based on the motion put forward by the Russian Federation. Created, as part of the Partnership for Peace program, Centre co-ordinates NATO the partner states' actions in the area of the Euro-Atlantic in a reaction to natural disasters and those caused by a man. All those tasks are conducted in a close co-operation with the UN Office for the Co-ordination of Humanitarian Aid (UN OCHA), which

has been entrusted with a superior role in co-ordinating international actions during disasters. Since 2001 EADRCC has played a role in co-ordinating the reaction of the countries to a terrorist attacks with the use of chemical, biological or radiological weapons, as well as the activities managing the consequences of those events.

The Military Concept for Defence against Terrorism and the Partnership Action Plan Against Terrorism were taken in Prague Summit in 2002. The Terrorist Threat Intelligence Unit – (TTIU) was created in Istanbul Summit in 2004.

In accordance with a strategic concept of defence and safety of the members of the NATO treaty, the Alliance takes on itself obligation to prevent crises, terrorist threats, managing conflicts and stabilising post-conflict situations, including closer co-operation with international partners, especially United Nations Organisation and European Union.

The European Council on the 25 march 2004 accepted the Declaration and a Plan, being an annex to the Declaration, on combating terrorism. One of the most important provisions is an acceptance of, so called, solidarity cause. It envisages, that in case of the terrorist attack on any EU member state, all remaining members will mobilise every available means, they regard as appropriate (including military ones), to help the state in trouble. In the declaration, there is a reference also to the European Security Strategy of December 2003, in which the terrorism was regarded as one of the most serious dangers to the international security.

The 2010 Strategic Concept threats include the proliferation of nuclear weapons, terrorism, cyber attacks and key environmental and resource constraints. NATO has adopted a holistic approach to crisis management, envisaging NATO involvement at all stages of a crisis. NATO will therefore engage, where possible and when necessary, to prevent crises, manage crises, stabilize post-conflict situations and support reconstruction.

At the EU level there has been Monitoring and Information Centre (MIC) established, which is available and capable of an immediate reaction, 24 hours a day, and also serving the member states and the European Commission to react to dangers. The Centre serves 31 states (27 EU states and Croatia, Lichtenstein, Iceland and Norway).

More over, the Common Emergency Communication and Information System (CECIS) is used in the Crises Situations in order to enable the communication between MIC and the contact points in the member states as well as sharing by them the information and managing them.

The Schengen zone countries use Schengen Information System – SIS, and the access to the system is in the possession of the police, consular offices and Border Guard together with Customs Offices and it enables the verification of people during the border control as well as during the control within the country.

The Counter-Terrorism Coordinator (CTC), the Counter Terrorism Group (CTG) and the Joint Situation Centre (SITCEN) were appointed to fight terrorism in the European Union.

At the EU level there have been independent institutions: Europol, Eurojust and Frontex.

Europol (European Police Office) is the European Union law enforcement agency that handles the exchange and analysis of criminal intelligence in preventing terrorism and serious international crime in order to raise the safety within the entire Europe. Europol commenced its full activities on 1 July 1999.

Eurojust (The European Union's Judicial Cooperation Unit) is a body established in 2002 to stimulate and improve the co-ordination of investigations and prosecutions among the competent judicial authorities of the European Union Member States when they deal with serious cross-border and organized crime.

Frontex (Franch: *Frontières extérieures)* is a European Agency for the Management of Operational Cooperation at the External Borders of the Member States of the European Union. Frontex started to be operational on October 3, 2005 and is headquartered in Warsaw.

On the wide international forum, INTERPOL has been used (188 countries) – the police organisation of the EU countries and those from outside the Union, e.g. Belarus, Russia, Ukraine. The co-operation takes place as part of the liaison officers network of the Polish Police operating in such EU countries as France, Holland, Germany, Great Britain, and countries outside the Union, i.e. Belarus, Russia, Ukraine. They have a direct access to the police data bases (lost and wanted persons, dactyloscopic cards, DNA profiles, stolen vehicles and documents, etc.).

3.3 *National institutions*

In Poland combating terrorism is dealt with by the following institutions (Strategy of National Security for the Republic of Poland 2007):
- at the strategic level: Government Centre for Security (GCS), National Crisis Management Team (NCMT), Interdepartmental Team for the Terrorist Threat (ITTT), Internal Security Agency (ISA), Intelligence Agency (IA), Military Intelligence Service (MIS), Military Counter-Intelligence Service (MCIS), Police Head Quarters (PHQ), Border Guard Head Quarters (BGHQ), Chief Inspector for the Financial Information (CIFI), Customs Service (CS), State Fire Service Head Quarters (SFBHQ), Government Protection Bureau (GPB), Military Police Head Quarters (MPHQ), General Staff of the Polish Armed Forces, National Atomic Energy Agency (NAEA), Civil Aviation Authority (CAA), Polish Air Navigation Services Agency (PANSA);
- at the operational level: Counter-Terrorist Centre (Makarski 2010, CTC 2012). The CTC operates in a twenty-four-hour system, 7 days a week. It comprises, apart from officers of the Internal Security Agency, seconded officers, soldiers and employees of e.g. the Police, the Border Guard, the Government Protection Bureau, the Foreign Intelligence Agency, the Military Intelligence Service, the Military Counterintelligence Service and the Customs Service. They carry out tasks within competences of the institution which they represent. Furthermore, together with the Counter – Terrorist Centre actively cooperate other bodies which participate in the system of anti-terrorist protection of the Republic of Poland such as: the Government Centre for Security, the Ministry of Foreign Affairs, the State Fire Service, the General Inspector of Financial Information, the General Staff of the Polish Armed Forces, the Polish Military Gendarmerie etc. The essence of the CTC functioning system is coordination of the information exchange process between the antiterrorist protection system participants, enabling implementation of the common procedures for reacting at the occurrence of one of the four categories of the defined threat:
  - terrorist threat occurring outside the Polish boarders but affecting the security of the RP and it's citizens,
  - terrorist occurrence taking place on the Polish territory affecting the security of the RP and it's citizens,
  - information obtained about a potential threats that may take place on the Polish territory outside the Polish boarders,
  - information obtained concerning laundering money or transferring financial resources that may be a proof of financing the terrorist activities.
- at the tactical level: special units, services and institutions answering to the Internal Affairs Minister (IAM), National Defence Minister (NDM), Financial Minister (FM), Minister of Transport, Construction and Maritime Economy (MTCME), special units.

Intelligence and counter-intelligence tasks of the special services (ISA, IA, MIS, MCIS) concern recognising and counteracting the internal and external occurrences, which threaten the state's interests. The essential role of the special services relies on obtaining, analysing, processing and conveying to the correct authorities, the information, which can be of considerable significance for the state security in every aspect, as well as on the pre-emptive informing about the potential and existing threats to the country. Special attention is given to preventing and counteracting the terrorism, protecting the defence and economic capabilities of the country, being a condition for it's international position. The special services ensure the counterintelligence protection of the country, especially in respect to the functioning of the main elements of the critical infrastructure, including transport networks.

The superior objective of the Police actions is to serve the society by effectively protecting the security of people and property as well as maintain security and public order. The Police prevent crime and criminogenic phenomena, including those of a cross-boarder character, co-operating with other guards, services and state inspections as well as the police of other countries together with the international organisations. The Police are prepared also for providing a wider support to missions conducted by other state and none-governmental entities, as well as the Armed Forces of RP. Preventing and effective response to the cases of organised crime should remain the care of not only the Police, but also other services and departments. Polish police should actively participate and initiate solutions at the international institutions of the police co-operation, such as Interpol or Europol, and develop the network of its liaison officers, actively representing Polish police outside the country.

As the priority actions, should be regarded the co-operation of the Police law enforcement bodies, Internal Security Agency with the Anti-Terrorist Centre, for the purpose of eliminating the terrorist and criminal events.

The superior objective of the Boarder Guard is an effective protection of the state borders and controlling the boarder traffic in accordance with the interests of the state security interests. It's special role is to protect one of the longest land sections of the external boarder of, both the European Union and NATO, as well as the responsibilities resulting from signing the Schengen Treaty by Poland.

It is necessary to continue and improve co-operation between the Border Guard, the Police and other services. The monitoring and migration control of the foreigners on the entire territory of the country is an important area requiring the intensification of the co-operation of these services, which should also have a preventive character. Both the Police and the Border Guard, as part of the statutory activities, ought to constantly monitor the threats of a terrorist character, co-operating in that, with other services.

The superior objective of the State Fire Service is recognising threats, preparing and carrying out the rescue actions. The SFS possesses the capability to immediately respond in cases of a threat to life and health, environment and property as well as in the cases of the extraordinary threats, disasters and natural disasters.

4 CONCLUSIONS

The terrorist attacks threats to the transport infrastructure are quite significant world-wide, and so far this trade attracted about 24% of all attacks carried out.

According to intelligence agencies, criminals and terrorists are planning to disrupt transport infrastructure in different countries all over the world because it is the open area that gather large numbers of people at predictable times in predictable places.

The tasks referring to protecting people and transport infrastructure while allowing transport systems to operate efficiently and effectively, should become a national and international priority. The solution requires global initiatives and close co-operations of transport operators, police, security and other international organizations, including the UN, NATO and European Commission.

The state services of the Republic of Poland are well prepared to recognise, prevent and combat terrorist threats. The most important is the co-operation of all services as well as an early recognition of the threat and not allowing the terrorist attack to take place (prevention).

More over, it would be advisable to prepare guide books concerning the behaviour of people in the crises situations, especially about the symptoms of the terrorist attack, being prepared, and the ways the people should behave in such cases.

REFERENCES

Aleksandrowicz, T. R. 2008. International terrorism, Warsaw.
Coburn, A. and Paul, M. and Vyas, W. and Woo, G. and Yeo W. 2012, Terrorism Risk in the Post-9/11 Era. A 10-Year Retrospective. Risk Management Solutions, Inc. Newark, USA.
Counter Terrorism Centre of the Internal Security Agency http://www.antyterroryzm.gov.pl/portal/eng/265/621/COUNTER__TERRORIST_CENTRE_CTC.html
Commission Implementing Decision 2011/453/EU of 13 July 2011 adopting guidelines for reporting by the Member States under Directive 2010/40/EU of the European Parliament and of the Council (notified under document C2011/4947). OJ of the EU, L 193/48, 49 of 23/7/2011.
Directive 2010/40/EU of the European Parliament and of the Council of 7 July 2010 on the framework for the deployment of Intelligent Transport Systems in the field of road transport and for interfaces with other modes of transport. OJ of EU, L 207 , 06/08/2010 P. 0001 – 0013.
Encyclopaedia PWN. 1982. PWN, Warsaw.
Global terrorism index. Capturing the Impact of Terrorism for the Last Decade. IEP. Dec 4th 2012. http://www.visionofhumanity.org/wp-

content/uploads/2012/12/2012-Global-Terrorism-Index-Report1.pdf

Makarski, A. 2010. Genesis, activities, rules and experiences after one year functioning of Anti-Terrorist Centre. Review of Internal Security, No 2. Anti-Terrorist Centre, Warsaw. http://www.abw.gov.pl/portal/pl/273/642/Przeglad_Bezpieczenstwa_Wewnetrznego_nr_2_2_2010.html,

Mandate 453 EN (2009). Standardization mandate addressed to CEN, CENELEC and ETSI in the field of information and communication technologies to support the interoperability of co-operative systems for Intelligent Transport in the European Community. Brussels, 6th October 2009. DG ENTR/D4.

Mirowski, W. 1996. Study on Polish country infrastructure. The equipment areas of social infrastructure. T. III. PAN Instytut Rozwoju Wsi i Rolnictwa, Warszawa.

NATO 2010. Strategic Concept for the Defense and Security of The Members of the North Atlantic Treaty Organization, Adopted by Heads of State and Government in Lisbon, Nov. 2010. http://www.nato.int/lisbon2010/strategic-concept-2010-eng.pdf.

Olechów, B. 2002. Terrorism in after bipolar Word. Adam Marszałek, Toruń.

Pawłowski, A. 1994. Terrorism in Europe of XIX & XX century. Lubuski Komitet Upowszechniania Prasy, Zielona Góra.

Regulation (EEC) No 1108/70 of the Council of 4 June 1970 introducing an accounting system for expenditure on infrastructure in respect of transport by rail, road and inland waterway. OJ L 130, 15.6.1970, p. 4–14 .

Regulation (EEC) No 2598/70 of the Commission of 18 December 1970 specifying the items to be included under the various headings in the forms of accounts shown in Annex I to Council Regulation (EEC) No 1108/70 of 4 June 1970. OJ L 278, 23.12.1970, p. 1–5.

Regulation (EC) No 851/2006 of 9 June 2006 specifying the items to be included under the various headings in the forms of accounts shown in Annex I to Council Regulation (EEC) No 1108/70. OJ L 158, 10.6.2006, p. 3–8

Rydzkowski, W. and Wojewódzka-Król, K. (red.). 2008. Transport. PWN, Warszawa.

Schmid, A. 1988. Political Terrorism: A New Giude to Actors, Authors, Concepts, Data Bases, Theories and Literature. New Brunswick, NJ 1988.

Schmid, A. and Jongman, A. 1988. Political terrorism. SWIDOC, Amsterdam.

Strategy of National Security for Republic of Poland, Warsaw, 2007.

The act from 16 April 2004 – Penal Code (Kodeks Karny) and others acts. OJ. 2004, No 93, pos. 889.

The report of Aon Global Risk Consulting, http://forsal.pl/wiadomosci/polska/341600.html http://www.aon.com/poland, 2008.

The report of UNIC from 10.02.2012, http://www.unic.un.org.pl/terroryzm/definicje.php,

Terrorist Attacks and Transport Systems. Institute of Transportation Studies at the University of Kalifornia. 2010.http://www.uctc.net/access/28/AccessComment/TerroristAttacksandTransportSystems.pdf

Terrorist attacks mapped around the world. 2012. Institute for Economics & Peace, New York. 13 December 2012. http://www.economist.com/blogs/graphicdetail/2012/12/daily-chart-0

US LRN. 2008. Light Rail Now Project. http://www.lightrailnow.org/features/f_lrt_2008-06a.htm

Zeng, D. and Chawathe, S. and Wang, F.Y. 2007. Protecting Transportation Infrastructure. ITS Department. IEEE Intelligent Systems, July/August 2007.

Terrorism and Piracy
STCW, Maritime Education and Training (MET), Human Resources and Crew Manning, Maritime Policy,
Logistics and Economic Matters – Marine Navigation and Safety of Sea Transportation – Weintrit & Neumann (Eds)

Maritime Piracy Humanitarian Response Programme (MPHRP)

A. Dimitrevich
Maritime Piracy: a Humanitarian Response Programme in CISB region, Ukraine

V. Torskiy
Maritime Piracy: a Humanitarian Response Programme, Ukraine

ABSTRACT: In recent years nearly 5000 seafarers have been hijacked and detained for months often in appalling conditions, while thousands of others have been the victims of a pirate attack. Given these numerous concerns, a pan industry alliance of ship owners, unions, managers, manning agents, insurers and welfare associations (maritime, labour, faith or secular) has come together to establish the "Maritime Piracy: a Humanitarian Response Programme" (MPHRP). The objectives of this programme address the three phases of "pre-, during and post incident", with the aim of implementing a model of assisting seafarers and their families with the humanitarian aspects of a traumatic incident caused by a piracy attack, armed robbery or being taken hostage.

Despite numerous diplomatic, military and other initiatives by many governments and governmental agencies, particularly in the Indian Ocean, and the protective measures and other actions adopted by ship owners, ship-managers and their representatives, ships are regularly attacked and seafarers put at risk as they go about their legitimate business in international waters. The period of detention for those hijacked has increased to an average of more than seven months; there are many more cases of brutal treatment, abuse and torture and lately, most regrettably, several fatalities, including the apparent murder of hostages, while others have been used as "human shields". Many other seafarers, even though their ship may not have been hijacked, have found themselves under armed attack and may have also been subject to a harrowing time locked in a citadel until released.

While acknowledging the actions of governments, the United Nations and the International Maritime Organization, the shipping industry has recognized that more needs to be done to support seafarers and their families. In recent years nearly 5000 seafarers have been hijacked and detained for months often in appalling conditions, while thousands of others have been the victims of a pirate attack. Every day of the year more than 100,000 seafarers experience anxiety while sailing in, or towards, piracy infested waters. Their families share these worries, often with a feeling of helplessness.

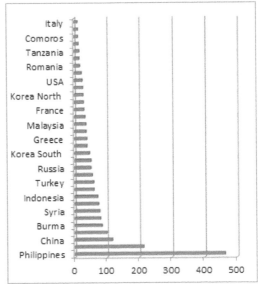

Figure 1. AFFECTED BY PIRATES' INCIDENTS. By Nations (IMB Statistics 2011)

Seafarers, obviously, play a pivotal role in any piracy incident and the appropriate preparations are integral to their well-being, as well as ultimately to that of their families, and to the overall outcome of the incident.

Given these numerous concerns, a pan industry alliance of ship owners, unions, managers, manning agents, insurers and welfare associations (maritime, labour, faith or secular) has come together to establish the "Maritime Piracy: a Humanitarian Response Programme" (MPHRP).

The programme has been built around:
- a task group of multi-disciplined, international experts,
- extensive fact finding and feedback gained from firsthand meetings and interviews with seafarers and families worldwide, including many with firsthand experience of attacks and hijackings,
- advisory groups on industry practices and procedures, pre-deployment piracy training and the skills required of responders, and
- the advice and assistance of a project steering group.

In its first phase the programme has developed:
- "good practice" guides for use by shipping companies, manning agents and welfare associations to support both seafarers and seafarers' families through the three phases of a piracy incident; pre-departure, the crisis and post-release/post-incident,
- associated training modules,
- an international network of trained first-responders with appropriate skills within partner and associated organisations,
- access to a network of professional aftercare,
- a 24 hour seafarers' international helpline

The objectives of this programme address the three phases of "pre-, during and post-incident", with the aim of implementing a model of assisting seafarers and their families with the humanitarian aspects of a traumatic incident caused by a piracy attack, armed robbery or being taken hostage.

To implement a model for assisting seafarers and their families with the humanitarian aspects of a traumatic incident caused by piracy attack, armed robbery or being taken hostage to be offered as an integral part of the Emergency Response Procedures of shipping companies and manning agencies in cooperation with partners involved in seafarers' welfare (maritime, union, faith or secular based agencies), company representatives and other bodies as appropriate.

The outcomes are anticipated to include:
- Guidelines on "good practice" for companies and seafarer welfare organisations on supporting seafarers and their family members through the three phases of a piracy incident from pre-deployment, during the crisis and post release/post incident.
- The development and implementation of relevant training modules.
- The development of an international network of trained first-responders with appropriate skills within partner and associated organizations.
- Access to a network of professional aftercare.
- The availability of a seafarers' telephone helpline.

A resource to collect appropriate research/information and to make this available where further advice or assistance is sought.

MPHRP has established regional offices and appointed its representatives in Ukraine, India and Philippines – in countries where greatest number of seafarers were affected by pirates' incidents.

In Ukraine representatives of Maritime Community has organized MPHRP National Working Group aimed to adapt and implement MPHRP guides and trainings for seafarers' benefits within the country taking into account peculiarities of National law, mentality and traditions. Thus Odessa National Maritime Academy is active participant of the programme which have started joint course of lectures and trainings for its students with MPHRP Director in CISB region.

REFERENCES

IMB Statistics 2011

Chapter 5

Maritime Policy and Global Excellence

Maritime Policy and Global Excellence
STCW, Maritime Education and Training (MET), Human Resources and Crew Manning, Maritime Policy,
Logistics and Economic Matters – Marine Navigation and Safety of Sea Transportation – Weintrit & Neumann (Eds)

Problem of Vessels "Escaping" from Under Flag by Case of Estonia and Some Possible Measures for Rising of Attractiveness of Them for Ship-owners

A. Alop
Estonian Maritime Academy, Tallinn, Estonia

V. Senčila
Lithuanian Maritime Academy, Klaipeda, Lithuania

ABSTRACT: The merchant shipping was during centuries and is now one of the most global business industries because of its transnational nature. Ship-owners enjoy freedom to choose freely their business place and flag state for their vessels and use such advantages busily for endeavouring the most suitable conditions for their businesses. Second half of XX – beginning of XXI century can be characterised by strong competition between so-called conventional countries and flag of convenience countries for having vessels sailing under their own flags. The amount of merchant ships sailing under Estonian flag was decreasing drastically during last 10-15 years and now it's close to zero. Situation in Lithuania and Latvia is not so unfortunate today, however there are number of specific problems in these countries as well. The authors of this presentation have intention to investigate backgrounds and reasons of such failures using method of comparative analysis.

1 INTRODUCTION

A transnational nature of merchant shipping and high level of its globalization cause the multinational character of competition amongst shipping companies around world. In fact, today not only companies but maritime countries themselves compete in field of marine transportation. Several countries are interested in bringing as many ships as possible under their flags. Thus, the attractiveness of their ship registers for ship-owners depends to large extent on state's economic policy, especially taxation regulations for the shipping and ship-management companies.

The flag state problems were coming forward especially during second half of XX – beginning of XXI century. That period can be characterised by strong competition between so-called conventional countries and flag of convenience countries for having vessels sailing under their own flags. The flag of convenience countries flatter the ship-owners to register ships under their flags by offering lower taxes and allowing employing of seafarers from so-called cheap labour force countries. Such state of matters lead for inequitable competition; however several regulations of European Commission advocated for EU countries to support local shipping and ship-owners by number of measures, mainly using tax allowances.

Last 25-30 years can be exemplified in general as period of "awakening" for conventional states governments that have taken unprecedented measures for retaining their national fleets or, furthermore, bringing vessels back under the national flags. Unfortunately there are still examples of so to say "sleeping" governments among maritime nations.

2 ANALYSIS OF SITUATION

2.1 *EC regulations on support measures for shipping*

Around 90 % of goods move in world trade using maritime transport. The world merchant fleet consists of more than 50,000 vessels that are registered under more than 150 state flags. About one million of seafarers are employed on them. Maritime transport is very important part of EU economy as sea shore areas in Europe contribute approximately 40 % of GDP and nearly 90 % of overseas trade of EU.

The European countries have started in latter 1980s to recognise the insufficient competitiveness of vessels that sail under their flags. The European common measures or regulations for improving of situation were non-exist then, thus a number of

Community's maritime countries invented their own steps and forced internal regulations to contribute the ship-owners and shipping companies. Despite of these means a strict competition with third countries was not relented and in 1989 the European Commission specified first guidelines on this topic with purpose to homogenize measures of various EU countries.

These guidelines were rather ineffective and decreasing of EU countries' fleets continued. After reconsidering of guidelines new set was published in 1997 that clearly determined possible state aid for shipping in EU [*Community guidelines on state aid to maritime transport (97/C 25/05)*]. As from that time a number of countries leaded by Greece and Netherlands not only implemented the tonnage tax but also used other possibilities brought on guidelines for taxation of shipping companies and ship-owners. Guidelines 1997 were complemented in 2004 [*Commission Communication C(2004) 43 – Community guidelines on state aid to maritime transport (2004/C 13/03)*].

In 2002 there was a questioning and statistical analysis carried out in EU member states to get a feedback on guidelines' measures to be used in guidelines 2004. It showed that in countries where measures envisaged by guidelines 1997 were applied, the amount of ships registered under their flags increased annually in average by 0.4 %. So it was affirmed that application of guidelines 1997 stopped the decrease of ships number under Community flags, thus the goal set by Commission was achieved, at least to certain extent. Therefore the guidelines 2004 follow the principles of guidelines 1997 and set up the aim of the state aid as furtherance of Community fleets without harming the economic interests of other member states, i.e. the principles of non-deformed competition between Community member states must not be suffered.

According to the guidelines the state aid shall be transparent and must lead to achieving the common national goals and at the same time to contribute to return of ships under Community flags and payment of taxes within Community. In addition, the ships must cope with safety requirements including working conditions on-board.

The guidelines 2004 include a number of fiscal and social measures to improve competitiveness of ships under Community flags. The guidelines call for "to improve the fiscal climate for ship-owning companies, including, for instance, accelerated depreciation on investment in ships or the right to reserve profits made on the sale of ships for a number of years on a tax-free basis, provided that these profits are reinvested in ships" (EC, 2004). The system of replacing the normal corporate tax system by a tonnage tax is another measure of a State aid. 'Tonnage tax' means that the ship-owner pays an amount of tax linked directly to the tonnage operated. The tonnage tax shall be payable irrespective of the company's actual profits or losses.

"Such measures have been shown to safeguard high quality employment in the on-shore maritime sector, such as management directly related to shipping and also in associated activities (insurance, brokerage and finance). In view of the importance of such activities to the economy of the Community and in support of the objectives stated earlier, these types of fiscal incentive can generally be endorsed" (EC, 2004).

The following action on employment costs should be allowed for Community shipping:
- reduced rates of contributions for the social protection of Community seafarers employed on-board of ships registered in a Member State,
- reduced rates of income tax for Community seafarers on-board of ships registered in a Member State.

However, the Commission has been reluctant to approve such schemes as subsidies for fleet renewal since they tend to distort competition except for areas where they form part of a structural reform leading to reductions in overall fleet capacity. Nevertheless, other investment aid may be permitted, if in line with the Community's safe seas policy, in certain restricted circumstances to improve equipment on board of vessels entered in a Member State's register or to promote the use of safe and clean ships. Regional aid for maritime companies in disadvantaged regions, which often take the form of investment aid to companies investing in the regions, may only be permitted where it is clear that the benefits will accrue to the region over a reasonable time period.

In conclusion it should be noted that assortment of measures and their range is very different state by state and depends on a lot of political, financial, and other factors. Even so it's necessary to avoid situations where the state aid is accompanied by distort competition between member states in field of subsidies.

2.2 *The present situation in Estonia*

Since Estonia re-independence in 1991 the Ministry of Economic Affairs and Communications of Estonia worked out and developed several leading documents in field of transportation. First such thorough document with name "Transport Development Plan 1999-2006" was approved by Estonian Government in the March 9th, 1999. The Annex 3 of it was dedicated to needs of maritime economics. It was declared that systematization of maritime economics and working out the new trends will be on important place in Estonian maritime activities at period of 1999-2006. Firstly it should bear on shipping industry and taxation policy regarding to seafarers. It was planned to work out during period 1999-2006 the package which should consist of:

- regulation for unification and redistribution of navigation fees by reducing of pilot dues and rise of navigation fee;
- working out and implementation the remission of social tax for seafarers employed in international navigation;
- the taxation privileges for ship-owners;
- the revising of personal income tax system for seafarers;
- working out the possible excise relief system.

Unfortunately, none of abovementioned taxation privileges measures was actually worked out in 1999-2006, far from the stage of putting them into practice.

The next strategic document worked out by the Ministry of Economic Affairs and Communications of Estonia that attended to problems of shipping was "Transport Development Plan 2004-2013" approved by Estonian Government in the January 24th, 2007. In this document the Estonian authorities avowed that significant decrease of vessels under Estonian flag during latest years exerts substantial negative effect to the Estonian economies, first of all taking into account decrease of employment and number of working places for seafarers, as well as a reduction of tax yield from marine industries and mariners. The other problem which vessels' "escaping" affects indirectly is decrease of maritime freight capacities and number of working places in ports and in other shore structures and undertakings.

The development plan declared that State will butt in development of transportation sector through purposeful subsidiaries, investments, purchasing of services, development of legal acts and making the supervision. One of tasks was the applying of the State shipping aid system in Estonia during 2006. Differently from the previous development plans, "Transport Development Plan 2004-2013" was succeeded by a practical output, namely by program of Enterprise Estonia for subsidy of Estonian flag flying merchant ships.

The Estonian Government admitted in 2007 the decision to support the Estonian ship-owners through abovementioned program of Enterprise Estonia. The target group was the companies owning or operating by bare-boat charter the merchant ships entered into Estonian ship register. The object for compensation was a social security tax paid by crew members of such vessels.

Unfortunately, the funds for this aid were allocated only once, in 2006. The source of foundation and regularity of subsidies were not determined in law, so ship-owners had no guarantees of stability and longstanding for compensation of their expenses. Taking into account extremely bureaucratic procedures it is not surprise that attitude of ship owners to the described system was,

to put it mildly, very inert. As result, we can read in "Transport Development Plan 2004-2013" only one remark regarding to subsidies invented in 2006: "The initially developed system did not sufficiently empower itself. The new system should be worked out in frame of a new development plan".

The latest development plan in the field is "The Estonian maritime policy 2011-2020" what was approved by Estonian Parliament in 2012. This document sets up five main priorities divided on eleven goals. The first goal of the first priority is to obtain the internationally competitive merchant fleet under Estonian flag. For this purpose a number of measures and actions were planned in the document.

According to these plans the summarised expenses of operation of Estonian ships must be approximately the same as those for Estonia's main competitors (mainly the countries around Baltic Sea). For achieving of this goal the additional analysis of EC guidelines 2004/C 13/03 shall be carried out making clear the possibilities for application of measures envisaged in guidelines. The shipping companies should have assuredness on long-time maintenance of condition and rules, so they will be interested in investments to ships and bringing them under Estonian flag. The development plan gives the following long-term perspectives: at least 35 ships with GT 500 and more under Estonian flag in 2020, at least 50 vessels in 2030.

The main activities for achieving of these goals stated in the document are:

- the analysis of possibilities for implementation of benefits for ship-owning companies;
- the analysis of applicability of giving the larger than before warranties to ship-owners;
- the revision of tariff policy (navigation and harbour dues, light dues);
- the examination of principles on using of foreign labour on vessels;
- strike up the transnational agreements in area of social security.

By opinion of authors the formulation of abovementioned activities are quite superficial and need more explicity for achieving of stated goals erected, especially taking into account importance of these goals and wide scope of problems concerned.

Regardless of number of development plans worked out in last decades there are not an adequate instruments for conducing to maritime economics and applying of state aid measures seems to be insufficient what lead to:

- passing away merchant fleet from under Estonian flag;
- decrease and sometimes disappearance of some economic sectors bound up to shipping industry (Dmitrijeva 2012).

2.3 Labour expenditure of seafarers

In general, the labour expenditure of seafarers is on third, sometimes even on second place in total structure of ship articles of expenditure, for what reason their impact on profit-and-loss results in companies is significant and the ship-owners' wish to decrease them as much as possible for maximization of company profit is understandable.

Table 1. The fiscal incentives for seafarers-residents in countries-members of EEA ((Shipping Industry Almanac 2011, Latvian legislation).

Country	SSP*	IT**
Belgium	An employer has a complete fiscal immunity; employee has it partly	Remission of tax for seafarers employed on vessels sailing under EEA flags
Estonia	An employer can apply for shipping subsidy for sums exceeding two minimum salaries	Not applicable
Netherlands	40% of computed SSP may be deducted	40% of computed SSP may be deducted
Ireland	Not applicable	Tax to be paid from sums exceeding €6,350 per annum
Italy	No SSP for seafarers-residents	No IT for seafarers-residents
Greece	Not applicable	IT rate is 3% less for ship officers and 1% less for ratings
Cyprus	No SSP for seafarers	No IT for seafarers
Latvia	To be paid from sums exceeding 2.5 minimum salary for ship officers and 1.5 minimum salary for ratings	To be paid from sums exceeding 2.5 minimum salary for ship officers and 1.5 minimum salary for ratings
Luxemburg	Not applicable	Not applicable
Malta	The rate for seafarers is reduced by 15%	The rate for seafarers is reduced by 15%
Norway	Two subsidy schemes fox tax refund	Two subsidy schemes for tax refund
Portugal	No SSP for seafarers-residents	No IT for seafarers-residents
France	Not applicable	No IT for seafarers-residents employed on-board more than 183 days per year
Sweden	The ship-owner can apply for shipping subsidy that covers SSP and IT	
Germany	40% may be returned	Not applicable
Finland	No SSP for merchant ships crew members	No IT for merchant ships crew members
Great Britain	One level lower tax rate for seafarers	No IT for salary earned at sea
Denmark	No SSP and IT completely or partly depending on residency and ship registration	

* Social security payment. **Income tax

In 2012 the income tax (IT) rate 21% and social security payment (SSP) rate 33% were in effect for all residents in Estonia. The fiscal incentives for seafarers-residents in countries-members of European Economic Area (EEA) are shown in Table 1.

It can be seen from Table 1 that almost every EEA country being active in shipping applying some measures for deduction of labour expenditure of ship-owners using the reducing of tax rates. The only exceptions are Estonia and Sweden where ship-owners have to apply for subsidies, and Luxemburg where fiscal incentives for seafarers are absent.

2.4 Tonnage tax

In countries where there is an interest for further development of shipping, the governments besides the measures for reducing of labour expenditure for ship-owners may apply other fiscal measures, with purpose to improve competitive ability of ship-owners and in this way to raise the attractiveness of country in question for them.

One of popular measures is so-called *tonnage tax* that may be described as alternative method for calculation of income tax for shipping undertakings. If the usual method of tax accounting is a revenue-based, then the method for accounting of tonnage tax is not revenue-based but it is based on total ship tonnage owned by ship-owner independently of annual revenue of company. In fact, the tonnage tax was applied still in 15th century because difficulties with identify of real makings of ship-owners (Grapperhaus 2009). Today it is in using rather as stimulus for bringing of more ship-owners as residents to country (Marlow 2002).

During last 15-20 years the number of EEA countries where the method of tonnage tax was applied raised significantly. The tonnage tax scheme was applied firstly in Greece in 1957 and next in Cyprus in 1963. Further it expanded in 1990s to Netherlands, Norway and Germany and at beginning of 21st century to Great Britain, Denmark, Spain, Ireland, Belgium, France, Finland, and Poland (Shipping Industry Almanac 2011).

In accordance with research work of G. Batringa (2010) there are three types of tonnage tax models in Europe:
- the Greek model that was taken into usage in 1957;
- the Dutch model that was taken into usage in 1996 and based on EC guidelines;
- the Norwegian model that based on EC guidelines as well despite the fact that Norway is not EU member country.

The Greek model is now in use in Greece, Cyprus and Malta. In Greece this model is obligatory for all ship-owners who have their revenues from shipping and their vessels must be registered under Greek flag. The same for Malta – vessels must be registered to Malta register and sail under Malta flag.

According to Greek model a tonnage tax for merchant ships should be calculated on base of summary gross tonnage of fleet that has to be multiplied by certain coefficient, after that it has to be corrected by coefficient for vessels age. For example, for the 5 years old vessel with gross tonnage 23,000 and net tonnage 20,000 we will have the tonnage tax approximately €13,000 per year (Dmitrijeva, 2012).

The Dutch model is in use, besides Netherlands, in Belgium, Denmark, France, Germany, Ireland, Italy, Poland, Spain, and Great Britain. According to that the tonnage tax may be used only in connection with merchant shipping activities. The taxable commercial profit is based on tonnage of vessels, not on profit-and-loss statement of company. The profit calculated in such way has to be taxed by usual income tax rate for undertakings.

The some particulars for calculation of tonnage tax by Dutch model are following:
- vessels are divided to four groups by scale of tonnage;
- the tonnage tax is based on net tonnage;
- the calculated profit to be taxed according to ordinary rules of taxation in country.

In Belgium there are five vessels' groups for taxation, whereat there is a special extra low taxation rate for vessels with net tonnage exceeding 40,000.

Using the data from previous example, i.e. 5 years old merchant vessel with gross tonnage 23,000 and net tonnage 20,000 it may be calculated that by Dutch model the tonnage tax in this case will be approximately €8,451 per year (Dmitrijeva, 2012).

The Norwegian model is in use now in Norway and in Finland. According to that the base for tonnage tax is the net tonnage of vessel and there are three taxation groups in Norway and four taxation groups in Finland. Besides, there are rebates for environment-friendly vessels in Norway scheme.

Making calculations for the same sample-vessel it is found that tonnage tax calculated by Norwegian model is €12,695 per year (Dmitrijeva, 2012).

From all the calculations is clear that difference between tonnage taxes calculated by three models is rather insignificant, especially between Greek and Norwegian model. Furthermore, it should be noted that in financially successful years the tonnage tax for shipping company may be in times less than the ordinary income tax for undertakings.

2.5 *The advantages and shortcomings of tonnage tax*

As it was mentioned previously, the main goal of governments who enable the ship-owners to choose between ordinary taxation model and tonnage tax model, is to create shipping business environment providing real advantages for all shipping companies who have intention to entry into shipping business or continue it successfully. The tonnage tax scheme has a number of advantages comparing to ordinary taxation system but there are some imperfections as well.

The next advantages may be named:
- *simplicity* – the simple and clear calculation methods should provide the high quality results, for shipping companies this means reduction of efforts and expenditures in course of completion of tax return annually;
- *increased cash flows* – low effective tax rates improve the competitiveness of shipping companies;
- *confidence* – thanking to foreknowledge of taxes level the companies must not make money reservations on their accounts and/or put forward the tax payments;
- *flexibility* – more freedom for companies making financial decisions;
- *clearness* – the tax position of company more understandable, the company is more attractive for business partners and investors;
- *conformity and better competitiveness* – the conditions for internal and external partners are more harmonized (19 countries of EEA are applying tonnage tax system);
- *employment and education* – the increasing number of vessels under flag of country and shipping companies as residents raise demand for domestic specialists in maritime and consequently for training and education institutions;
- more *stability* in income to state budget – as rule, earnings on base of tonnage tax are less than those by ordinary tax but fixed taxes provide the stabile income to budget independent on changes in economic situation.

Regardless of long list of advantages there may be remind some disadvantageous consequences for undertakers who have decided to choose tonnage tax as tax system for their shipping companies, especially in period of economic recession. They are:
- *too high level of taxes* – in period of economic recession the turnover and income of company may fall remarkably, in case of negative annual results a tonnage tax becomes the loss-increasing component;
- *inevitability* – tonnage tax payment is obligatory for ship-owner independent on state of return of his company and on changing in economic situation. Thus, for business with annual loss and a high operating costs in previous years, a fixed payment liability is an inevitable additional cost, especially for ships with big gross tonnage;

- prescheduled exit from tonnage tax regime is *expensive* – coming back to ordinary taxation system usually goes together with complicated transition rules and penalty sanctions which are hardly managed;
- possible accompanying *additional obligations* – in legislation of several countries the tonnage tax system is regarded as favourable taxation, this brings together a number of additional obligations, for instance the obligation to create the fractional reserves or obligation to employ seafarers graduated from domestic education and training institutions.

Taking into account above listed arguments every shipping company should before entering the tonnage tax system to investigate deeply its own financial and economic situation and possible development scenarios in company as well as in world economy making clear the possible advantages and avoiding unexpected problems.

3 THE POSSIBLE SOLUTIONS AND RECOMMENDATIONS

3.1 *The factual output of fiscal measures in selected countries*

For estimation of long-time impact of applying of tonnage tax and measures for decreasing the labour expenditure the dynamics of ships moving to and from under flag for four countries was investigated: Norway, Denmark, Netherlands and Latvia.

Norway. The tonnage tax was enforced in Norway in 1996. The particular feature of it was initially the taxation of dividends payable by shipping companies according to state law regulations. Additionally the tax rebate from taxes paid by seafarers was enabled.

After establishment of tonnage tax the amount of vessels under Norwegian flag increased quickly – from about 100 in 1997 up to approximately 450 in 2011, in 2009 this number was nearly 500 thanking to changing in legislation in 2007 by that the ship-owners were exempted from dividends taxation.

Denmark. The tonnage tax was enforced in Denmark in 2001 whereby the dividends are not taxable. The shipping companies can use tonnage tax system for owned or bare-boat chartered vessels. It is possible for time chartered vessels as well if they constitute not more than 75 % of total fleet tonnage of company.

The income tax rebate for seafarers' salary is in effect in Denmark from 1988 whereby the ship-owner can request for income tax reimbursement once a year or ones a quarter depending on agreement with tax office.

In fact, the number of vessels under Danish flag has not changes in 1992-2011 and stays about 600

units but total tonnage of fleet has raised from 5 million up to approximately 12 million.

Netherlands. In Netherlands the tonnage tax was enforced in 1996, the dividends are not taxable. The tonnage tax system may be used only for owned or bare-boat chartered vessels. The ship-owners can reduce taxes from labour expenditure by 40 % immediately and to pay to tax office reduced taxes. This allows him to use more money in cash flows.

By CIA – the World Factbook 2011 there was 744 vessels registered into Dutch ship register whereat in 2006-2011 327 ships were registered as new ones.

Latvia. The tonnage tax was enforced in Latvia in 2002. By Latvian ship register the number of merchant vessels under Latvian flag decreased in 2003 from 30 by 8 vessels but in 2004 this number has recuperated but started to decrease again in 2008.

By CIA – the World Factbook 2011 the number of vessels under Latvian flag was 11 vessels in 2011 compared to 21 vessels in 2010, so the decrease was significant. At the same time there was 79 companies based in foreign countries on Latvian capital having 79 vessels in 2011 under foreign flags.

On the base of information for these four countries it may be concluded that the enforcing of tonnage tax only is not enough for improving of situation with bringing vessels under state flag. The taxation of income is important expenditure category but not as important as aggregate costs connected to labour expenditure of seafarers. For stimulation of shipping development in a single country the combination of several measures directed to achieving of optimal state aid is essential; the stability of these measures must be guaranteed by legal acts.

3.2 *The possible developments in Estonia*

Shipping subsidy system has been applied in Estonia from 2007. That envisaged the growth of merchant fleet under Estonian flag by 1 vessel per year. In fact this goal was not achieved. The opinion of Ministry of Economic Affairs and Communications is that change in taxation of seafarers and shipping companies are necessary. The measures envisaged by preliminary plan may be described as follow:
- approximately 80 % of paid social security tax may be attested as qualifying for aid measures;
- refund of up to 50 % of income tax for seafarers;
- concluding of long-term (for example 10 years) agreements between ship-owners and subsidy payer what would give more confidence to ship-owners.

The other possibility of bringing to Estonia investors and returning ship-owners and their vessels to under Estonian flag is to create attractive environment not only for investments but for optimal using of profit earned. For this purpose the establishment of tonnage tax instead of regular income tax seems to be right solution.

The taxation system of Estonia is quite liberal and friendly for investors already now. There is no corporate income tax in Estonia, but income tax on paid out dividends. Thus, it may be rather difficult to convince investors to bear to expenses some fixed sum of taxes as tonnage tax if they have an existing taxation system where choice the first liability to pay taxes arises only when profit is shared in form of dividends. However the authors of this paper are sure that as the main goal of businesses is earning profit to investors, it is possible to regulate the corporate income taxation of particular shipping companies in such way that this kind of activity will be more attractive for investors.

4 CONCLUSIONS

Estonia and other Baltic States had always geographically and historically very good prospects to be successful maritime nations: there are more high qualified people who had obtained a good maritime education and seafaring experience; there are own education and training institutions in the field; there are ports and cargo flows moving through them but at the same time there are the serious problems with optimal using of them as well. For example a number of ships under Estonian flag are close to critical.

A lot of maritime nations in world including the EEA countries have taken in use a number of measures for providing of competitive ability for their shipping companies and keeping merchant fleet under their flag and ship-owners as their residents.

In Estonia such measures are actually absent or are not sufficient and/or not effective. In opinion of authors of this paper the fundamental changing and resolute steps in this field are necessary. The analysis carried out shows that government and decision-makers has began to understand the importance of problems and the certain measures have worked out and have undertaken by them. The livening up of these activities may be expected in next 5-10 years.

REFERENCES

Batrinca, G. 2010. Considerations on Introduction of Tonnage Tax System in the European Union. *Maritime Transport & Navigation Journal* Vol. 2, No 2.

CIA. 2012. The Factbook 2011.[https://www.cia.gov/library/publications/the-world-factbook/geos/countrytemplate_en.html].

Commission Communication C(2004) 43 - Community guidelines on State aid to maritime transport. [http://eur-lex.europa.eu/LexUriServ.do?uri=CELEX:52004XC0117(01):EN:NOT].

Dmitrijeva, E. 2012. *The Analyses of Shipping-related State Aid Measures in Estonia and Change of Taxation regulations as One of Means to Improve the Situation.* Tallinn: Estonian Maritime Academy.

Grapperhaus, F.H. 2009. *Taxes Through the Ages*. Amsterdam: IBFD.

Marlow, P.B. 2002. *Ships, Flags and Taxes*. London: LLP.

Maritime Policy and Global Excellence
STCW, Maritime Education and Training (MET), Human Resources and Crew Manning, Maritime Policy,
Logistics and Economic Matters – Marine Navigation and Safety of Sea Transportation – Weintrit & Neumann (Eds)

Analysis of Safety Inspections of Recreational Craft in the European Union. A Case Study

J. Torralbo & M. Castells

Universitat Politecnica de Catalunya (UPC), BarcelonaTech, Spain

ABSTRACT: Inspections of pleasure boats in Spain can be carried out by collaborating entities of inspection, and these entities must have the previous authorization by the administration. This authorization allows these entities performing effective inspections and technical controls of pleasure craft. Recreational crafts are subjected to surveys that are based on the registration list and hull construction material. In addition the required safety equipment of the recreational boat depends on the distance that the recreational boat is authorized to navigate.

The aim of this paper is to analyze and compare the types of survey / inspections to be carried pleasure craft (non-commercial use), periodicity and required safety equipment in some member states of the European Union. A case study of Spain is presented. From the results obtained, we can determine if there is a lack of coordination in this area and, if required, indicate the need to unify a common pattern in inspections and surveys of recreational boats in the EU.

1 INTRODUCTION

The recreational craft industry covers boats of a certain length designed for sports and leisure purposes. These are high-value and very movable products, intended mainly for end consumers and with a relatively long life-cycle, which means that they are often on the market for a long time. The recreational craft industry has attracted the interest of the EU Commission because of its impact on the environment and its economic significance in the European Union. To implement EU-wide initiatives in this sector, the Commission has legislated on recreational craft, which are boats of any type, regardless of their means of propulsion, between 2.5 and 24 metres hull length. This EU legislation (Directive 94/25/EC, as amended by Directive 2003/44/EC) includes a number of exceptions and derogations.

In order to promote sustainable development, the European legislation on recreational craft also introduced standard requirements regarding user safety, as well as exhaust and noise emissions. As a result, this European legal framework has removed disparities among Member States, while facilitating free competition across the Union and trade with foreign countries.

Although Directive 94/25/EC establishes safety issues, there is a lack of coordination and equivalence among the EU countries according to the survival and safety equipment compulsory for recreational crafts.

2 CASE STUDY: SPAIN

With the approval of the Spanish legislation (*Real Decreto 1434/1999*), the government authorizes collaborating entities perform the inspections and surveys that must be submitted in pleasure craft registered in Spain, regardless of the means of propulsion. These entities can survey recreational crafts with a hull length between 2.5 and 24 meters, designed and aimed for recreational and sports purposes (list/register sixth and seventh), and not allowed to board more than 12 passengers.

The ship's register in Spain is done by a lists, which will be registered all ships, boats and floating structures. The sixth list is intended for sporting or recreational craft that are operated for commercial gain. The seventh list is for non-profit pleasure craft or fishing boats unprofessional.

All yachts have a certificate of seaworthiness. This certificate is a document attesting that the craft is fit to sail and in which is shown information such as the name and port of register of the vessel, its technical features and navigation area.

Table 1. Navigation areas based on distance of operation. Source: Own based on FOM/1144/2003

Area	Distance of operation
Area 1	Unlimited
Area 2	Up to 60 miles
Area 3	Up to 25 miles
Area 4	Up to 12 miles
Area 5	Up to 5 miles
Area 6	Up to 2 miles
Area 7	Protected waters in general

2.1 Technical Inspections

Certificate of seaworthiness have expiry date and its validity is determined by the registration list (sixth or seventh), the length and the material of the hull. Tables 2 and 3 detail the types of inspections to be carried out in recreational boats according to the registration list and material of the hull:

Table 2. Type of technical inspections for List 7th depending on the length and hull material. Source: Own, based on RD 1434/1999

Inspection	Frequency	List 7
Regular	Every 5 years	6-24m
Intermediate	Between 2nd and 3rd year	15-24m 6-24m/wooden hall
Additional	In case of repairs or modifications	
Extraordinary	Under request of the Maritime Administration	

Table 3. Type of technical inspections for List 6th depending on the length and hull material. Source: Own, based on FOM/1144/2003

Inspection	Frequency	List 6
Regular	Every 5 years	2.5-24m
Intermediate	Between 2nd and 3rd year	6-24m
Additional	In case of repairs or modifications	
Extraordinary	Under request of the Maritime administration	

It is important to note that boats that have less than 6 meters in length and registered in the 7th list, should not perform periodic inspections and certificate of seaworthiness will include the phrase "No Expiration."

Figure 1. Hull inspection in a shipyard

The collaborating entities of inspection (approved by the Spanish government) can carry out periodic, intermediate, additional and extraordinary inspections.

The periodic and intermediate inspection of the boat must be performed in a shipyard (dry) and floating. The inspection for boats less than 7 meters in length can be only performed in a shipyard (dry), provided that it's possible to start the engine of the boat.

According to the type of boat (motor / sail), all surveys are made considering the following points: hull and equipment, engine and auxiliary machinery, mast and rigging, electrical installation, radio communication equipment, survival and safety equipment, fire extinguishing equipment, nautical material, and navigation lights and anchoring equipment.

The owners and / or user of recreational craft are responsible for keeping up to date surveys and inspections. Emphasize that the inspection aims to ensure personal safety and preventing marine pollution accidents. Therefore, it is of vital importance to navigate with valid certificate of seaworthiness. In addition to the security issues indicated, to have the current certificate can avoid the possibility of being fined by the administration.

2.2 Compulsory survival and safety equipment

Compulsory survival and safety equipment will be determined by the navigation area of the boat. The equipment for area "1" is most complete and is reduced to lower areas being the minimum equipment for the area "7". Table 4 shows the compulsory survival and safety equipment according to the navigation area:

Table 4. Compulsory safety equipment. Source: Own, based on FOM/1144/2003

Equipment	Navigation area						
	1	2	3	4	5	6	7
Liferaft	X	X	X				
Lifejacket	X	X	X	X	X	X	X
Lifebuoy	2	1	1	1			
Parachute flare	6	6	6	6			
Hand flare	6	6	6	6	3	3	
Buoyant smoke	2	2	1	1			

Table 4 shows that in areas "1", "2" and "3" is compulsory boarding a liferaft. All liferafts must be reviewed annually by an authorized station. In the case of a new liferaft revision, it must be made before two years from the date of manufacture.

A SOLAS (International Convention for the Safety of Life at Sea, 1974) liferaft means that complies with the requirements set by the International Convention and also indicates that manufacturing has been approved by the Spanish government. The SOLAS liferaft in area "1" must carry inside an emergency package type A SOLAS and in the case of a liferaft for areas "2" and "3" the emergency package must be type B SOLAS. Both packages contain emergency supplies to survive at sea, as first aid, food and water among others.

Figure 2. Container liferaft with hydrostatic release.

Regard to the compulsory pyrotechnics on board (parachute flare, hand flare and buoyant smoke), pointed out the need to control their expiration date. Due to its danger, it is recommended stow them in a dry and safe place. All the pyrotechnics, according to SOLAS, must be stored in a hydroresistant box and must indicate brief instructions or clearly diagrams explaining how to use them.

Table 5 provides details of the means of fire extinguishing and bilge drainage equipment:

Table 5. Compulsory fire extinguishing and bilge drainage equipment. Source: Own, based on FOM/1144/2003

Equipment	Navigation area						
	1	2	3	4	5	6	7
Portable extinguishers (depending on length)	X	X	X	X	X	X	X
Portable extinguishers (depending on power)	X	X	X	X	X	X	X
Fire buckets	2	2	1	1			
Scoop	2	2	2	1	1	1	
Extractor fan	X	X	X	X	X	X	X
Bilge pump	2	2	2	1	1	1	1
Fixed installation	X	X	X	X	X	X	X
Gas detector	X	X	X	X	X	X	X

Engines that use gasoline as fuel must install an extractor fan that works for suction and discharge directly from the outside. This extractor must renew completely the air of the engine compartment and fuel tanks in less than four minutes. Before starting the engine, the extractor must be switch on to ensure the renewal of the air and avoid a possible explosive atmosphere.

The engines that use gasoline are equipped with a fixed fire extinguishing installation. This installation has to be activated manually from outside of the engine compartment. Automatic fire extinguishers are only allowed when the engine compartment is not accessible.

Table 6 shows other survival and safety equipment compulsory according to the navigation area:

Table 6. Other survival and safety compulsory equipment compulsory equipment Source: Own, based on FOM/1144/2003

Equipment	Navigation area						
	1	2	3	4	5	6	7
Rudder emergency	X	X	X	X	X	X	X
Mooring line	2	2	2	2	2	2	2
Boat hock	1	1	1	1	1	1	1
Oar	1	1	1	1	1	1	1
Inflator	X	X	X	X	X	X	X
Repairing set	X	X	X	X	X	X	X
First aid outfit	X	X	X	X	X	X	X

Table 7 provides details of the navigation equipment required depending on the navigation area:

Table 7. Navigation compulsory equipment compulsory equipment. Source: Own, based on FOM/1144/2003

Equipment	Navigation area						
	1	2	3	4	5	6	7
Anchor lines	X	X	X	X	X	X	X
Navigation lights and shapes	X	X	X	X	X	X	X
Compass	2	2	1	1			
Speedometer	1	1					
Sextant	1						
Chronometer	1						
Dividers	1	1					
Course protractor	1	1					
Rule of 40 cm	1	1					
Binoculars	1	1	1	1			
Nautical charts	1	1	1	1			
Fog horn	1	1	1	1	1	1	1
Bell	1	1	1	1			
National flag	1	1	1	1	1	1	1
Flag code	1	1					
Waterproof torch	2	2	1	1			
Ship's log-book	1						
Heliograph	1	1	1	1	1	1	1
Radar reflector	1	1	1	1			
Signals code	1	1	1	1	1	1	1

Finally, radio and navigation equipment is required in all area navigation as follows:

Area 1:
- VHF with DSC class A (SOLAS)
- MF / HF with DSC or LES(Land Earth Station - SOLAS)
- RLB (Manual and automatic)
- NAVTEX (SOLAS)
- Portable VHF adapted to GMDSSM
- SART 9 GHz (Search And Rescue radar Transponder - SOLAS)

Area 2:
- VHF with DSC
- RLB (Manual and automatic)
- Portable VHF (GMDSSM or submersible IPX7) or SART 9 GHz (Search And Rescue radar Transponder)

Area 3:
- VHF with DSC
- RLB (Radio Locator Beacon -Manual and automatic or manual only)

Area 4:
- Fixed VHF

Area 5:
- Fixed or portable VHF

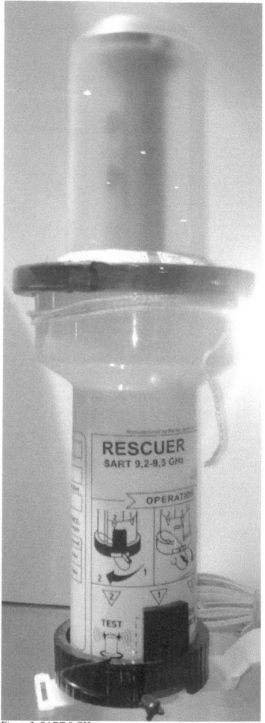

Figure 3. SART 9 GHz

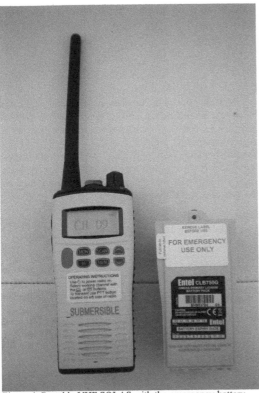

Figure 4. Portable VHF SOLAS with the emergency battery.

Table 8. Mandatory equipment in France. Source: Own, based on French Division 241.

Equipment	Navigation area		
	Basique	Côtier	Hauturier
	X (>50N)	X (>100N)	X (>150N)
Navigation lights	X	X	X
Manual bilge pump	X	X	X
Ladder	X	X	X
MOB device (>4.5 kW)	X	X	X
Fire fighting device	X	X	X
Towing device	X	X	X
Sea anchor	X	X	X
Lifebuoy light	X	X	X
Lifeline	X	X	X
National flag	X	X	X
3 Hand flare		X	X
1 Lifebouy		X	X
Fog horn		X	X
Heliograph		X	X
MOB system		X	X
Steering compass		X	X
RIPA		X	X
Signal code		X	X
Navigational charts		X	X
Harness and H. line			X
Liferaft			X
3 parachute flare or VHF with DSC			X
2 Buoyant smoke or VHF with DSC			X
Weather Device aboard			X
Dividers			X
Sextant or GPS			X
Rule for navigation			
Bearing compass or GPS			X
Lighthouse notebook			X
First aid outfit			X
Ship's log-book			X
Tide book (not in Mediterranean)			X

3 COMPARISON STUDY WITH OTHER MEMBER STATES

3.1 *France*

In France, from April 15, 2008 is in force Division 240 and this normative applies to all ships less than 24 meters in length.

This legislation establishes safety and rescue equipment that is compulsory according the navigations distance allowed. So this can be basic - *basique* (navigation to two miles from a shelter), coast - *côtier* (up to 6 miles from a shelter) or ocean navigation - *hauturier* (navigation over 6 miles from the coast).

Some of the notable aspects of the French legislation would be that the VHF is not required for recreational craft. But is highly recommended to equip the boat with a VHF. In the case of installing a VHF, the skipper must have a restricted certificate of GMDSSM. It is also not compulsory to install a beacon neither an electric bilge pump.

Regarding inspections for the French case, the first inspection must be performed three years after the purchase of the boat. The next inspection should take place in a period not exceeding 36 months, and at least two inspections should be carried out within a period of five years. For boats boarding more than 12 people in addition to the crew, the maximum time between inspections is 12 months.

3.2 *United Kingdom*

Pleasure craft of less than 13.7 metres in length are not covered by any statutory requirements as far as lifesaving or fire fighting equipment is concerned.

At 13.7 metres in length and over they are, however, obliged to comply with the Merchant

Shipping (Life-Saving Appliances for ships other than ships of Classes III to VI (A)) Regulations 1999 and the Merchant Shipping (Fire Protection: Small Ships) Regulations 1998 respectively. According UK non-passenger ship classification, these vessels are classified as being Class XII in these Regulations. It is compulsory for Class XII vessels to carry Life-Saving Appliances and Fire Protection equipment.

Table 9. Mandatory equipment in United Kingdom. Source: Own, based on www.rya.org.uk

Equipment	For vessels of 13.7 m in length and over			
	>3'	3-20'	20-150'	>150'
Lifejacket	X	X	X	X
Lifejacket lights		X	X	X
Life raft		X (B)	X (B)	X(A)
Lifebuoys and lines	2	2	2	2
Flares				
Training manual	Containing instructions and information on the life-saving appliances provide in the vessel and their maintenance			
Lifesaving signals	A copy of the table "Life-saving signals and Rescue methods, SOLAS 1" or "Life-saving signals and rescue methods, SOLAS 2".			
Maritime radio	Capable of transmitting and receiving, appropriate to the area of operation.			
Ladder	X	X	X	X
Fire extinguisher	X	X	X	X
Fire buckets	Not less than 2 with lanyards			

According to the Survey and Certification Policy Instructions for the Guidance of Surveyors by the Maritime and Coastguard Agency (MCA), for Pleasure Vessels (Non-Commercial) are required the following certification: If certified for > 15 persons and on international voyages, an International Sewage Pollution Prevention Certificate, if registered: Certificate of Registry, which will normally be valid for a period not exceeding 5 years and the Certificate of Measurement.

Smaller boats may be registered on the Small Ships Register (SSR) if they are to be used abroad and they should display a number preceded by the letters 'SSR'.

For commercial boat, the Small Commercial Vessel and Pilot Boat (SCV) Code, also known as the harmonised code, states that the hull, shell fittings, external steering and propulsion components of the vessel should be examined out of the water at intervals not exceeding 5 years. The Certifying Authority may stipulate a lesser interval in consideration of hull construction material or the age or the type and service of the vessel. A certificate is to be valid for not more than five years. Every boat has to be inspected by an RYA (*Royal Yachting Association*) Inspector at least once during its first three years of operation, and be completely resurveyed and certificated every five years

For non-commercial recreational craft, Boat Safety Certificate (BSC) will be issued for a 4-year period, after which it must be renewed.

3.3 *Italy*

In Italy the *Capitanerie di Porto – Guardia Costiera* according to the *Decreto 5 ottobre 1999 n. 478 "Regolamento recante norme di sicurezza per la navigazione da diporto" (G.U. del 17.12.1999)* establish the survival and safety equipment compulsory in recreational crafts. The survival and safety equipment is related with the distance of the boat is allowed to navigate.

Table 10. Mandatory equipment in Italy. Source: Own, based on Capitanerie di porto - Guardia Costiera

	No limitation	Less than 50 miles	Less than 12 miles	Less than 6 miles	Less than 3 miles	Less than 1 mile	Less than 300 meters	Inland waters
Liferaft	X	X	X					
Lifejacket	X	X	X	X	X	X		X
Lifebuoy	1	1	1	1	1	1		1
Light buoy	1	1	1	1				
Buoyant smoke	3	2	2	2	1			
Compass and deviations table	X	X	X					
Watch	X	X						
Barometer	X	X						
Binoculars	X	X						
Nautical charts								
Instruments for charts: dividers, course protractor, rules,etc.	X	X						
Hand flare	4	3	2	2	2			
Parachute flare	4	3	2	2				
First aid outfit	X	X						
Navigation lights	X	X	X	X	X			
Fog horn / Bell	X	X	X	X	X			
Navigation equipment (LORAN, GPS)	X	X						
VHF	X	X	X					
Radar reflector	X	X						
EPIRB	X							
Bilge pump / Scoop	X	X	X	X	X	X		
Fire extinguisher	X	X	X	X	X	X		

In Italy the first inspection of the recreational craft must be performed between the eighth and tenth year (depending on the design and construction category A, B, C or D). After the first inspection, the security certificate -*Certificato di sicurezza*- has an expiration date of five years, so the next inspection should take place in a period of five years.

4 CONCLUSIONS

Since June 1998, it has been a requirement that all new boats offered for sale within the EU comply with the Recreational Craft Directive (RCD). This indicates that the craft fulfils certain essential criteria concerning safety and other associated matters. The boat must display a CE mark together with a plate detailing the maximum payload and operational limits. Local Authority Trading Standards officers have the responsibility to ensure that CE-marked craft comply with the RCD circulation of recreational craft in the EU.

Recreational craft marketed in the EU must comply with harmonised technical safety and environmental requirements and meet a number of administrative obligations defined by Directive 94/25/EC, as amended in 2003. These safety and environmental requirements address the design and construction of the craft, and set limit values for their exhaust and noise emissions.

Although Directive 94/25/EC establishes safety issues, from the results obtained in the above sections, we can state that there is no coordination and equivalence among the EU countries according to the survival and safety equipment compulsory for recreational crafts.

In the case of the countries analysed, we note that each country establishes its classification of pleasure boats, required different safety equipment and types and frequency of the mandatory inspections are also different.

The diversity of criteria of the topics discussed states that it would be necessary to establish mechanisms to unify some aspects among countries of the European Union like:

- Classification of pleasure boats.
- List of mandatory safety and survival equipment according to classification of pleasure boat.
- Periodicity of inspections and items to be inspected.

To conclude, we can state that there is a lack of coordination in this area and further research is necessary to unify a common pattern in inspections/surveys and survival and safety equipment of recreational boats in the EU.

REFERENCES

[1] Arrêté du 11 mars 2008 modifiant l'arrêté du 23 novembre 1987 relatif à la sécurité des navires - Division 240. Division 241, France.
[2] Directive 94/25/EC of the European Commission.
[3] Directive 2003/44/EC of the European Commission.
[4] International Convention for the Prevention of Pollution from Ships, 1973 and Protocol of 1978 (MARPOL 73/78).
[5] International Convention for the Safety of Life at Sea (SOLAS), 1974.
[6] International Regulations for Preventing Collisions at Sea (COLREGs), 1972.
[7] Merchant Shipping (Vessels in Commercial Use for Sport or Pleasure) Regulations 1998; SI 1998/ 2771 as amended, and
[8] Merchant Shipping (Small Workboats and Pilot Boats) Regulations 1998, SI 1998/1609.
[9] MERCHANT SHIPPING- REGISTRATION- The Merchant Shipping (Registration of Ships) Regulations 1993
[10] Orden FOM/1144/2003, de 28 de abril, Spain.
[11] Real Decreto 1434/1999, Spain.
[12] Decreto 5 ottobre 1999 n. 478 "Regolamento recante norme di sicurezza per la navigazione da diporto" (G.U. del 17.12.1999), Italy.
[13] Walliser J., Piniella F., Rasero J.C., Endrina N.: Maritime Safety in the Strait of Gibraltar. Taxonomy and Evolution of Emergencies Rate in 2000-2004 Period. TransNav - International Journal on Marine Navigation and Safety of Sea Transportation, Vol. 5, No. 2, pp. 189-194, 2011

Maritime Policy and Global Excellence
STCW, Maritime Education and Training (MET), Human Resources and Crew Manning, Maritime Policy,
Logistics and Economic Matters – Marine Navigation and Safety of Sea Transportation – Weintrit & Neumann (Eds)

Stakeholders' Satisfaction: Response to Global Excellence

R.A. Alimen, V.B. Jaleco, R.L. Pador & M.C. Sequio
John B. Lacson Foundation Maritime University-Molo, Iloilo City, Philippines

ABSTRACT: The study investigated the stakeholders' satisfaction on the competencies and performances among marine engineering graduates in the maritime university in response to global excellence. The rescarchers employed descriptive quantitative-qualitative method of data collection. Respondents of this study were manning officers, crew managers, training officers, recruiting officers, and shipping owners of different shipping/manning agencies based in Manila, Philippines. The study entailed personal interviews with the respondents and accomplished the data-gathering instrument titled "Stakeholder Satisfaction Survey" conducted last December 2009. Based on the quantitative analysis of data, competencies and indicators in terms of communication, professionalism & trustworthiness, communication, discipline, loyalty, consistency of performance, leadership skills, honesty, industry, social responsibility, initiative, and inter-personal competency were identified. Qualitative data from the interviews and answers to open-ended questions were utilized to enhance and further discuss the quantitative results of the present study to substantiate the data to achieve excellence in maritime education.

1 INTRODUCTION

It is proper as the premier and only maritime university in the country conducts assessment study on the competency and performance exhibited by the marine engineering graduates. As supported by the statements mentioned in the study of Japos (2010) in his words "Performance appraisal allows the employees to determine their strengths and weaknesses as basis for improvement and the management to design appropriate interventions."

Through this study, the administration obtained the necessary information to determine the strengths and weaknesses of the graduates. This was done by conducting appraisal survey or stakeholders' satisfaction survey. The stakeholder research approach was anchored on the premise that organizations must concern themselves with the demands of multiple constituents (Freeman, 1983; Jones, 1995; Donaldson & Preston, 1995; Moldoveanu, 2003).

Experiences and perceptions of groups of people who have vested interests in the services delivered by the organization – customers, employees, strategic partners, and special-interest groups were looked into. 'Stakeholder satisfaction' is often used to represent the views of these groups, and a common approach to its measurement is to focus on the concept of 'satisfaction' either as an exogenous variable or as a construct based on various attributes of satisfaction. It is defined as the "critical investigation of the experiences and views of sets of people who have vested interests in the products and services delivered by an organization."

2 CONCEPTUAL FRAMEWORK OF THE STUDY

The conceptual framework of the study was anchored on Three Hundred Sixty (360) Degrees Feedback cited by Newstrom (2007) but modified by the researchers to suit the need of the time to address global excellence in maritime education. The framework was considered as a process of systematically gathering data on personal skills, abilities, and behaviors from a variety of sources such as (a) managers, (b) peers, (c) subordinates, (d) customers, (e) clients. The different perspectives were examined to see where the problems exist in the eyes of one or more groups. The 360-degree feedback framework was used in this study as assessment approach to get information of one's need for change. The framework was considered by the researchers to obtain multiple data and information (both positive and negative) that were used in performance improvement that university considered to enhance teaching-learning delivery to achieve global competence in maritime industry. To further understand the conceptual framework, the diagram was shown below.

What is the level of satisfaction among marine engineering graduates in terms of different competencies such as: (a) communication, (b) professionalism and trustworthiness, (c) discipline, (d) loyalty, (e) consistency of performance, (f) leadership skills, (g) honesty, (h) industry, (i) social responsibility, (j) initiative, and (k) inter-personal competence?

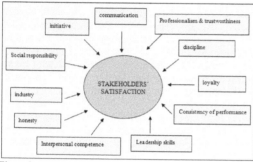

Figure 1.

3 STATEMENT OF THE PROBLEM

This study determined the level of satisfaction among company partners- - crew managers, personnel managers, training directors based in Manila, Philippines last December 2009.

Specifically, the following questions were advanced:

1 What is the stakeholders' level of satisfaction of the performance of marine engineering graduates employed in the international vessels/ships when classified according to the different categories such as: (a) management level, (b) operational level, (c) engine ratings, and (d) engine cadets?

2 What is the level of satisfaction among marine engineering graduates in terms of different competencies such as: (a) communication, (b) professionalism and trustworthiness, (c) discipline, (d) loyalty, (e) consistency of performance, (f) leadership skills, (g) honesty, (h) industry, (i) social responsibility, (j) initiative, and (k) inter-personal competence?

3 What suggestions do stakeholders have for performance improvements?

4 THEORETICAL FRAMEWORK OF THE STUDY

The theoretical framework of the present investigation is the Performance Prism (PP). The theory is used to measure stakeholder value. Cranfield University originated the utilization of the Performance Prism as an innovative performance measurement and performance management framework of the second generation.

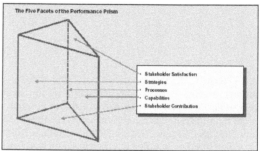

Figure 2. The Performance Prism (Cranfield University)

The underlying framework of the present study was based on the workability of the Performance Prism (PP) as the belief that for organizations aspiring to be successful in the long term must have a clear picture of who their stakeholders are and what they want. This clearly indicated in essence the interrelationships among the facets in the Performance Prism. The proponents of this framework suggest that for a performance to possess quality, the process should start not from the strategies but from the stakeholders and basically on what they want. In a similar manner, the theoretical framework of the PP has found its way on how this study has been theorized.

5 RESEARCH DESIGN

This study employed the quantitative-qualitative method of analyzing the obtained data from the different company-partners of JBLFMU-Molo, Iloilo City, Region 6 conducted last December 2009. The data were obtained by utilizing the following components: (1) survey personally administered by the researchers, (2) personal interview with the stakeholders (crew and personnel managers, training officers, and HRD heads of different company-partners of JBLFMU-Molo, Iloilo City, and (3) analysis of the results of the survey using descriptive statistical method such as mean, frequency count, and percentage. Qualitative data from the interviews were also used to enhance the quantitative results of the study.

6 THE PARTICIPANTS

The respondents of this study were Greek, Japanese, Norwegian, Singaporean, Italian, German, and American. There were fifty two (52) respondents interviewed for this study. The distribution of the graduates employed in the international vessels was presented in Table 1. Graduates who had obtained management levels were 6 (12%), graduates who had operational levels were 6 (12%), graduates who belonged to engine ratings were 20 (38%), and engine cadets respondents were 20 (38%). To further illustrate the distribution of graduates who were assessed by the company-partners, the data are shown in the table below.

Table 1. Distribution of the Respondents Classified According to Levels

Level	f	%
A. Management Level	6	12
B. Operational Level	6	12
C. Engine Ratings	20	38
D. Engine Cadets	20	38
Total	52	100

7 DATA-GATHERING INSTRUMENT AND STATISTICAL TOOLS

The data-gathering instrument in this research was the "Stakeholder Satisfaction Survey" which consisted the following areas of competencies: (a) communication, (b) professionalism and trustworthiness, (c) discipline, (d) loyalty, (e) consistency of performance, (f) leadership skills, (g) honesty, (h) industry, (i) social responsibility, (j) initiative. These areas were applied to the different levels of seafarers such as: engine ratings (electricians, fitters, oilers, and wipers) and engine cadets. The data-gathering instrument had rating scales of 1 to 10, which were arranged in ascending manner.

8 RESULTS OF THE STUDY

The results of the present study revealed that the level of stakeholders' satisfaction is "satisfied" with mean scores of 7.55, 7.48, 7.47, and 7.38 in all levels such as (a) management level, (b) operational level, (c) engine ratings, (d) engine cadets.

Table 2. According to the Different Categories

Level	Mean	Description	Rank
A. Management Level	7.48	Satisfied	2.0
B. Operational Level	7.38	Satisfied	4.0
C. Engine Ratings	7.55	Satisfied	1.0
D. Engine Cadet	7.47	Satisfied	3.0

Legend:

Scale	Description
7.76 – 10.00	Very Satisfied
4.51 – 7.75	Satisfied
2.26 – 4.50	Dissatisfied
1.00 – 2.25	Very Dissatisfied

As to the management level, the competencies were perceived by the stakeholders' as "very satisfied." These competencies were (a) hardworking, (b) consistency of performance, and (c) social responsibility. The other competencies were all "satisfied" according to customers' satisfaction.

Table 3. Competencies of Graduates in the Management Level

Category	Mean	Description	Rank
Hardworking	8.0000	Very Satisfied	1.0
Consistency of Performance	7.8333	Very Satisfied	2.5
Social Responsibility	7.8333	Very Satisfied	2.5
Professional & Trustworthy	7.6667	Satisfied	4.5
Leadership Skills	7.6667	Satisfied	4.5
Discipline	7.3333	Satisfied	7.0
Honesty	7.3333	Satisfied	7.0
Initiative	7.3333	Satisfied	7.0
Communication Skills	7.1667	Satisfied	9.5
Loyalty	7.1667	Satisfied	9.5
Interpersonal Competence	7.0000	Satisfied	11.0

Table 4. Competencies of Graduates in the Operational Level

Category	Mean	Description	Rank
Leadership Skills	7.6667	Satisfied	1.5
Discipline	7.6667	Satisfied	1.5
Honesty	7.5000	Satisfied	4.0
Social Responsibility	7.5000	Satisfied	4.0
Interpersonal Competence	7.5000	Satisfied	4.0
Loyalty	7.3333	Satisfied	7.0
Consistency of Performance	7.3333	Satisfied	7.0
Hardworking	7.3333	Satisfied	7.0
Professional & Trustworthy	7.1667	Satisfied	9.5
Initiative	7.1667	Satisfied	9.5
Communication Skills	7.0000	Satisfied	11.0

Table 5. Competencies of Graduates (Engine Ratings)

Category	Mean	Description	Rank
Hardworking	7.9000	Very Satisfied	1.0
Loyalty	7.8500	Very Satisfied	2.0
Honesty	7.8000	Very Satisfied	3.0
Consistency of Performance	7.7000	Satisfied	4.0
Discipline	7.6000	Satisfied	5.0
Leadership Skills	7.5000	Satisfied	6.0
Interpersonal Competence	7.4000	Satisfied	7.0
Social Responsibility	7.3500	Satisfied	9.0
Initiative	7.3500	Satisfied	9.0
Professional & Trustworthy	7.3500	Satisfied	9.0
Communication Skills	7.2000	Satisfied	11.0

The stakeholders' satisfaction on Table 6 revealed the stakeholders' satisfaction of engine ratings as 'very positive' on loyalty, while the following competencies such as: (a) initiative, (b) discipline, (c) social responsibility, (d) honesty, (e) hardworking, (f) internal competence, (g) consistency of performance, (h) leadership skills, (i) communication skills, and (j) professional and trustworthy were "fairly positive." Furthermore, the quantitative results revealed that "loyalty" ranked first. The "initiative" was next, followed by "discipline," "social responsibility" and "honesty." The last competency was "professional & trustworthy."

Table 6. Competencies of Graduates (Engine Cadets)

Category	Mean	Description	Rank
Loyalty	8.0000	Very Satisfied	1.0
Initiative	7.7500	Satisfied	2.0
Discipline	7.6000	Satisfied	3.0
Social Responsibility	7.5000	Satisfied	4.5
Honesty	7.5000	Satisfied	4.5
Hardworking	7.4500	Satisfied	6.0
Interpersonal Competence	7.4000	Satisfied	7.0
Consistency of Performance	7.3500	Satisfied	8.0
Leadership Skills	7.3000	Satisfied	9.0
Communication Skills	7.1500	Satisfied	10.5
Professional & Trustworthy	7.1500	Satisfied	10.5

As to the operational level, the data in Table 4 revealed that the level of stakeholders' satisfaction on competencies of graduates is "satisfied" but the first two competencies noted were (a) leadership skills and (b) discipline. The communication skill was the last competency based on the stakeholders' satisfaction of different shipping/manning companies.

The result on the stakeholders' satisfaction when the respondents were classified according to engine ratings. The first competency is "hardworking," followed by loyalty, and followed again by honesty with "very satisfied" according to the different shipping/manning managers, crewing managers, training directors/personnel. The other competencies that had "satisfied" level of satisfaction were (a) consistency of performance, (b) discipline, (c) leadership skills, (d) interpersonal competence, (e) social responsibility, (f) initiative, and (g) professional & trustworthy.

The stakeholders' satisfaction revealed the stakeholders' satisfaction of engine ratings as 'very positive' on loyalty, while the following competencies such as: (a) initiative, (b) discipline, (c) social responsibility, (d) honesty, (e) hardworking, (f) internal competence, (g) consistency of performance, (h) leadership skills, (i) communication skills, and (j) professional and trustworthy were "fairly positive." Furthermore, the quantitative results revealed that "loyalty" ranked first. The "initiative" was next, followed by

"discipline," "social responsibility" and "honesty." The last competency was "professional & trustworthy."

The "very positive to fairly positive" stakeholders' satisfaction indicates the quality of education and training demonstrated by graduates of maritime university. The results also signified the attainment of the objectives of the University to provide competent and qualified graduates to the global maritime world. The University needs to monitor their competitiveness in order to further improve the stakeholders' level of satisfaction as well as to remain competitive as a major supplier of seafarers in the global maritime market. Interview results were processed and the stakeholders' satisfaction was further reinforced by the statements derived from the interviewees. Based on the responses derived from the interview questions, the stakeholders' view the JBLFMU graduates' performance was perceived to be satisfactory. Table 7 highlights the responses to the interview questions.

Table 7. Answers to the Interview Questions on

Answers to the Interview Questions on Suggestions/Recommendations for future training/curriculum innovation to further equip JBLFMU-Molo Graduates

"Most seafarers have poor personal growth and had not initiated to strive for personal enhancement on board ship. Very few had become good leaders because most of them develop technical skills but not interpersonal competence. To this effect most of them as well became ineffective fathers and parents. How can school help/supplement this to develop their preparedness to become better officers and individuals. It is my great hope that the school can develop a curriculum that would tackle this problem. A class that would tackle family, life growth, understanding of how one needed this aspect in life to be better individuals and leaders."

"Best students in training not only in academics."

"Curriculum-core subject: philosophy, psychology, foreign language, sociology, and anthropology."

"ROTC for discipline and basic human actual inter-personal performance (Japanese & Koreans). You should encourage basic military training for cadets due to marine profession should not be included in the banning of mandatory ROTC for college students. Koreans are trained and required to serve in Korean military for 2 years. Japanese are rapidly nationalistic. So we have to train our cadets in this respect."

* The answers are not edited to capture authentic data and candid reactions/comments from the respondents.

Table 8

Answers to the Interview Questions on Comments/Remarks Regarding the Education and Training of JBLFMU-Molo Graduates Employed in the International Ships/Vessels.

"The seafarers and cadets engine that we presently employed who are graduates of JBLFMU-Molo have shown impressive technical competence and attitude. In view of this observation I perceived that education and training of these cadets received from the school is good."

"Shy, they should also interact or mingle with other marines."

"Give importance to referral, Good Cadets! Problem is communication."
"Loyalty is always the problem."
"No interaction, officers are OK....satisfied and got promotions."
"Generally they are OK...."
"Officers are OK"
"Fear to transfer to other companies"
"They are generally good. However as basis for their 3 years academic program, CMO 013 it is considered limited. That is why we have adopted 1 year experience scheme onboard of all types of vessels. So we have them retrained and re-educated in-house."
"Improve, refresh, and enhance computer-based assessment..."
"Communication is OK..."

---

* The answers are not edited to capture authentic data and candid reactions/comments from the respondents.

9 CONCLUSIONS

In response to the demands of global competition and the increasing use of knowledge to create products and services, institutions have been moving toward a form of assessment that organizes graduates into competitive individuals and dedicated members of the workforce. Thus, this study has been one of those strategies to measure academic excellence and institutional efficiency. The study yielded that most of the competencies of graduates were "satisfied" as assessed by the different crewing managers/ship owners/ training directors/ HRD heads in relation to the graduates' educational and work performances. Likewise, the respondents suggested a more careful look at the suggestions/recommendations to reveal an equally difficult challenge to the institution that has gained a good reputation in the field of seafaring industry.

10 RECOMMENDATIONS

Based on the findings of this study, the researchers arrived at the following recommendations:

In this vein, the university has to look into further personal enhancement of the graduates, curriculum, basic military training skills, technical competence, communication skills, and issue on loyalty that turn out better results as far as content, education, character building, and training are concerned. This way, stakeholders' suggestions can be addressed and graduate quality can be assured. Research design like this one must be continuously done to elicit issues and queries about an educational performance and thus feedbacks can be taken as challenges for more improved educational and institutional reforms. Direct and specific feedback scheme may be done to stimulate administrators and other stakeholders concerned to act, address lapses, and definitely improve institutional output.

REFERENCES

Australian Quality Council. (2002). Australian Business, Excellence Framework 2002, AQC, Sydney.

Bayley, S. (2001). 'Measuring customer satisfaction: comparing traditional and latent trait approaches using the Auditor-General's client survey,' Evaluation Journal of Australasia, vol. 1, no. 1, March 2001, pp. 8–18.

Brooks, M., Milne, C., and Johansson, K. (2002). Using Stakeholder Research in the Evaluation of Organizational Performance.

Chennell, A.F. et al. (2000). 'OPM: a system for organizational performance measurement', Proceedings of Performance Measurement: Past, Present and Future, University of Cambridge, Cambridge, 19–21 July.

Commonwealth Department of Finance and Administration (2000). The Outcomes & Outputs Framework: Guidance Document, Canberra, November.

Donaldson, T. and Preston, L. (1995). The Stakeholder Theory of the Corporation: Concepts, Evidence, and Implications. Academy of Management Review 20 (1):65-91.

Evan, W., & Freeman, R.E. (1993). A Stakeholder Theory of the Modern Corporation: Kantian Capitalism. In Ethical Theory and Business. Tom Beauchamp and Norm E Bowie. Englewood Cliffs, CA, Prentice Hall.

Fletcher, A., J. Guthrie, P. Steane, G. Roos and S Pike. (2003). Mapping stakeholder perceptions for a third sector organization. Journal of Intellectual Capital 4(4): 505 – 527.

Freeman, R. E. (1984). Strategic Management: A Stakeholder Approach. Boston: Pitman. The Seminal Work in Stakeholder Research.

John-Cramer, M & Berman, S. (2003). Reexamining the Concept of Stakeholder Management. Unfolding Stakeholder Thinking. Volume 2. London.

Jones, T. (1995). Instrumental Stakeholder Theory: A Synthesis of Ethics and Economics. Academy of Management Review 20 (2): 404-437.

Kaplan, R. & Norton, D. (1996). The Balanced Scorecard: translating strategy into action, Harvard Business School, Boston, Massachusetts.

Mitchell, R. K., B. R. Agle, and D.J. Wood. (1997). "Toward a Theory of Stakeholder Identification and Salience: Defining the Principle of Who and What really Counts." Academy of Management Review 22(4): 853 - 888.

Naisbitt, J. & Aburdene, P., Megatrends 2000: Ten New Directions for the 1990's, Avon Books: New York, 1990.

Newstrom, J.W. (2007). Organizational Behavior: Human Behavior at Work. McGraw Hill. Avenues of America, New York, USA.

Phillips, R. (2003). Stakeholder Theory and Organizational Ethics. San Francisco: Berrett-Koehler.

Rowley, T.J. & Moldoveanu, M. (2003). When Do Stakeholders Act? An Interest and Identity-Based Model of Stakeholder Mobilization. Academy of Management Review 28 (2): 204-219.

Savage, G. T., T. W. Nix, Whitehead and Blair. (1991). Strategies for assessing and managing organizational stakeholders. Academy of Management Executive 5.

Singer, P. & Daar, A. (2001). Harnessing Genomics and Biotechnology to Improve Global Health Energy. Science 294: 87-89.

Chapter 6

Baltic Sea Logistic and Transportation Problems

Baltic Sea Logistic and Transportation Problems
STCW, Maritime Education and Training (MET), Human Resources and Crew Manning, Maritime Policy,
Logistics and Economic Matters – Marine Navigation and Safety of Sea Transportation – Weintrit & Neumann (Eds)

"Ground Effect" Transport on the Baltic Sea

Z.T. Pagowski, K. Szafran & J. Kończak
Institute of Aviation, Warsaw, Poland

ABSTRACT: Ground effect technology is recognized from the late sixties. Problem with wide expanding of this technology named WIG (Wing in Ground effect craft (WIG)) or ekranoplane was connected with stability and control systems, aerodynamics of landing and take off also sea state. Different civil and military configurations were tested, but the first Ground Effect Craft named Airfish 8 produced by Airfoil Development Gmbh (AFD) was certified in late 90-ies by Germanischer Llyod (+100 A0 WIG – A , WH 0,5/1,5 EXP) . Currently, it is now owned and developed by Wigetworks of Singapore and testing of Hoverwing 20 is expected. Last developments in the aviation field indicates new role in maritime transport of WIGs The second generation of ground effect craft sizes up to 200 seats in the future is planned. Authors recommend to discuss proposition of new generation of WIGs on Baltic sea transport similar to effect introduced in European General Aviation by project of EPATS and SAT Rdmp.

1 "GROUND EFFECT"

"Ground effect" is known like aerodynamic phenomena observed when an airplane is landing, and pilots feel it from start of aviation like floating or landing on a cushion of air. First theoretical works was known from 1920 (A.E. Raymond Ground influence on Airfoils T.N. No 67 NACA 1921 or C. Wieselberger Wing resistance near the Ground T.M. no77. NACA 1922).

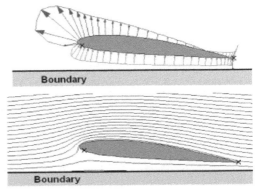

Figure 2 Surface Pressure and flow field in Ground Effect acc [8]

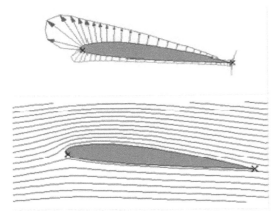

Figure 1. Surface Pressure and flow field out of Ground Effect acc [8]

Ground effect practically does not exist when a plane flies more than one wingspan above the surface and increase with speed of aircraft named WIG (Wing in Ground effect craft (WIG)) or ekranoplane. Ground effect technology is recognized well from the late sixties mainly in Germany (Hanno Fischer + Dr Lippisch developed the X-113) and Soviet Union (Rostislav Alexeiev and his all known Central Hydrofoil Design Bureau) with all known Caspian Monster or next generation WIG named the Lun for high-speed attacks. Idea of this type of aircraft is still attractive to commercial and military operators in few counries.

Source: http://www.combatreform.org/wig.htm
Figure 3. High speed attack ekranoplane Lun

The integration and competitiveness in the maritime sector technology named WIG (Wing in Ground effect craft (WIG)) or ekranoplane is promising idea. Different civil and military configurations were tested, but the first Ground Effect Craft named Airfish 8 produced by Airfoil Development Gmbh (AFD) was certified in late 90-ies by Germanischer Llyod (+100 A0 WIG − A , WH 0,5/1,5 EXP) . FS8 was produced under license of AFD by Flightship Ground Effect Pte, Cairns,Australia.

Currently, it is now owned and developed by Korean company Wingship Technology Corp. (WST Corp.) of Singapore. In October 2011 WST Corp. have finalized WIG craft WSH-500 Type A, with 47 seats + 3 crew. Tests confirmed 73knots in "Ground Effect" mode with good stability, however is not known level of future production.

Last developments in the aviation field indicates new role in maritime transport of WIG's The next generation of WIG's is planned in Russia (BE-2500), Korea (WSH 1500), USA (Boeing Pelican), China (Hubei type1) and Australia (Dragon clipper or Hoverflight 90) etc.

Source: AFD AirfoilDevelopment GmbH
Figure 1. Flightship 8

2 CONSOLIDATION OF TRANSPORT

The real internal market for all types of transport in Europe does not yet exist. The research concerning the relationship between maritime and other type of transport and their consolidation is needed to reduce

main emission factors like CO, HC, NOx, PM and improve economy of all type of transport, especially in Poland, where maritime passenger movements and is on very low position in Europe. Also direction of movements (by country of departure or destination) for purpose to Gdansk is limited only to Sweden.

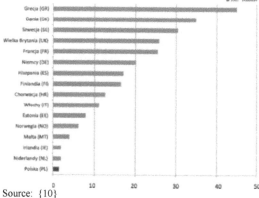

Source: {10}
Figure 2. Passenger movements in Europe (Poland is on last position)

Acc. preliminary analysis for Balticc sea by AFD Airfoil Development GmbH using WIG technology decreasing of time of trip and cost efficient maritime transport between main ports is expected.

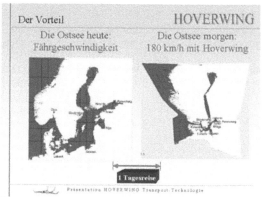

Source: Hanno Fisher
Figure 3. WIGs vision in Europe

Similar point of view focuses on the Highly Customer Oriented and Time, and Cost Efficient Air Transport System was prepared by the EPATS (European Personal Air Transportation System) and its continuation SAT Rdmp (Small Air Transport Roadmap) which considered "the wider use of small aircraft, served by small airports, to create access to more communities in less time" but was not compatible with WIG also. Deliverables of EPATS indicate that cumulative distribution of different type

222

of transport and aircraft is possible.It seems that additional compatibility testing of both types of transport are necessary with participation of experts from aviation and maritime transport, including big and small passenger traffic incorporating small airports and seaports. Actually the total number of passengers passing 1200 sea ports of Europe is estimated at level of 396 million every year and at Europe's airports - over 1.600 million and still grow, especially in large units, which concentrate 85% of traffic. EPATS or SAT Rdmp indicates that is possible to modify system of transport and flow of traffic reducing pilot workload and improving safety without increasing the acquisition cost thanks to small aircrafts. Like effect of these projects is consortium of few European partners with proposition of futuristic amphibious plane with novel type of propulsion but not using system of ground effect.

Source: EPATS
Figure 3. Hubs, regional and General Aviation airports

First calculation of type of WIGs on the Baltic sea needs formation of this type of transport vision for all European marine area. Below, taking under care special situation on Baltic sea in Institute of Aviation was prepared initial vision of WIG for special tourist or business trip with 4 and 50 seat WIG bus prepared by J.W. Kończak using classical calculations for this type of aircraft. This vision incorporate using of this type of WIGs like transport medium for number of person small business trip or patrol – rescue also special touristic season marine bus.

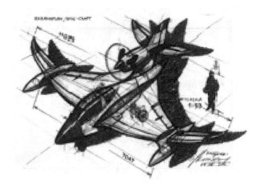

Source: J.M. Kończak
Figure .4. Vision of small 4 person WIG

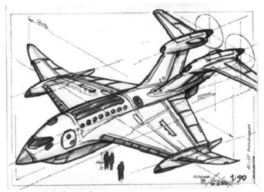

Source: J.M. Kończak
Figure 5. Vision of 50 person WIGbus

This vision is one of the possible WIGs now prepared by specialists, artists or producers of aircrafts in few options presented below.

Source: [4}
Figure 6. Russian BE-2500

Source: http://www.combatreform.org/wig.htm
Figure 7. Boeing Pelican

Source: [6]
Figure 8. Howerwing 20

Source: [5]
Figure 9. Ekranoplane-yacht

Source: [13}
Figure 10. Ekranoplane "Mobula"

3 CONCULSION

The Ground effect transport on the Baltic sea requires a new vision based on the development of the multidirectional point of view on marine and air transport including technical infrastructure and airplane and organizational solution also for small seaports.. The planned expansion of all type of transport will include probably analysis of existing transport systems and its strategies and interaction connected with analysis of futuristic concepts. Authors thinks that its paper will be next element stimulating discussion firstly in the Baltic region sea.

REFERENCES

[1] Climate change and the freight industry ,2012 http://www.fta.co.uk/policy_and_compliance/enviroment/logistics_carbon_reduction_scheme/climate_change_and_freight.html
[2] FS 40 Dragon Clipper http://www.export61.com.au/companydetail.asp?id=0003058406521063&pid=193&cid=80
[3] EPATS http://www.epats.eu/index.htm
[4] Evolution of the ekranoplane Be-2500 Neptun 2022 http://www.export61.com.au/companydetail.asp?id=0003058406521063&pid=193&cid=80
[5] Ekranoplane Ekranoplane- a yacht that can fly http://wordlesstech.com/2011/01/23/ekranoplane-a-yacht-that-can-fly-video/
[6] Getting more strategic lift http://www.combatreform.org/wig.htm
[7] Hanno Fisher ,Flightship 8, AFD Airfoil Development GmbH,2010
[8] Halloran M. O'Meara S. Wing in Ground Effect Craft Review, The Sir Lawrence Wackett Centre for Aerospace Design Technology Royal Melbourne Institute of Technology Contract Report CR-9802
[9] Hoverwing HW-20, Flyer operationHW-20, February 2011
[10] Jones Brett L. Experimental investigation into the aerodynamic ground effect of a tailless chevron-shaped ucav Ensign, USNR June 2005
[11] Maritime ports freight and passenger statistics http://epp.eurostat.ec.europa.eu/statistics_explained/index.php/Maritime_ports_freight_and_passenger_statistics
[12] Passenger Traffic Study http://ec.europa.eu/transport/modes/rail/studies/doc/2003_passenger_trafic_2010_2020_en.pdf
[13] SAT – Rdmp http://www.epats.eu/SATRdmp/index.htm
[14] Strange vehicles Mobula http://www.diseno-art.com/encyclopedia/strange_vehicles/mobula.html
[15] WSH 1500- specification data http://wingship.webs.com/wsh1500.htm

Baltic Sea Logistic and Transportation Problems
STCW, Maritime Education and Training (MET), Human Resources and Crew Manning, Maritime Policy,
Logistics and Economic Matters – Marine Navigation and Safety of Sea Transportation – Weintrit & Neumann (Eds)

Methods and Models to Optimize Functioning of Transport and Industrial Cluster in the Kaliningrad Region

L. Meyler, S. Moiseenko & S. Fursa
Baltic Fishing Fleet State Academy, Kaliningrad, Russia

ABSTRACT: The concept of economic development of the Kaliningrad region provides high rates of growth of gross regional product (GRP). Transport being interlinks between spheres of production and consumption has a significant impact on GRP by accelerating goods distribution activity and reduction in transportation costs. It can be achieved through the use of modern transport technologies, logistics, improve the technical level of vehicles and equipment, the quality of services. In this regard, organization and functioning of the transport and industrial cluster (TIC) is an objective factor in economic development of the region. TIC of the region can consist of industrial companies, enterprises of the transport sector, as well as of educational and financial institutions and government bodies. This paper presents a demo version of optimization of functioning of the transport and industrial cluster that can be considered as a simplified simulating model of the cluster operation.

1 INTRODUCTION

The Kaliningrad region is the most western region of the Russian Federation. It has no common borders with the rest territory of Russia but is connected by the Baltic Sea. The region borders with Lithuania (200 km) on the North and the East and with Poland (210 km) on the South. The length of the sea coast is 140 km. The area of the region is 15,100 km². Thus the region has a half-exclave geographical status (Figure 1).

Figure 1. Geographical location of the Kaliningrad region

The status of the Kaliningrad region is unique both in geographical and socio-economic aspects. It complicates interaction between the Kaliningrad region and other Russian regions and creates a lot of problems related to such status. On the other hand the Kaliningrad region has the westernmost position in Russia and therefore closeness to the industrial developed countries of Europe which can be potential market outlets and sources of investments. In this connection the economy of the region is more vulnerable to changes in the international economic environment. It was confirmed during the world economic crisis in 2008-2009. Figure 2 illustrates the dynamics of the GRP in 2007 - 2011 and the forecast of growth rate up to 2017 in accordance with the project of socio-economic development of the Kaliningrad region in the medium and long term.

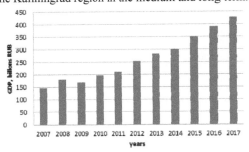

Figure 2. GRP dynamics in the Kaliningrad region

GRP growth will be caused primarily by development of production industries, including the automotive cluster, growth in construction, creating tourist and sports clusters in the region, the successful implementation of large investment projects. The base of the regional economy is production of vehicles, electronic and optical equipment and food products. It is planned to create several industrial zones in the Kaliningrad region in order to place newly industrial enterprises there. The forecast of total amount of investment in the economy of the region is at least 703 billion rubles in 2013-2017. The Kaliningrad region is the only free economic zone in Russia (the special economic zone since 1996) which is identical to its borders with the territory of the administrative region of the Russian Federation. The exceptional customs and tax privileges for the Kaliningrad region are objectively due to the fact that all industrial and trade enterprises in the region have additional transport costs associated with the need of cargo transit in the direction from Kaliningrad to the rest part of the Russian Federation through territories of Lithuania and Belarus. Also there are significant customs costs associated with long customs clearance procedures and the frequent downtime of cargo on the Lithuanian-Russian border.

2 TRANSPORT AND LOGISTICS COMPLEX OF THE KALININGRAD REGION

Transport and logistics complex of the Kaliningrad region plays an important role in the socio-economic development of the region (Kalashnik O. 2011). Stable and efficient functioning of the transport complex is essential condition for high rates of economic growth and effective integration of the region in the Russian and world economy. The transport complex combines rail, sea, river, air, road transport modes and applicable infrastructure facilities (roads, railway stations, ports and an airport, border crossing). Dynamics of the major indicators of cargo transportation in 2006-2011 is presented in Table 1.

Table 1. Cargo transportation in the Kaliningrad region

Mode of transport	Annual transportation, mln t					
	2006	2007	2008	2009	2010	2011
Railway transport	19,1	19,4	18,5	12,7	15,5	14,7
Road transport through the border crossings	1,99	2,19	2,16	1,88	2,22	2,3
Cargo handling by the port complex	15,2	15,6	15,4	12,4	13,9	13,4

Sections of the Kaliningrad railways are included in the two branches of the Trans-European Transport Corridors (Figure 3): №1A (Riga-Kaliningrad-Gdansk) of the route № 1 "Via-Baltica" (Helsinki – Tallinn – Riga – Kaunas – Warsaw) and № 9D

(Kaunas-Kaliningrad) of the route №9 (Kiev – Minsk – Vilnius – Kaunas – Klaipeda) cross the Kaliningrad region. More than 80% of goods are imported to the region by these routes. It gives an opportunity to attract freights between Europe and Russia and also through the Trans-Siberian Railway from the countries of Asia to Europe as well as freights of a corridor 'North – South' transported from ports of the Caspian Sea to the countries of Scandinavia and Northern Europe. It is possible to say that the Kaliningrad region is integrated in the European transport system taking into consideration some transport-technological aspects.

The combination of two kinds of the rail gauge: narrow (European) with 1435 mm wide and broad (Russian) with 1520 mm wide from Kaliningrad to the station Braniewo (Poland) and from Chernyakhovsk to the station Skandava (Poland) just makes the uniqueness of the railway.

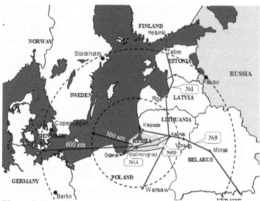

Figure 3. Branches of international transport corridors passing through the Kaliningrad region

Density of public railways in the Kaliningrad region is 9 times higher than on the average in Russia and the capacity is more than 25 million tons.

The regional road network has the length of 4,660.2 km and provides a transport link with the Baltic, St. Petersburg, Finland, and Poland, Germany and several other countries. The density of roads is 303 km per 1000 square km which is significantly higher than the same average parameter in Russia.

The "big port" of Kaliningrad consists of the Sea Commercial, Fishing, River ports and several bulk and container terminals located along the Kaliningrad sea canal. Distance from the port to the capitals of neighboring states - Vilnius, Riga, Minsk, Warsaw, Berlin, Copenhagen and Stockholm is 400 to 650 km, and to the biggest foreign ports on the Baltic Sea - from 400 to 700 km. Total design capacity of all terminals is 39.0 million tons of various types of cargo per year.

According to the materials of the IV Baltic Transport Forum (Materials...2012), existing resources of the regional transport complex are used at present in the following proportions: rail about 50%, port about 35%.

The Kaliningrad region has a system of internal waterways (bays, navigable rivers and canals). It allows to sail from the Baltic Sea to the Vistula Lagoon, along the Pregel and Dame rivers to the Curonian Lagoon, or passing it through the canals from the Dame river to the Neman river. These waterways of the region have access to the inland waterways in the neighboring countries.

3 PREREQUISITES FOR THE FORMATION OF REGIONAL TRANSPORT AND INDUSTRIAL CLUSTER

3.1 Cluster approach

The above information shows that the Kaliningrad region as a half-exclave territory has conditions to form the territorial-industrial complex, that is a group of interconnected in industrial relations and geographically concentrated enterprises. According to (Bulatova N. 2005) the high spatial intensity of industry at the limited territory favours that the structural elements of the complex are closely related to each other and transport as one of the elements has a special role for the maintenance of these relationships. A logistic support of enterprises of the territorial production complex will increase the efficiency of its elements and management of the complex. According to analysts, integration of industry and transport reduces delivery costs by 20 - 30%. It will lead to technological improvement of goods exchange, restructuring elements of goods distributional and transport networks, development of transport processes at a better level. Therefore it is necessary to form such an institution that is based on cooperation and interaction between industrial and transport sectors in the region and can be realized using the cluster approach. The cluster approach in the transport and industrial sectors can improve the functioning and development of transport and industrial and economic systems in the region by means of:

– reduce transaction and transportation costs due to geographical closeness of enterprises;
– providing better access to raw materials, components and services;
– increase in innovative activity;
– increase the level of training;
– enhanced the spread of knowledge and technologies;
– lobbying for common interests of companies of the cluster, etc.

The concept of a cluster appeared relatively recently in the economy. In spite of the impressive growth of research on this topic, a cluster theory still remains uncertain in terms of content and terminology. There is no generally accepted way to identify clusters and the very concept of 'cluster' is treated differently. In particular (Porter M. 1990), the cluster is considered as a group of geographically neighboring interconnected companies and associated organizations, acting in a certain area, characterized by community of activities and complementary to each other. The other cluster definition (Jacobs D. & De Man A. P. 1996) is given as a geographical or spatial association for economic activities which suggests horizontal and vertical relationships, the common technology use, the presence of 'core', sustainable collaboration. According to (Bergman, E. M. & Feser E. J. 1999) the cluster is a group of businesses and organizations for which this membership is an important element in enhancing the individual competitiveness. The cluster as a whole is completed with transactions of buying and selling, general technologies, customers, channels of distribution and labor markets. And one of recent definitions (Voynarenko M. 2003) is: the cluster is a territorial branch-wise voluntary association of business organizations that work closely with academic institutions and local authorities to improve the competitiveness of their products and growth in the region. The author mentioned above (Bulatova N. 2005) describes the clusters as geographically concentrated groups of interconnected companies, specialized suppliers, service providers, firms in related industries, as well as organizations connected to their activity.

3.2 Conditions for cluster realization in the Kaliningrad region

Transport as the serving industry strongly depends on the state of the industry in the Kaliningrad region (Meyler & Fursa 2010). TIC where transport acts as a link between local industrial and service companies can give a possibility to make the regional economy more effective and competitive (Stepanov A. & Titov A. 2008). Thus, it can be argued that the formation of clusters in the enclaves and exclaves territories is one of the most effective methods of economic development in these regions.

Above given classification of clusters allows to identify a group of companies which should be focused on determining or correcting the strategy for the region. All these definitions can be applied to the existing Kaliningrad region status. Which branches of the regional economy can be organized as a cluster? An analysis of regional specialization is used to determine the cluster of the Kaliningrad region with coefficients of jobs localization as criterion for assessment (Porter M. 2003). The basis of this approach is the historical experience of cluster formation, indicating that the cluster can form a company of basic industries in the region. The choice is due to the relative availability of information and the prevalence of the method. These coefficients reveal specialization of the regional branches of industry by comparing the regional industrial structure with such structure of the country. The branch of industry is considered as specialized one in the region when the coefficient is more than 1.0. The higher value of this coefficient corresponds to a greater concentration of the industry branch in the region. It is considered that the use of a cluster theory is irrational if the value of the coefficient is less than 0.75.

$$LQ_{EMP} = \frac{E_{ig}}{E_g} \Big/ \frac{E_i}{E} \quad (1)$$

where LQ_{EMP} = jobs localization coefficient of employment in the industry branch; E_{ig} = the number of employed people in the industry branch i in the region g; E_g = the total number of employed people in the region g; E_i = the number of employed people in the industry branch i in the country; and E = the total number of employed people in the country.

Values of jobs localization coefficient for such industry branches as agriculture and fishery, mining operations, production, building industry, trading and transport were calculated for the period of 2007 – 2010 (Meyler et al. 2012). Two of branches have the largest number of jobsites in the Kaliningrad region: production (1.12 ÷ 1.32) and transport (1.14 ÷ 1.24).

Also coefficient of enterprises and organizations localization LQ_{ENT} was studied for the same period. The coefficient LQ_{ENT} is an indicator of attractiveness some branches of the regional industry. The coefficient is defined as:

$$LQ_{ENT} = \frac{N_{ig}}{N_g} \Big/ \frac{N_i}{N} \quad (2)$$

where N_{ig} = the number of companies and organizations in the industry i in region g; N_g = the number of companies and organizations in the region g; N_i = the total number of enterprises and organizations in the industry i in the country; and N = total number of enterprises and organizations in the country.

The largest values of the coefficient have transport (1.49 ÷ 1.53), agriculture and fishery (1.35 ÷ 1.49) and production (1.26 ÷ 1.34).

It is possible also to study the dynamics of changes in financial results in order to evaluate the priority sectors of the economy for the formation of the cluster. One of the main indicators characterizing the operation of enterprises is the net financial result (profit minus loss) – the final financial result. Table 2 illustrates values of this indicator for 2007 – 2011 on data of official statistics (http://kaliningrad.gks.ru). Table 2 shows that enterprises of transport and production industries were the most susceptible to the 2008 financial crisis. The positive economic effect on the transport after the crisis appeared only in 2011.

Correlations between branches using the Pearson coefficient are found according to these data (Table 3). Calculation of the Pearson correlation coefficient suggests that the variables x and y are normally distributed. The coefficient ranges from -1 to 1 and the higher it is the more direct relationship between variables.

Table 2. Net financial result of enterprises in the Kaliningrad region

Economy branch	Annual net financial result, Rubles				
	2007	2008	2009	2010	2011
Agriculture	153.5	33.7	268.8	66.9	228.9
Fishery	-263.3	-292.7	1247.8	1178.6	2473.8
Production	1101.6	-6239.9	-698.1	4429.4	357.5
Trading	716.7	651.4	352.2	1023.2	1019.1
Building industry	422.3	399.5	415.7	545.3	196.5
Transport	631.4	-3147.6	-7557.9	105.1	1139.9
Total	11872	-1653	3074	22420	12410

In general, the formula for calculating the correlation coefficient is:

$$r_{xy} = \frac{\sum (x_i - \bar{x}) \cdot (y_i - \bar{y})}{\sqrt{\sum (x_i - \bar{x})^2 \cdot \sum (y_i - \bar{y})^2}} \quad (3)$$

where x_i = values taken by the variable x; y_i = values of the receiving variable y; $\bar{x}$ = average for x; and $\bar{y}$ = average for y.

Table 3. Pearson correlation coefficient

	Agri-culture	Fishery	Produc-tion	Trading	Building industry	Trans-port
Agriculture	1	0,59	0,20	-0,33	-0,51	-0,30
Fishery	0,59	1	0,42	0,39	-0,51	0,11
Production	0,20	0,42	1	0,50	0,30	0,44
Trading	-0,33	0,39	0,50	1	-0,16	0,89
Building industry	-0,51	-0,51	0,30	-0,16	1	-0,19
Transport	-0,30	0,11	0,44	0,89	-0,19	1

As a result there is a strong relationship between economic efficiency of transport and trading (r_{xy} = 0.89) and production (r_{xy} = 0.44) companies. Meanwhile, despite the strong dependence of the above branches still it is possible to observe the imperfect coherence between all transport stakeholders, starting with the main consumers of transport services - large industrial enterprises. Consequences of such situation are the losses of enterprises due to downtime of vehicles and products, increased transit time, loading and unloading, etc. There are more than 50,000 organizations in the above branches in the Kaliningrad region at the end of 2011 and some of them are closely connected to the transport and interested in joint development. One solution to the problem is the inclusion of all parties in the transport chain in the TIC of the Kaliningrad region.

4 APPROACHES TO THE CLUSTER DESIGN AND OPTIMIZING ITS OPERATION

4.1 *TIC is a complex production - economic system*

TIC is considered as a complex production - economic system. The effectiveness of such system depends on the level of its internal organization and the interaction with the environment. In this regard the current problem is the optimal design of the cluster and optimizing its operation. As a methodological basis of designing and organizing the cluster functioning a methodology of a systematic approach, systems theory, methods of operations research are used. The main tasks that need to be solved to ensure effective operation of the cluster are determined according to the objectives and main activities of enterprises. The complex of tasks consists of monitoring and analysis of the market, information management, optimization of transport service of production and marketing, optimization of inventory management, traffic management and production costs, investment and risk management (Moiseenko S. 2010).

A demo of TIC optimization is presented below. The proposed demo can be considered as a simplified simulation model of the cluster. The use of such models both on the design stage and during operation of the cluster will significantly increase the effectiveness of the design and management decisions of all elements and objects in the system and TIC as a whole. A cluster which combines enterprises of fishing industry and transport will be considered. At the first stage of its development the cluster includes: the core - the production of fish and main objects - port, means of transport (road, rail and sea), fishing fleet, barrel - tare factory, fishing gear factory, educational and research institutions. Schematic logistic diagram of TIC functioning includes the following processes: fish catch - transportation of raw materials to the port - handling of ships at the port - transportation of raw materials to fish processing factory - production of finished fish products - products transportation to consumers.

4.2 *Model of the optimization task of TIC activity*

Let's consider the problem of optimization of TIC in the fishing industry. The objective function of the cluster is focused on maximizing production of commodity products (or profit). Then other objects included in the cluster will organize its activities in accordance with the needs of production. Of course, it should not be an obstacle to solve the problems of the own development of the facilities as independent production - economic systems. Let's suppose that a fish processing factory can produce products of different assortment of different types of raw materials (fish). Also the basic technical and economic characteristics of the production lines, wholesale unit price of commodity outputs, as well as resources constraints are known. Then the mathematical model of the task of production optimization (Wagner G. 1973) of fish products on maximum commercial outputs in cost criterion can be written in a linear form.

The objective function:

$$\sum_i \sum_j C_{ij} X_{ij} = \max \tag{4}$$

with constraints:

$$\sum_i \sum_j K_{ij} X_{ijk} \leq \sum_i \sum_j Q_{ij} \tag{5}$$

$$\sum_i \sum_j X_{ij} \leq \sum_i \sum_j \Pi_{ij} \tag{6}$$

$$\sum_i \sum_j N_{ij} X_{ij} \leq T_p \tag{7}$$

$$X_{ij} \geq 0; i = 1,2,...I; j = 1,2,...J \tag{8}$$

where X_{ij} = value of fish products of i assortment made from j kind of fish; K_{ij} = coefficient of j kind of raw materials consumption for fish products of i assortment; Q_{ij} = amount of raw material; Π_{ij} = capacity of a technological processing line for fish products of i assortment made from j kind of fish; N_{ij} = time rate for fish products of i assortment made from j kind of fish; C_{ij} = wholesale price of a ton of commercial fish products of i assortment made from j kind of fish; and T_p = labor recourses of the factory, man-hour.

Let's assume that the initial data for the calculation of the optimal production plan are given in Table 4.

Table 4. Data for the calculation of the optimal production plan

X_{ij}, t	C_{ij}, thousands rubles/t	Π_{ij}, t/day	K_{ij}	Q_{ij}, t/day
X_1	20	60	1.0	
X_2	40	40	1,3	100
X_3	80	20	1,8	

229

It is possible to determine the optimal production plan for the next day of the fish factory work inserting data of Table 4 into the mathematical model and solving the task of linear programming. The optimal plan is as following $X_1 = 12$ t; $X_2 = 40$ t; $X_3 = 20$ t. The value of the objective function is 3440 thousands rubles. To ensure the uninterrupted work process of the factory the raw material should be delivered "just in time" which is not always feasible. The problem of stocks management arises in this connection.

4.3 Stocks management model

An important task of production is maintaining a reasonable stock of production resources or component parts to ensure continuous of the productive process. Traditionally (Mirotin L. 2002), the stock is regarded as inevitable costs. It leads to costly shutdowns of production when the stock level is too low and to "necrosis" of capital when it is too high. The problem of stocks management is to determine the level that balances the two mentioned extreme cases.

An important factor (Mukhin V. 2002; Sulitskiy V. 2004) in determining the formulation and solution of the stocks management problem is that the demand for stored reserves (per unit time) can be either deterministic (fairly well-known), or the probabilistic one (described by a probability distribution). In case of fish products the demand and constant supplies of raw materials consumed uniformly, because production capacity is constant. Thus, the deterministic static model of stocks management will be used, that is each next order is made at regular intervals time. Also it is important for interaction with the fleet in the fishery (Moiseenko S., Meyler L. & Semenkov V. 2011), because it is necessary to synchronize the supply of goods to the coastal fish-production factory with an optimal plan of storage and release of fish products (Moiseenko S. & Meyler L. 2011). Such task is important taking into account interaction between a maritime transport ship and the factory in the frame of this example of TIC.

The simplest model of optimization (Bellman R. and Kalaba R. 1969) of the current stock of raw materials is considered here. It allows to improve efficiency of the fish processing factory and commercial enterprises selling fish products. The model is designed in the following situation: a fish processing factory plans to manufacture a certain range of products within a fixed period of time (one month, for instance). This process requires a certain amount of fish raw material. It is necessary to simulate the factory operation in a way that will minimize the total cost of raw materials storage. We used the following initial conditions in this model:

- one product or one product group stocks are planned only;
- stocks level is reduced uniformly according to the manufacturing of products in accordance with the plan;
- the demand in the planning period is fully defined in advance;
- costs of stocks management consist of the costs of delivery and storage only.

Let's assume that total costs depend on the value of one delivery which is denoted as q. Thus, the optimal stocks management reduces to finding the optimum value of one delivery is q_0. Then other parameters of the model, namely: the number of deliveries n_0, the optimal time interval t_{so} between two consecutive deliveries, minimal (theoretical) total cost Q_0 can be calculated. The following designations for the predetermined model parameters are introduced: T = full time period for which the model is designed; R = the whole volume (the total demand) of raw materials for time T; C_1 = storage cost of one unit of raw materials per unit of time; C_s = costs of a raw materials lot delivery.

Let's denote the total cost of stocks is denoted as Q that is the objective function. The task of modeling consists in designing the objective function $Q = Q(q)$. The total cost will consist of the costs of delivery and storage of goods. The known formulae (Wagner G. 1973) are used for calculations.

Total costs of the storage of the current stock will be equal to $C_1 T q / 2$, i.e. cost for storage per goods unit - the "average" current stock product. As stated above, the stock level is reduced uniformly by uniformly produced sales, i.e. if at the initial moment of the stock it is equal to q, then at the end of the period of time t_s, it was equal to 0. Total costs for the delivery of the goods will be equal to: $C_s R / q$, that is a product cost of one lot of goods delivery to the number of deliveries $n = R/q$. Then the total cost of current stocks management will be:

$$Q = Q_1 + Q_s = \frac{C_1 T q}{2} + \frac{C_s R}{q} \to \min \qquad (9)$$

i.e. the objective function Q is a nonlinear function of q, changing in the range from 0 to R and trends to minimize. The solution of the task proceeds by well-known scheme.

$$Q'(q) = 0; \quad q_0 = \sqrt{\frac{2C_s R}{TC_1}} ; \qquad (10)$$

$$Q''(q) = \frac{2C_s R}{q^3}; \quad Q''(q_0) > 0; \qquad (11)$$

Thus, the optimal size of one delivery is:

230

$$q_0 = \sqrt{\frac{2C_s R}{C_1 T}} \; ; \tag{12}$$

The optimal average current stock:

$$\frac{q_0}{2} = \sqrt{\frac{C_s R}{4C_1 T}} \; ; \tag{13}$$

The optimal number of deliveries:

$$n_0 = \frac{R}{q_0} = \sqrt{\frac{C_1 RT}{2C_s}} \; ; \tag{14}$$

The optimal interval between two deliveries:

$$t_{s0} = \frac{T}{n_0} = \sqrt{\frac{2C_s T}{C_1 R}} \; ; \tag{15}$$

Optimal (theoretical) costs are:

$$Q_0 = \sqrt{2C_1 C_s RT} \; ; \tag{16}$$

Let's assume that the fish processing factory plans to deliver and to process fish with the total value R = 50,000 tons per a year (T = 12 months). The cost of one raw materials lot delivery C_s = 650,000 USD, and the storage of one ton of fish costs C_1 = 320 USD per a year. The value of Q can be calculated according to the above formulae. Thus, the optimal size of a delivery $q \approx 4500$ t; the optimal number of deliveries $n_0 \approx 11$; the optimal interval between two deliveries $t_{s0} \approx 33$ days; optimal costs $Q_0 \approx 4\,560,700$ USD.

It should be noted that the conditions of this problem are in mainly the idealized ones. In practice it is not always possible to adhere to the theoretical parameters of the stocks management model. For example, in the considered problem, it is found that the optimal size of delivery is about 4,500 tons. But it may be that the total required amount of raw material is not satisfied in the case of a failed fish catching. It means that the optimal size of delivery has to be changed. Therefore it is important to define such limits of a change that does not lead to a significant increase in total costs.

4.4 Organizing delivery of finished products to consumers

It is also important to solve the task of organizing delivery of finished products to consumers, for instance, within the city or the region in the frames of interactions between road or rail transport and the factory in the above mentioned TIC. Using a solution of the known "a commercial traveler's task", as the objective function can be considered to minimize the route network or minimizing service time of clients/consumers.

For example, it is considered the network shown in Table 5, where the number 1 is the factory and numbers 2 - 5 are consumers. The distance between destinations (it is given in each cell, km), the average speed of traffic (given in each cell in parentheses, km/hour) on each network segment, i.e. network nodes and time of loading and unloading operations (it is taken 30 min for each client) are known.

Using an algorithm for solving the "a commercial traveler's task" it is possible to find the optimal route of the consumer service according to criteria of minimizing the distance traveled in the network and/or time.

Table 5. Matrix of network: the factory - consumers

	1	2	3	4	5
1		8 (40)	9 (50)	6 (30)	6 (40)
2	8 (40)		7 (30)	11 (50)	6 (40)
3	9 (50)	7 (30)		9 (40)	10 (50)
4	6 (30)	11 (50)	9 (40)		11 (50)
5	6 (40)	6 (40)	10 (50)	11 (50)	

Thus for the suggested example of the network the best route through the points 1-5-2-3-4-1 by minimizing the traveled distance is 34 km and time of customer service is 194 min. The route through points 1-2-3-5-4-1 by minimizing service time will take service time of 181 min and the distance will be 49 km.

The decision regarding the choice of the route will be taken by a supplier.

5 CONCLUSION

For enterprises of the Kaliningrad region it is actual to solve transportation problems associated with both delivery of raw materials and components and export of finished goods from the region. It is necessary to consider the fact that the region has a unique location as a semi-exclave of the Russian Federation which has no common borders with the rest territory of the country.

Statistic data show an increase in industrial production in the region and its investment attractiveness in recent years. In this regard establishment of a regional transport and industrial cluster gives opportunities for joint activities of transport and production.

The proposed demo as one of the options for the transport and industrial cluster can be considered as a simplified simulation model for its optimal functioning.

The use of such models both at the design stage and during operation of the cluster will significantly increase the efficiency of the design/management decisions of all elements/objects of the transport and industrial cluster as a whole. This model has been implemented in the development of the project of diversification of the Kaliningrad maritime fishing port and interaction with its associated companies of sea, road and rail transport.

REFERENCES

Bellman R. & Kalaba R. 1969. *Dynamic programming and modern control theory.* Moscow: Nauka. (in Russian)

Bergman, E. M. & Feser E. J. 1999. *Industrial and Regional Clusters: Concepts and Comparative Applications.* Web Book in Regional Science. (Regional Research Institute, West Virginia University).

Bulatova N. 2005. *Formation of a regional industrial and transport complex of a cluster type.* Saint-Petersburg: Publishing house of the Polytechnic University. (in Russian).

Jacobs D. & De Man A. P. 1996. Clusters, industrial policy and firm strategy: a menu approach. *Technology Analysis and Strategic Management* 8: 425 - 437.

Kalashnik O. 2011. Kaliningrad region transit potential and its place in the transport system of Europe: Advantages and Challenges. *Maritime logistics in the global economy. Current trend and approaches, Proc. of the Hamburg intern. conf. on logistics (HICL 2011), Hamburg, September 2011*: 313 – 326. Hamburg: E. Schmidt Verlag GmbH & Co.

Materials of the IV Baltic Transport Forum. Kaliningrad, September 2012. http://www.baltic.konfer.ru/

Meyler L. & Fursa S. 2010. The cluster theory of economic development and its application in the regional port complex. *Maritime Industry, Transport and Logistics in the Baltic Sea Region States: Proc. 8th intern. conf.: Kaliningrad, 27-29 June 2010*: 201 – 206. BFFSA Publ. house. (in Russian).

Meyler L., Moiseenko S. & Volkogon V. 2011. Diversification of the seaport activity as an effective post-crisis strategy. *Proc. Europ. conf. on shipping, intermodalism and ports. Maritime transport: Opportunities and threats in the post-crisis world, Chios, Greece, June 2011*: Rep.02-03.

Meyler L., Moiseenko S. & Fursa S. 2012. Current problems of maritime transport complex development in the Kaliningrad region. *WCTRS SIG2 Conference at the University of Antwerp/TPR, 21-22 May 2012.*

Mirotin L. 2002. *Logistics: management of cargo transport - logistical systems.* Moscow: Jurist. (in Russian)

Moiseenko S. 2010. *Designing transport and logistics systems.* Kaliningrad: BFFSA Publ. house. (in Russian).,

Moiseenko S. & Meyler L. 2011. Optimal management of fleet relocation at deep-sea fishing grounds. *Proc. 12-th Annual Gen. Assembly .IAMU AGA12. (Green ships, Eco shipping, Clean seas), Gdynia, Poland, June 2011*: 197-208

Moiseenko S., Meyler L. & Semenkov V. 2011. Organization of fishing fleet transport service at ocean fishing grounds. *Maritime logistics in the global economy. Current trend and approaches, Proc. of the Hamburg intern. conf. on logistics (HICL 2011), Hamburg, September 2011*: 193-203. Hamburg: E. Schmidt Verlag GmbH & Co.

Mukhin V. 2002. *Study of management systems. Analysis and synthesis of management systems.* Moscow: Ekzamen. (in Russian)

Porter M. 1990. *Competitive Advantage of Nations.* New York: Free Press.

Porter M. 2003, The Economic Performance of Regions. *Regional Studies.* 37: 549 - 578.

Sulitskiy V. 2004. *Methods of statistical analysis in the management.* Moscow: Delo. (in Russian)

Socio-economic development of the Kaliningrad region in the medium and long term. Project. 2012. http://economy.gov39.ru/ (in Russian).

Stepanov A. & Titov A. 2008. Theory and practice of port clusters development, *Exploitation of maritime transport* 2(52): 11-14 (in Russian).

Voynarenko M. 2003. Cluster technology in business development, integration and investment attraction. http://www.unece.org/fileadmin/DAM/ie/wp8/documents/voynarenko.pdf).

Wagner G. 1973. *The fundamentals of operations research,* Moscow: Mir. (in Russian).

Baltic Sea Logistic and Transportation Problems
STCW, Maritime Education and Training (MET), Human Resources and Crew Manning, Maritime Policy,
Logistics and Economic Matters – Marine Navigation and Safety of Sea Transportation – Weintrit & Neumann (Eds)

Safety Culture in the Baltic Sea: A study of Maritime Safety, Safety Culture and Working Conditions Aboard Vessels

F. Hjorth
Kalmar Maritime Academy, Linnaeus University, Sweden

ABSTRACT: This paper explores the safety culture aboard vessels trading in the Baltic Sea. The common denominator for vessels trading in the Baltic Sea is frequent port visits, coastal voyages, voyages in dense trafficked areas and small crews. In a safety culture values, attitudes, competence and behavioral patterns are four important factors. The maritime safety management system aims towards a proactive and evolutionally safety culture. This study has an ethnographic perspective and totally eleven vessels participated in the study. Collected data has been analysed through a model and the study reveals that there is a need to discuss and change the safety culture, in large as on the single vessel in shipping today. As well as a need for further studies of how the safety culture can be improved and a need for education aimed towards system thinking, organizational theory and safety culture. The study was funded by the Swedish Mercantile Marine Foundation.

1 INTRODUCTION

Ship management is a complex business; many parameters combine to allow the operations to function satisfactorily, from maintenance of the vessel, equipment, organisation aboard and ashore to financial performance of the vessel. All this together forms a socio-technical system which covers the whole spectra of shipping and maritime operations. It is therefore vital to use a socio-technical perspective to deal with both small and large areas and to explain the different activities and their mutually coupling.

Several different studies have attempted to shed a light on different perspectives and show problems which affects short sea shipping. According to the UK Marine Accident Investigation Branch, (MAIB, 2004) a lack of proper lookout and tight working schedules is not unusual in short sea shipping and goes even further and implies that there is a systematic error in shipping. Furthermore the Swedish and Norwegian Maritime Administrations have revealed similar problems as the MAIB, fatigue and lack of a lookout can be a contributing factor in the majority of the accidents that occur in short sea shipping and that working hours at several occasions exceeds 90 hours per week and that they are not recorded according to the rules (Lindquist, 2003, 2005; Sjöfartsdirektoratet, 2004). Even the report

Seafarers fatigue (SIRC, 2006) reveals that the records of workinghours are not according to the truly worked hours and that the records are not completed according to the rules. Further a study in Sweden shows short sleep periods such as 6 hours for watchkeepers at sea on most ships in the study. Sleep periods which leads to continuous lack of sleep, which ultimately can cause decreased alertness and fatigue, thereby worsening the maritime safety. (Lützhöft et al, 2007)

Vital parameters that define a safety culture are values, attitudes, skills and behaviour, (Reason, 1997). Based on the problems mentioned above there is a need to examine the safety culture in the whole industry as well aboard the vessels. That is since the culture aboard affects how the individual crew member is acting as well as the room for manoeuvre for the individual. Culture aboard is a shared culture that the crew members are involved in building, maintaining and also forced to relate to. It is built on old inhabited patterns that meets the new technology and new requirements such as the International Safety Management System (ISM) or newly implemented administrative obligations imposed by authorities, shipping companies and charterers. That several studies reveals shortcomings in watchkeeping, tight workhour schedules and workhour schedules not kept truthfully reveals a shortcoming in the state of values, attitudes, skills and behaviour patterns in the short sea shipping. It is a sign of deficiencies in the safety culture that prevails aboard and in shipping as a whole.

1.1 Research Question

The introduction of the ISM shall, according to the Code, seek to support and encourage the creation of a safety culture in shipping and further pointing out that the attitude, values and beliefs as three important aspects of promoting a good safety culture. According to the guidelines for ISM shall the certifying administration measure the effectiveness of the safety management system aboard the vessel and in the company (IMO, 2010, 2.1.14).

ISM thus points to the importance of a good safety culture and demonstrates that it is the effectiveness of the safety management system that should be measured. Therefore to examine the safety culture in shipping on multiple socio-technical levels is of vital interest. If it is possible to define a safety culture and if so, how is the safety culture within the vessel and in the short sea shipping in the Baltic Sea?

1.2 Objectives

The study's main objective is to investigate safety in the Baltic Sea trade, the safety culture aboard the individual ship and of the individual crew member affected by the surrounding factors such as the shipping company, segment and maritime safety throughout. How does such culture affect safety aboard as well as the single crew member's perspectives on risk and safety in their work aboard? Even though the main focus is on the single vessel and the single crew member, the study also focuses on surrounding areas that affects the vessel and the crew. This is because the vessel and shipping should be viewed from a systems perspective, where the socio-technical system interacts to promote safe work.

2 METHOD

A study on vessels with a focus on safety culture aboard means that we are studying a part of a system. Shipping in general, as a global industry, could be viewed as a system in its entirety, where the system contains several components, subsystems, that both relates to each other and also acts with and against each other. The interaction and the relationship between people, systems and technology can be divided in mental-, social- and technical systems, (Moe, 1996). A further investigation of the system in a system theoretical perspective means we have to study how the subsystems interact with each other, how they take account of each other and where they adopt different roles in order to fulfil different functions. In a way we can see the system as a circle, where systems assume a circular embossed relationship where concepts are woven into each other and become dependent of each other.

A system, such as the maritime, separates to the surrounding but not more than events on the surrounding affects the system. In this way it is difficult to see what is causing events in a system, the events outside or in the system. System and subsystems becomes self-regulating, auto poetic, in its form, in which the system and its subsystems create their own identity and their own relationship to the environment, (Luhmann, 1995; Moe, 1996). Studying this requires, first, that we can generate a holistic approach but even a grasp of the parts. We need to use multiple tools to investigate the auto poetic system in its entirety. We must be able to both manage and analyse human actions and strategies and the organisation's actions against both to the individual and to other organisations in the system. Methodologically, we adapt based on where in the system and subsystems we are in our investigation. The human in the system is a part which adapts, influences and orients itself based on a systemic context. To understand its behaviour, we must adopt an interpretive approach, in accordance with Mead´s view in *Mind, Self, and Society,* (Mead, 1972).

In the ethnographic study, based on the anthropological discipline and method, the researcher spends a long time in the environment they want to explore and creates an opportunity for in-depth and more detailed descriptions of the observed pattern. The whole purpose of the ethnographic method is to create a so thick and full description of the culture under study, (Geertz, 1973). Traditionally there have been a number of studies, (Gouldner, 1954; Powell et al, 1971) in which ethnography has been used as a method to describe an organisations culture and in addition there are studies, (Weick & Sutcliffe, 2001) which also use an ethnographic method in control and identification of cultural elements which affects the safe and reliable operations. .

In terms of literature on safety culture studies the use of questionnaires are the dominant method of choice. Since this studies aim is to capture an image of the safety culture and at the same time present the complex system in which this culture are placed the statistic method of questionnaires is rather obsolete. The ethnographic method gives a much fuller account of the culture of an organisation than surveys can do, (Abbott, 2003). The big question is the validity of the ethnographic description, accuracy of the description is judged by the credibility of that description to insiders who live in the culture and, at the same time, to outsiders who are trying to understand it, (Schein, 2004).

Ethnographic method is much more than just being there, on site at exact that moment. Even though access to the site for the study is central for the success of the data collection phase, access isn't only physical access but much more important the access to the social subsidence on the site, (Hammersley & Atkinson, 2005). To be able to take part of the individual experiences of their situation and how they experience the safety culture I have used interviews and observations. Interviews linked to observations in different contexts have been used to form separate pictures of the same phenomena. With this method a wide range of material have been collected that on different forms describes the same context and event, which gives strength to the collected data.

The material collected consisted of scripts from interviews, observation and field notes during the visit to the eleven vessels. It was grouped together and read at first for each vessel, then for each profession and last grouped in themes according to the research themes. The interpretations are divided into three degrees, close to the hermeneutical spiral, where each reinterpretation turns the interpretation an extra turn and increases the complexity of interpretations, (Fangen, 2005).

2.1 Field studies

The data collection field work was done during the summer and fall of 2007. It was done in two main categories, visit aboard and interviews with shore based personnel. A total of 31 people were interviewed in the study, see Table 2.1.

Table 2.1. Interviews in the study

Aboard interviews		Shore based interviews	
Master	9	Inspectors	2
Mate	6	Company staff	5
Seafarer/Chef	2	Union staff	2
Seafarer/Engine	2		
Seafarer/Deck	2		
Chief Engineer	1		

Totally for the study 11 vessels did participate. Common for all vessels was that they sailed with two nautical officers aboard sharing the watchkeeping responsibility. In addition, manning varied depending on flag and trade. Normal manning was three seafarers in addition to the watchkeepers. The vessels were engaged in coastal trade with short voyages with up to two days between port visits. Only one of the vessels did have dedicated engine personnel aboard. Seafarers doubled up in responsibility for deck, catering or engine duties. Nationality of crews shifted after the flag on the ships visited. They consisted of people with

Swedish, Polish, Russian, Latvian, German, Finnish, Ukrainian, Filipino and Portuguese nationality. A clear picture of how international even coastal shipping in the Baltic Sea is.

3 SAFETY CULTURE, A MODEL FOR ANALYSIS

To work out a possible method to evaluate and analyze the safety culture is a multifaceted work based on that the shipping and the work aboard is influenced by a variety of factors. The safety culture provides a dual role, to prevent accidents with the ship and accidents with the single crew member and therefore the impact aboard are at different levels. Rasmussen (1997) suggests that the entire socio-technical system is involved in risk management, thereby affecting the safety culture.

Depending on the wherein the system, what level, we study it requires different methods and disciplines to succeed. Rasmussen's levels of the socio-technical system are (1) Authorities, (2) Rule Creation organizations, (3) the Company, (4) leadership, (5) staff and (6) The workplace with its technical equipment (Rasmussen, 1997). All of these components work together. Sometimes counteracts and thus affect the culture that arises and is created a workplace or in the wider sphere of its activities. Translated to the shipping and the present paper, we can distinguish three categories that affect the safety culture of the individual seafarer, (1) the system of shipping in general, (2) the segment, and (3) the ship, see Figure 3.1.

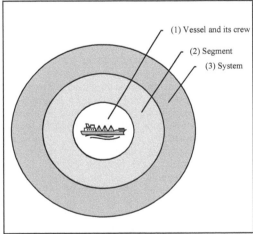

Figure 3.1. Levels for analysis (Hjorth, 2012)

Based on these three categories and Rasmussen's levels different parts of the system affects in different ways, it is first and foremost (1) the ship and its crew, the safety culture that is accepted on board and where the crew together, collectively agree on what is accepted and not accepted. This is done based on several aspects and from different angles, depending on the individuals in the crew and the ship's physical characteristics. An older experienced crew member is likely to affect more than a young inexperienced, the experience and knowledge of the tacit communication are important. The safety culture aboard is influenced not only by the crew and their shared agreements but also by the surrounding systems, which are in turn influenced by other surrounding systems.

The ship's crew is also affected by (2) the segment the vessel operates within. Is it a vessel in short sea shipping or in ocean shipping, is it a dry cargo or a tanker etc. These conditions affect how the work is carried out aboard. Typical segment-specific characteristics are crew size, type of ship, the type of trade and type of cargo. Clearly, the crew sizes are a strong influence on safety culture, but also the differences in ship management's influence on the crew and ship. It can be the difference between a large shipping company with a large shore organization and a small shipping company in which the shore organization is minimal or sometimes non-existent.

As a holistic perspective is all (3) shipping in the background. The impact may be from the special conditions which shipping operates within, regardless of sub-sector, flags, ship or trade. There is an international regulatory framework, the ship is outside the direct physical reachability when traveling between ports, the ship is constantly moving in order to carry out its mission, the crews come and go and we have a tradition inherited by the particular conditions with floating workplaces and homes wide apart. In addition to these three levels are the whole global community and influence from there as well but it is so differentiated and diverse that it is difficult to substantiate on the basis of the individual ship.

With the themes finalized as shown in Figure 3.1, System - Segment – Ships we need a model, an analytical model to analyze the themes. A possible model for further work will be to analyze the themes based on Reason's four elements vital for a safety culture, (1) reporting culture, (2) just culture, (3) flexible culture and (4) learning culture and the Westrum´s expanded classification affecting the safety culture, (Reason, 1997; Westrum, 1992).

Since the study is based on the crew's experience of the phenomenon of safety culture the focus will be how the crew is affected and adapts depending on the outsider's influence on the work aboard. Reasons four elements translated in to Westrum extended model provides the analytical factors. To create a reporting culture it needs information, feedback and that liability issues are resolved. A just culture requires clarity regarding responsibility and

cooperation, for the culture to be seen as flexible it needs feedback and collaboration and finally for to culture to be a learning culture it needs feedback, follow-up and development, (Reason, 1997; Dekker, 2007).

Further it is on the classes Westrum and Reason worked out the discussion will be based and further relate the safety culture to. In the model there are six factors that lay the base for the analysis in the result, how information is shared, feedback is passed on, how responsibility is shared, co-operation is dealt with, the process involving follow-up and how development is handled.

Table 3.1. Model for analysis based on Westrum, (1992) and Reason, (1997) (Hjorth, 2012)

Analysis Factors	Pathologic	Reactive	Bureaucratic
Information	Information is hidden	Information is misunderstood	Information may be ignored
Feedback	Messengers are "shot"	Messengers are ignored	Messengers are tolerated
Responsibility	Responsibility are shirked	Responsibility is personal	Responsibility is compartmentalized
Co-operation	Bridging is discouraged	Bridging when demanded	Bridging is allowed but neglected
Follow-up	Failure is covered up	Failure is punished	Organization is just and merciful
Development	New ideas are actively crushed	New ideas are ignored	New ideas create problems

Continuation of table 3.1: Model for analysis based on Westrum, (1992) and Reason, (1997) (Hjorth, 2012)

Analysis Factors	Proactive	Generative
Information	Information is shared	Information is actively sought
Feedback	Messengers are controlled	Messengers are trained
Responsibility	Responsibility is controlled	Responsibilities are shared
Co-operation	Bridging is natural	Bridging is rewarded
Follow-up	Failure is investigated	Failure causes inquiry
Development	New ideas are shared	New ideas are welcomed

The five different safety cultures behave in different ways. The pathological culture does not care about safety, so long as we do not get caught. This means that everything is done with a focus on everything to be hidden away and all new is potentially dangerous. In the reactive safety culture safety is important, and as soon as an accident occurs, very much is done. A work that seeks to protect the organization from environmental influences is started and the person at fault is hung out to demonstrate decisiveness, and new rules are developed to prevent a similar accident to happen again. We use our experience to determine our future behavior.

At the next level of culture, the bureaucratic, security monitoring system has been developed to handle all hazards postulated to occur. This means that our own horizon is the limit of what we can anticipate. The system is important and all information and events that fit into the system, as has been predicted to take place can be treated. Events and information that does not fit into the system creates major problems and could be neglected.

The penultimate cultural level, proactive, means that the safety management system is so constructed that it contains elements which try to capture potential safety problems before they occur. For example, by actively process the incident and deviation reports, and to share information. The last and highest safety cultural level, the generative, means that the entire organization is permeated by a safety think, which goes beyond just the safety management system. Vulnerabilities are searched actively in all operations in order to find them before they occur. All events are processed and investigated in order to obtain more reliable operations; safety permeates the entire organization and comes before anything else. Everything, except safety, is up for discussion aiming towards continuous improvement.

A clear distinction between the different cultural levels is that the confidence in the individual and the organization increases the further up in the stair we go. In the lower levels, pathological, reactive and bureaucratic, the individual and the organization are monitored, controlled and have a little confidence. Threats of retaliation are used, and different systems to control behavior and that individual do what is expected of them. At the highest level the individual and the organization is relied up on with the confidence and trust that they have a motivation to work mature and trustworthy with the safety system.

4 RESULTS

4.1 Safety culture within the system of shipping

Shipping as a system places itself on the border between the pathological and the bureaucratic state, i.e. the reactive stage. Most would surely oppose it, especially due to the spirit of the ISM. And it is also correct, in terms of ISM and its regulatory framework fits better into the development stage. The problem with ISM is that it has only scratched the surface. When it actually comes down to a critical examination, the ISM has probably not affected the shipping industry as much as was intended, yet. ISMs SMS manuals tend rather to become shelf warmers on the bookshelf than a natural part of the daily work, which makes it more impressive on paper than the vital part of a safe work patterns as it is meant to be.

The reactive categories are denoted by that information is misunderstood, messengers ignored, accountability is personal, bridging only when required, mistakes are punished and new ideas are ignored. The problems facing the system are (1) lack of implementation of regulations, (2) lack of clarity in how incidents are handled by the authorities in the country, (3) globalization that impede cooperation and monitoring, and (4) separation from the outside world. Globalized shipping where actors from different parts of the world come together around the same ship and the same cargo means increased difficulty in handling information. Based on that the various players have a difference in culture and handles information differently. Regardless of who is better or worse at handling information. Just that multiple players are involved, from different nations, cultures, organizations and organizational cultures, implies that the dissemination of information becomes more difficult. Information stops and can at best only be ignored or maybe simply misunderstood.

Furthermore, the interchange ability of flags and seafarers might halt the vital communication and information flow. The substitution creates a lack of continuity, an uncertainty on the status and unhealthy competition. Especially the risk of losing their job or losing ships from the flag can create a fear of making demands and thereby only compliance with regulations. The special working condition found in shipping, the ship physically isolated from the outside world for long periods, means that the ability to control and verify the work aboard difficult. Although the specific situation regarding work and leisure aboard creates special conditions, the difficult of distinguishing work and spare time aboard may create a fatigue, which can hinder the will and ability to maintain a high level of information flow, feedback and collaboration.

The allocation of responsibility is a key element in the construction of a safety management system, in which each participant in the system clearly must know when and how mistakes and accidents are handled. According to Dekker (2007) there arises a dilemma for the individual, with the threat of prosecution hanging over them for mistakes and accidents that have occurred. A dilemma if the incident should be reported or not. Here is the problem, no one can predict how blame is shared or punished. Only that it sometimes happens and sometimes not. Mistakes are investigated but the responsibility is avoided, ignored, or worse, pinched on individuals.

Also the extensive regulatory flora in shipping is another example that there is a problem. Only IMO has produced a large number of conventions, rules, procedures and guidelines. The large amount of rules, procedures, advice and guidance should generate safe shipping. However, most indicators show that this is not so. For example, the number of accidents over time remains at a similar level, there are years that are better or worse, (Anderson, 2005).

But over a longer period, there is no major change. The big change depends probably more on shifts in the economy and not on the regulatory impact. It is also within the IMO structure as one of the major problems lie. IMO is working to build up a minimum acceptable standard for all, which means that they set the bar for the lowest permissible level. This means that a company or a flag that meets this minimum level meets an acceptable level. IMO is also, normally, working after a consensus process in which the desire is that everyone should be in agreement before any decision is taken. Given that, this need not only be detrimental but can also be seen as an advantage. Since the agreements are supported by the whole maritime world.

4.2 Safety culture within the segment

The problems facing the segment is (1) the low crew levels, (2) the lack of further training of the crew, (3) lack of support from state organization and (4) lack of inspection. As with shipping in large the spirit of the ISM was to promote the proactive approach. Nevertheless there are apparent shortcomings that rather place the segment in a reactive stage. The impact the introduction of ISM has hade has been very limited, more of another regulation which is not considered necessary and also awkward to use.

There are both positive and negative opinions on flag- or port- state inspections. There are examples of different inspectors focus on different things, they focus more on technology than on other issues, and that inspectors sometimes do not choose to note points made by crew members. The phenomenon indicates that the information available is ignored or hidden, which in turn suggests a reactive way, where the messengers are ignored and not included.

Based on crew levels and its consequences, we see a similar behavior, information is hidden or ignored, no accountability, and the messengers ignored. Why this happens is probably because the issue on crew levels is too sensitive, it is a *hot potato*, which was shown in the discussions for the revision of the STCW in Manila. Among ship-owners and some flag states there were a massive resistance to create mandatory rules on criteria's to determine the manning level a ship should have, (IMO, 2010). This demonstrates a bureaucratic behavior, where the information is ignored, or reactive, because the information to some extent is misunderstood.

The inherited traditions in the form of, *we can handle ourselves* or *we do not need help* mentality in an increasingly turned up the tempo may mean a

high risk of accidents, where more should be done by fewer people in less time. This is because staffing steadily decline and that lying in port also steadily reduced, (Kahveci, 2001; Hjorth, 2008). In an industry and a segment where the visibility to outsiders is minimal, since the access to the production platform, the vessel is not possible for most of the running time; there is no control mechanism that can control work functions aboard when the ship is at sea.

Furthermore, we also see how the feedback and follow-up are difficult as the reporting of deviations and incidents are not always in accordance with the ISM or SMS. One of the most important points of the ISM is the continuous improvement in the organization. This improvement should be documented in order to be verified by inspecting agencies and organizations. Other studies, (Jense, 2009), reveals a tactical action with reporting, something that also appears in the present study. A tactical action is either to report a moderate amount, that doesn't stand out as too many of too few, or to use reporting as a mean for extortion against higher organization. Feedback and follow-up is thereby lost, hidden and information and cooperation takes place only at the requirements.

Collaboration is about to jointly create an atmosphere which enhances the activity, a bridge between organizations, ships, departments, and finally individuals that provides a natural workflow. Where, it is seen natural to share experiences and to develop a common strategy to work proactively. If we look at the report as part of the cooperation, it only works when conditions outside forces a reporting to take place. When that requirement is not found, there is no natural incident reporting, so far.

Collaboration involves both to learn from each other's mistakes, but also to work for the same objective. Inspections should be seen as part of this collaboration as they can detect flaws which the crew or the company itself do not see and reporting to be seen as something that shows the vitality of the system. Much of this, including the development of the ISM, comes from above, which is not always beneficial to the actual development aboard a ship or in the segment. In a way, in terms of the study of the segment, one example is the ISM and SMS, which aboard more or less is seen as a paper tiger, and thus not something to use when the actual problems are to be solved, whether it be physical problems with work or more internal procedural matters.

The development in the segment is hampered by the general level of education is undersupplied. This should not be misinterpreted that the seafarers aboard are bad sailors or omissions in their practical profession. Aboard there are highly competent professionals, but there is no general targeted training in new areas, such as organization and systems thinking to manage a complex system that

ISM and SMS correctly or training in resource management and leadership. Having served as master aboard for 20 to 30 years creates a vast array of experience to use in their daily professional lives. But it also means that there is a risk that new ideas, technology and work methods are not received in the right way. Expressions such as; *We do what we have always done,* or; *So have we always done it,* is frequent used aboard. Development is about to welcome new ideas, not crushing them or at best, just ignore them.

4.3 *Safety culture within the vessel*

From the population being studied it is impossible to definitively declare a state of all vessels in the segment. This study on vessels can only be seen as an example. Nevertheless, the study shows that even vessels, as individuals, position itself in the reactive level. This is on the basis that (1) information available is ignored; (2) feedback tolerated but often ignored, (3) personal accountability, (4) cooperation only when required, (5) mistakes punished and (6) new ideas are ignored. On board the ship there is a significant lack of information. Especially notable is that the information is hidden, misunderstood or ignored. Information is rarely shared or actively sought, except when demanded from the outside, for example, during inspections. The information is there but will not be passed on in the system or if it is passed on stops at the next level in the system. There are several examples on how the crew aboard have tried to pass on information to shore based managers in the shipping companies regarding faulty equipment. It was clear how the information was ignored regarding the need for repair or exchange of the equipment at the shipping company. In these cases the crew used inspections as a mean to put pressure on the shipping company to overhaul the faulty equipment.

Furthermore, information about how the crew work on board is available to all parties, both the shipping office, inspection, whether it is inspection by authorities or classification societies and unions. The crew constantly and continually violate the required rest periods is not new news. In this case feedback from the crew aboard to the shipping company and other shore based organizations is ignored and the information hidden.

There is a certain mistrust between crews and those they communicate with on the shore side, which is not conducive to cooperation. The interviews maintained the important form of cooperation between crew and shore based personnel. The shortage occurs when information is not been taken seriously, misunderstood or ignored; this in turn inhibits cooperation when the crew needs support from its shore based colleagues at the most.

Watchkeeping aboard include a requirement for cooperation and accountability in the regulatory framework, but these are put aside aboard. There is a tacit agreement not to make contact with the master at night, although there is according to the regulations a must. This suggests that responsibility is avoided or neglected and that cooperation, bridging is resisted, not open officially but unofficially and in silence.

Risk is seen as a probability of physical injury caused by technical or other form of process, while the accident is what is the result when the risk passes from being likely to actual, (Beck, 1992). Constantly aboard there are risks, incidents and accidents. However, there is no direct discussion of the hazards or risks, a bit like Beck is on,

Where everything turns into a hazard, somehow nothing is dangerous anymore, (Beck, 1992).

Hazards are ever-present but currently they are not discussed. There is an unnoticed externalization of accidents and incidents, *it does not happen me, it happens only those botched* or the accident is so infantile stupid that it becomes ridiculous. This procedure allows the development aboard in preventing accidents to stop; or rather never get started. According to the interviewed the personal clumsiness cannot be stopped, with ISM system and its SMS manuals since the accident never depend on other factors than the purely personal clumsiness.

This externalization is on both personal accidents and ship accidents. Ship accidents always depend on something else, ice, the pilot, breakdown of machinery or just bad luck. The interviews showed that the accident never depends on the individual capacity, organization or factors that are modifiable aboard or in the shipping company. This suggests that mistakes are hidden, because they never come to the surface with the focus on external factors in accidents.

5 CONCLUSIONS

The study's main purpose to investigate the safety culture began with a question: Is it possible to define a safety culture? From my study and its results, the answer is, yes it is possible to define a culture of safety. The Westrum (1992) and of Reason (1997) extended model served as a starting point for the development of analytical factors and my own model, i.e., the analysis of factors which later was used to analyse the safety culture. The results of the study and the links given to previous studies with a similar theme show the model's usefulness well. The model has thus proved satisfactory. Now, let's remember that the study was not designed to test the model or models, instead trying to define the safety culture in a given area, with the help of a model. To really test a model fully requires extensive testing in different specific environments to really define its function, a more abstract modelling to test the model's function, (Abbott, 2003). The purpose was not that, instead using a model to define and explain the safety culture, which the analysis model could handle.

Barry Turner & Nick Pidgeon (1997) argues in the book *Man-Made Disasters* that accidents takes place by restricted the perceptions and beliefs of the organization, which is associated with which information is interpreted, how it is interpreted, by whom and where in the organization. Turner & Pidgeon calls this the *failure of foresight*, meaning that we have no foresight or improper planning due to restrictions in our perception. This is linked to the safety culture, which in its proactive and developmental level is proactive and forward-looking. In order to prevent the problems that might arise. If we compare this with the navigation that is proactive, i.e. to be able to navigate safely, we need to pre-build a mental model of the environment, how it can be developed based on the data we have to process. But we also need to pre-ensure that future measures are working, that they are both planned properly and that in the current situation can be used. This is a natural part of the navigation of the ship. What needs to happen is that this naturalness found in navigation of the ship is transferred to other parts, then more specifically the safety culture and how we work with the safety management system daily.

Failure of foresight is in link with Weick´s & Sutcliffe´s concept *collective mindfulness* one way to look at how we should work and provide further approach to the solution, (Weick & Sutcliffe, 2001). With *collective mindfulness* principles of performance and the ability of containment we work focused. Both proactively to process the data we have to work with, but also to contain the initial events, so they do not develop into major events beyond our control. But to take preventive and containment action requires a change of focus and awareness level of our work.

Reason´s parameters for the safety culture to function are attitude, skills, knowledge and behaviors of those who work with and in the safety culture (Reason, 1997). Without a high level of these parameters do the safety culture do not work in a proactive and developmental way. But this requires education and that everyone in the organization changes the focus. Instead of focusing on reacting to events in retrospect, we must learn to prevent incidents, which require a proactive individual, group and organization.

The results I have presented in this study clearly show that the safety culture is reactive. To lift it, primarily to a bureaucratic level and then be able to take it to the proactive and preferably all the way to development, requires that these parameters,

attitude, skills, knowledge and behavior is constantly evolving. It becomes clear from an examination of the ISM and its implementation, that there are many more factors that affect safety culture than only to introduce the ISM. There is a worn but useful expression, *commitment from the top*, as an ultimate guide to successfully implementing an advanced security management system that ISM actually is.

The adage *commitment from the top* means that it is in the overlying subsystems that one sets the standard of the underlying subsystem. Underlying subsystems will only follow what the overhead does. Is safety important, if it feels as if it is an honest and sincere approach and will to create a good safety culture. Then it will also be reflected in the underlying subsystems of the whole system - shipping.

REFERENCES

Abbott, A. (2003). Methods of discovery - heuristics for the social sciences. New York: Norton.

Anderson, P. (2005). ISM Code: A Practical Guide To The Legal And Insurance Implications, London: LLP.

Beck, U. (1992). Risk Society: Towards a New Modernity. London: Sage Publications.

Dekker, S. (2007). Just Culture - Balancing Safety and Accountability. Aldershot: Ashgate Pub Co.

Fangen, K.. (2005). Deltagande observationer (Participating observations). Malmö, Sweden: Liber ekonomi.

Geertz, C. (1973). Thick Description: Toward an Interpretive Theory of Culture. In: The Interpretation of Cultures (pp. 3-30). New York: Basic Books.

Gouldner, A.W. (1954). Patterns of Industrial Bureaucracy. New York: Free Press.

Hammersley, M., & Atkinson, P. (2005). Ethnography. London: Routledge.

Hjorth, F. (2008). Arbetstider och arbetsvillkor ombord på tvånavigatörsfartyg: en studie av fartyg i Östersjöfart med enbart befälhavare och endestyrman som nautisk kompetens ombord (Working hours and working conditions on board twonavigatorships: a study of vessels in the Baltic Sea trade with only the master and one officer with nautical skills board). Kalmar, Sweden: Högskolan i Kalmar.

Hjorth, F. (2012). Säkerhetskultur i Östersjöfart – En studie kring sjösäkerhet, säkerhetskultur och arbetsvillkor ombord på fartyg som trafikerar Östersjön. (Safety culture in the Baltic Sea trade - a study on maritime safety, security and working conditions on board ships operating in the Baltic Sea). Växjö, Sweden: Linnaeus University.

IMO (2010). ISM Code and Guidelines on Implementation of the ISM Code 2010, London.

Jense, G. (2009). ISM-koden: den svenska erfarenheten i ett globalt sjösäkerhetsperspektiv, (ISM-Code: the Swedish experience in a global maritime perspective). Växjö, Sweden: Växjö University.

Kahveci, E. (2001). Fast Turnaround Ships and Their Impact on Crews. Cardiff: Seafarers International Research Centre (SIRC).

Lindquist, C. (2003). Studie av visst statistiskt material med anknytning till ett urval av kollisioner och grundstötningar i vilka trötthet/sömn har konstaterats ha haft stor eller helt avgörande betydelse (Study of some statistical data related to a variety of collisions and groundings in which fatigue /

sleep has been found to have had a major or critical importance). Norrköping, Sweden: Sjöfartsverket.

Lindquist, C. (2005). Studie av vissa arbetsförhållanden och arbetstider på ett antal mindre lastfartyg (Study of certain working conditions and working hours on a number of small cargo ships). Norrköping, Sweden: Sjöfartsverket.

Luhmann, N. (1995). Social systems. Stanford, Calif.: Stanford University Press.

Lützhöft, M., Thorslund, B., Kircher, A., & Gillberg, M. (2007). Fatigue at sea - A field study in Swedish shipping. Linköping, Sweden: VTI (the Swedish National Road and Transport Research Institute).

MAIB (2004). Bridge Watchkeeping Safety Study. Southampton: UK Marine Accident Investigation Branch.

Mead, George, H. (1972). Mind, Self, & Society. Chicago: University of Chicago Press.

Moe, S. (1996). Sociologisk betraktelse – En introduktion till systemteori, (Sociological reflection - An Introduction to Systems Theory). Lund, Sweden; Studentlitteratur.

Powell, P.I., Hale, M., Martin, J., Simon, M. (1971). 2000 Accidents: A Shop Floor Study of their Causes, Report 21. National Institute of Industrial Psychology, London.

Rasmussen, J. (1997). Risk management in dynamic society: A modelling problem. *Safety Science*, 27(2/3), 183-213.

Reason, J. (1997). Managing the Risks of Organizational Accidents. London: Ashgate.

Schein, E. H. (2004). Organizational Culture and Leadership 2:nd. ed. San Francisco: Jossey Bass.

Seafarer's International Research Center (SIRC) (2006). Seafarers Fatigue. The Cardiff Research Programme, SIRC: Cardiff.

Sjøfartsdirektoratet (2004). Hviletid og navigasjonsulykker på 2-vaktskip (Rest hours and naviagion accidents on Two watch vessels). (No. 2004/18098). Oslo.

Turner, B. A., & Pidgeon, N. (1997). Man-made Disasters. Oxford: Butterworth-Heinemann Ltd.

Weick, K. E., & Sutcliffe, K. M. (2001). Managing the Unexpected: assuring high performance in an age of complexity. San Francisco: Jossey-Bass.

Westrum, R. (1992). Cultures with Requisite Imagination, In Wise, J.A., Hopkin, V.D., Stager, P. (Eds.), Verification and validation of complex systems: human factors issues. Springer-Verlag: Berlin, s. 401 – 416.

Baltic Sea Logistic and Transportation Problems
STCW, Maritime Education and Training (MET), Human Resources and Crew Manning, Maritime Policy,
Logistics and Economic Matters – Marine Navigation and Safety of Sea Transportation – Weintrit & Neumann (Eds)

Development of Logistics Functions in the Baltic Sea Region Ports. Case Studies

M. Grzybowski
Gdynia Maritime University, Gdynia, Poland

ABSTRACT: The article presents the investments in the chosen ports of Baltic Sea Region, whose aim was to improve the competitiveness through the development of logistics services. The investments were carried out by the boards of directors of the ports and supported financially by governments and cities. The examples from Russia, Baltic Sea Region countries, Scandinavia, Germany and Poland were presented. In order to develop the logistics functions of the regions, new ports and terminals are being constructed.

1 INTRODUCTION

Baltic Sea Region belongs to the fast developing areas. The rate was slowed down only by the recession in years 2008-2009. However, in 2008 the number of containers reloaded was record-breaking in many ports. After the decrease in 2009, the market restored in 2010-2012. For example the number of containers reloaded in DCT Gdańsk increased repeatedly thanks to the enlargement of the terminal and launching the connection with The Far East[19].

One of the basic causes of this occurrence was a need to prepare ports for competition challenge on the sea transport market. The analysis of logistic investments' market in the ports of Baltic Sea Region, based on the source materials, leads to the conclusion that most of governing bodies of ports, regions and countries was prepared to the challenge. The process of development was comprehensive and long-term. Not only the superstructure and infrastructure of the ports was modernized, but also new terminals and ports were constructed, for example: DCT Gdańsk, UST Luga, LNG Świnoujście Terminal and Norvikudden nearby Stockholm. The special areas for technological parks and logistics centre are created close to the ports. European Union and public funds are used to develop the infrastructure leading to the ports,

improving the accessibility both from land and sea. The development of intermodal transport is supported as well[20].

2 LOGISTICS INVESTMENTS IN SELECTED RUSSIAN PORTS

A number of logistic centers is being created around St. Petersburg. Itella Logistics St. Petersburg of 10 000 sq meters area, was located in Utkina Zavod. The centre has good railway connection to Moscow and offers a vide span of logistic services: assembling cargo, storing and transport services. In 2010 Finnish company SRV launched new A class logistics centre, warehouses of 120 000 sq meters area. The center's buildings are 12.5 m high[21]. Ahlers Logistic Center in St. Petersburg was created in order to provide services for French Groupe SEB, the manufacturer of household equipment (Rowenta, Krups, Tefal, Moulinex). Groupe SEB imports products from China on European and Russian market. Because the latter noted lately 20% growth per year, it was necessary to open logistic centre near the port. The task was given to Ahlers Logistic Center, which owns 25 000 sq meters[22] warehouses.

[19] Grzybowski M. (2012). Strategie rozwoju portów morskich w konkurencyjnym otoczeniu portów regionu Morza Bałtyckiego. Studia przypadków Gdańska i Gdyni, "Logistyka" 2012, nr 2, ISSN 1231-5478, s. 643-652.

[20] Grzybowski M. (2012). *Rozwój połączeń intermodalnych z portami regionu Morza Bałtyckiego, Adriatyku i Morza Czarnego. Studia przypadków* [w:] *Transport morski w międzynarodowych procesach logistycznych*, pr. zb. pod red. H. Salmonowicza, Wyd. Zapol Szczecin 2012, ISBN 978-83-7518-418-1, s. 35-57.

[21] Itella (2011). *Itella Annual Report 2010*, Helsiniki 2011, s.18-21.

[22] *Ahlers Logistic Center (St. Petersburg, Russia)*, http://www.ahlers.com/news/articles/2006/alc/, 2012-08-11.

Despite the recession the rate of investments in many Russian ports didn't slow down. The intensive works are carried out in the port UST Luga. Thanks to the investments, by the year 2018 the port will have reached reloading potential 180 millions tonnes. It is worth to notice that in 2001 around 11.8 millions tonnes of cargo were reloaded and in 2009 around 10.4 tonnes. The development of the ports is supported by major investments in road, railway and pipeline transport. Since 2009 the turnover of UST Luga port is dynamically growing. In 2011, the UST Luga port transshipped 22.7 millions tonnes of cargoes and 133,837 vehicles. In January-September 2012 stevedoring companies of Ust-Luga port handled 33.4 millions tonnes of cargo. It is two times higher than the same period last year. Total volume of vehicle handling reached 119,323 units and increased on 1.4 times[23].

Port **St. Petersburg** (60 millions tonnes in 2011) invests over 2.1 billions rubles in order to improve the access to the port and services provided for the ships. 500 millions rubles were assigned for builiding ro-ro terminal with around 1 million tonnes reloading capacity. JSC "Sea Port of Saint-Petersburg" invested 170 millions rubles into modernization of port infrastructure and technical re-equipment of the terminals within 9 months of 2012[24]. Viktor Larin, the director of BaltTehProm, announced on the conference *Port infrastructure of the North-West: from design to operation*: "**In Kalinigrad** (13.4 millions tonnes in 2011) BaltTehProm is in the process of building an universal port with container terminal of 880 000 TEU year reloading capacity with the possibility of providing services for ro-ro ships.". In 2013 the investment is going to be started, worth 20 millions euros. By the year 2017 the services for the ships will have started[25].

3 LOGISTICS INVESTMENTS IN THE MOST IMPORTANT BALTIC SEA REGION PORTS

In the first half of 2012 the ports of Baltic Sea Region (Lithuania, Latvia, Estonia) noted increase of reloading in comparison to the same period in the previous year. That put them on the positions just under the biggest Russian ports.

The port in Klaipeda reloads around 20% of Baltic Sea Region cargo. In 2011 the Klaipeda Seaport and Būtinges terminals handled 45.5 millions tonnes of cargo (13% more than in 2010). In the period of the January–September of 2012 the aggregated throughput of seaborne cargo in Klaipėda Port totalled 25.9 millions tonnes[26].

In 2010 498 meters of landing moors had been constructed for 21 millions euros. Next investments were planned, estimated on total cost of 47 millions euros. 680 meters of outworn moors were reconstructed for the cost of 11 millions euros[27]. In 2011 and 2012 investment projects were realized, which were described in strategic programs connected with EU structural programs. In order to develop multimodal transport, 1480 meters of railways for 7 millions euros were build and 812 meters were renovated for 1.1 millions euros. The governing body of Klaipeda Port assigned 31.6 millions euros to deepening works. 336 ha area located near the port was intended for building new logistic centre. A new passenger-cargo terminal is going to be constructed on 80-81 moor for 225 millions euros, supported by funds from EU program INTERREG IIIB. The new moor and pier is going to provide services for three vessels at once, the type of ro-ro, ro-pax, ferry or cruise[28].

The management of port of Tallin (36.5 millions tonnes in 2011) manages 5 ports: Muuga (located 17 km from Tallin), Tallinn, Paljassaare, Paldiski South oraz Saaremaa[29]. Muuga is one of the newest ports in the Baltic Sea Region (operating since 1986) and one of the most important in Estonia. It is equipped with terminals providing services to reload crude oil and its products, bulk cargo and containers, what meets all qualities of an universal seaport hub. The plan of the port's development contains building new moors, development the potential of the containers reloading and the connections with Moscow. The Port in Tallin (known also as the Old Port) is specialized in passenger services (ferries and cruise).

In 2010 the Group of Tallin ports have made investment worth 28.5 millions euros, in 2009 worth 55 millions euros, in 2008 worth 35 millions euros. The funds were assigned to build new logistics infrastructure and for the modernization of the ports,

[23] JSC "Ust-Luga" (2012). *Cargo turnover of Ust-Luga port had doubled for 9 months of 2012*, http://www.ust-luga.ru/pr/?s=news&id=408&lang=en, 2012-12-27
[24] JSC "Sea Port of Saint-Petersburg" (2012). *"Sea Port of Saint-Petersburg" transferred over 170 millions. rubles for the development of port infrastructure*.
[25] Kalinigrad Port (2012). *OOO BaltTehProm to put new transport industrial warehouse into operation in the Kaliningrad region in November 2012*,

[26] KLAIPEDA STATE SEAPORT AUTHORITY (2012). Within the period of 9 months of the current year 31,9 million tn. of seaborne cargo were handled in the Port. http://www.portofklaipeda.lt/news/65/579/Within-the-period-of-9-months-of-the-current-year-31-9-million-tn-of-seaborne-cargo-were-handled-in-the-Port/d,press 2012-12-28
[27] KLAIPEDA STATE SEAPORT AUTHORITY (2012). *Envelopes with submitted large scale of Port Capital Dredging and Widening Works Were Opened*, http://www.portofklaipeda.lt/en.php/news/news/envelopes_with_submitted_large_scale_of_port_capital_dredging_and_widening_works_were_opened_/13520, 2012-07-07.
[28] Port of Klajpeda (2012). *Activities of 2010*, http://www.portofklaipeda.lt/en.php/urgencies/activities_of_2010/124 89, 2012-10-10.
[29] Port of Tallinn (2012). www.portoftallinn.com, 2012-10-30.

managed by the board of port Tallin. In 2010 19.9 millions euros were assigned to build new moors and the development of the infrastructure in the ports. One of the biggest logistic objects put into service in 2010 is container terminal in Muuga port. Besides, part of the funds were assigned to purchasing new area as the preparation for further investments[30]. In 2011 the works focused on creating the new infrastructure, connecting the terminals of east part of Muunga port and the industrial park, which is on planning stage[31]. Port of Tallin participated in the project titled: „Port Integration: Multimodal Innovation for Sustainable Maritime & Hinterland Transport Structures". It is a three-year 1.47 million euro project, including a 1.11 million euro contribution from the European Regional Development Fund. The project has thirteen port and political partners from ten EU countries and Russia[32].

Latvia has three significant ports and four small ones. The port of Ventspils (28.5 millions tonnes in 2011) is nowadays a transit port, providing a wide range of services[33]. On the area of 30 ha Ventspils High Technology Park was located. It is a good place for industrial investments[34].

Riga is an universal port (34 millions tonnes in 2011). Most of the reloaded cargo, over 80% is transit from Russia and Belarus. It includes general cargo in containers, steel, wood, fertilizer and crude oil[35]. The new program of the port development was accepted in 2009 and will have been realized by the 2018. For 140 millions euros a new moor 1780 meter long, located on 65 ha area is going to be constructed, with seven landing places for ships[36]. In order to develop the multimodal transport by the end of 2021 the road and railway infrastructure will have been modernized and the new bridge, connecting port areas on Kundzin island, will have been constructed. By the year 2017 a new terminal providing services for passenger and ropax ships

will have been built, for the cost estimated on 50 millions euros[37].

4 LOGISTICS INVESTMENTS IN SELECTED SCANDINAVIAN PORTS

The group of ports Copenhagen-Malmö together with Malmö city built a new terminal and a new logistics centre Northern Harbour for 900 millions Swedish kronor. The investment was partially financed by Europan Union funds. Northern Harbour was launched after two years of works on the beginning of 2011. The new port allowed to increase Malmö port's turnover five times. It is the biggest investment in Scandinavian ports of the last years. It includes building new loading places by the moor of 1300 meters length. In the new port ramps for ro-ro ships were installed. Since March the port can provide services for three ferries (ro-ro ships) at the same time. In 2012 a new system of lighting and a new infrastructure IT, based on optic fibre was installed[38]. In 2012 the board of Copenhagen-Malmö ports signed an agreement with port Helsingborg in order to strengthen the position on transport market[39].

A new system Autogate was installed, which allows to control the movement in the port automatically. It is going to be supervised by the system of cameras and the computer system which is especially adjusted to the conditions of Northern Harbour. In Kopenhaga the terminal Prøvestenen was expanded by 18 ha and moors of 650 meters length. The new moor for passenger ships (1100 meters length, 60 meters width) will have been constructed by the year 2013 in the terminal Nordhavnen[40]. Since several years the administration of the port **Goeteborg** aims to improve the ecological transport through the development of multimodal transport connections. Thereby during the last decade the number of operators offering railway connections increased from 1 to 10, and the number of intermodal trains increased from 1 to 27 scheduled connections with the most important cities of Sweden and Norway. The port terminals provide services for 70 trains

[30] Port Of Tallin (2012). *The extension to the Eastern part of Muuga Harbour,* http://www.portoftallinn.com/muuga-development-plans, 2012-12-12.
[31] Port Of Tallin (2012). *AS Tallinna Sadam, Consolidated Annual Report for the Financial Year Ended 31 December 2010,* AS Tallinna Sadam, Tallinn 2011, s. 12.
[32] Port Of Tallin (2012). *The Single Window challenge for ports,* http://www.portoftallinn.com/news?art=261, 2012-10-09
[33] Ventspils Port (2013). http://www.portofventspils.lv/en, 2013-01-03.
[34] Ventspils High Technology Park (2013). www.vhtp.lv, 2013-01-03.
[35] Freeport of Riga (2013). http://www.rop.lv/en/about-port/facts-a-figures.html, 2013-01-02.
[36] Freeport of Riga (2013). Development of Infrastructure on Krievu Sala for the Transfer of Port Activities from the City Center, http://www.rop.lv/en/about-port/projects/1082-development-of-infrastructure-in-krievu-sala-for-relocation-of-port-activities-out-of-the-city-center.html, 2013-01-03.

[37] Freeport of Riga (2013). Development of Infrastructure on Krievu Sala for the Transfer of Port Activities from the City Center, http://www.rop.lv/en/about-port/projects/1082-development-of-infrastructure-in-krievu-sala-for-relocation-of-port-activities-out-of-the-city-center.html, 2013-01-10.
[38] Copenhagen Malmö Port (2013). *Three new terminals,* http://www.cmport.com/Corporate/Investments/, 2013-01-10.
[39] Grzybowski M. (2012). Skandynawski alians portów. Kopenhaga-Malmö, Helsingborg, „Polska Gazeta Transportowa". 20 czerwca 2012, nr 25, s. 6.
[40] Copenhagen Malmö Port (2013). New cruise-ship quay, http://www.cmport.com/Corporate/Investments/, 2013-01-10.

daily. The railway transport of the containers increased from 144 000 to 380 000 TEU[41].

The board of Stockholm port manages the ports in Stockholm, Kapellskär and Nynäshamn. Environmental Protection Agency given the board an permission to build a new port Norvikudden, located near Nynäshamn. On 44 ha space terminals providing services to containers transported by ro-ro vessels are going to be provided. A new moor is going to be constructed, 1400 meter length (16.5 meter length) with seven mooring places for ships. The terminal is going to be equipped with four or five container cranes. It is estimated that the terminal will be able to provide services for 300 000 TEU containers and 200 000 ro-ro cargo units yearly[42]. The terminal is going to be constructed by Hutchison Port Holdings. Container Terminal Nynashamn (CTN) is going to be located in Norvikudden near Nynashamn. It is going to occupy 25 ha, equipped with the moor of 800 meter length[43].

5 LOGISTICS INVESTMENTS IN LÜBECK AND ROSTOCK

There are five German merchant ports on Baltic Sea. The biggest ones are Lübeck and Rostock (together with Warnemünde), while Sassnitz/Mukran, Wismar and Kiel are coming next[44]. After the reunification, when German Democratic Republic joined the Federal Republic of Germany, the governments of Germany and Mecklenburg-West Pomerania started new programs (tax reliefs for investors) and assigned funds (partially coming from EU) to activate former industry areas in East Germany, including ports[45].

Rostock is second German Baltic port when it comes to cargo reloading (22.3 millions tonnes in 2011, including 11,7 millions tonnes of ferry cargo)[46]. The crucial investment in the development of logistics services was rebuilding of landing places no. 64 and 66. The board of the port assigned around 72 DM to adjust them to providing services for modern ferries. Today it is a landing place for ferries from Trelleborg[47]. Another important investment was the expansion of Warnow Ferry Terminal,

which costed 80 millions DM. Around 20 millions euro was assigned for building second passenger terminal (600 square meters surface), which was launched by P8 moor in the former shipyard Warnemünde[48].

The port of Lübeck is nowadays the biggest German merchant port on Baltic. In the second part of 90s LHG and the city assigned over 300 mln DEM to develop the logistic infrastructure of the port in order to adjust it for the needs of the clients. Thanks to it the import of paper was taken over from Kiel[49]. Federal and country funds were assigned to modernize the channel Lübeck-Elbe, electrification railway line and increasing the capacity of A1 from Lübeck to Hamburg and A20 to Rostock and Szczecin. For 15.3 millions euros a new modern terminal for intermodal transport was built. Terminal Skandinavienkai's surface was increased almost double, from 47 ha to 77 ha, and additional 35 ha was assigned for the development of logistics functions. On this area, for further 25 millions euros offices and warehouses for clients were constructed. In the middle of 2006 a new Seelandkai Terminal (18.5 ha area in the Lübeck Siems district) was put into service, with landing places for vessels operating in lo-lo system (one place) and ro-ro (two places)[50].

6 SELECTED LOGISTICS INVESTMENTS IN POLISH PORTS

Nowadays the strategic logistics investment in Polish pots is **building liquid natural gas terminal LNG in Świnoujście**. The contract was worth around 3 billiards zlotys. The terminal is going to allow to receipt around 5 billion cubic meters of natural gas yearly, which is 35% of the consumption in the country. The construction of the terminal allows to increase the receipt of the gas to 7.5 billions cubic meter yearly[51]. By the year 2013 the terminal will have been connected to the industrial network. By 30 June 2014 the object will have been put into operation[52].

[41] *Rail Services,* Port of Gothenburg, February 2011, s. 9-15.
[42] Stockholm-Nynäshamn Port (2013). Norvikudden, http://www.stockholmshamnar.se/en/Our-Ports/Nynashamn/Stockholm-Nynashamn-Norvikudden/, 2013-01-11.
[43] The World of Hutchison Port Holdings, Broszura HPH, April 2011.
[44] University of Turku (2010). Centre for Maritime Studies, Baltic *Port List 2009,* Turku 2010, s. 17-20.
[45] Association of German Seaport Operators (2011). Germany's Seaports 2011, Germany Trade & Invest, Berlin 2011, s. 12-14.
[46] Rostock Port (2013). Key Facts and Figures, http://www.rostock-port.de/en/rostock-port/key-facts-figures.html, 2013-01-12
[47] Rostock Port (2013). Investment, http://www.rostock-port.de/en/rostock_port/key_facts_figures/investment.htm, 2011-07-12.

[48] Rostock Port (2013). *Investment,* http://www.rostock-port.de/en/rostock_port/key_facts_figures/investment.htm, 2011-07-12.
[49] Shipgaz (2011). *Lübeck: Logistics the focus for growth,* http://www.shipgaz.com/old/magazine/issues/2002/03/vessel_0302.php, 2011-08-11.
[50] The LHG – Company (2012). http://www.lhg-online.de/, 2012-10-18.
[51] Ministerstwo Skarbu Państwa (2010). *Umowa z wykonawcą terminalu LNG w Świnoujściu podpisana,* informacja prasowa, , http://www.msp.gov.pl/portal/pl/29/11215/Umowa_z_wykonawca_terminalu_LNG_w_Swinoujsciu_podpisana.html?search=156318, 2010-07-15.
[52] Polskie LNG (2011). *W Świnoujściu powstaje gigantyczny „termos LNG",* http://www.polskielng.pl/biuro-prasowe/aktualnosci/wiadomosc/artykul/201137.html, 2011-08-09.

The East Pomerania Logistic Centre in Szczecin was built in port of Szczecin on the area of 20 ha. The centre was connected with road and railaway infrastructure. Gdynia Logistic Centre is going to be located nearby Baltic Container Terminal, Gdynia Container Terminal and Ferry Terminal which provides services to Stena Line Gdynia – Carlscrona connection[53]. Pomeranian Logistic Center in Gdańsk is located on the area of 130 ha near DCT Gdańsk. The investment was started by company Goodman. The construction of Logistic Centre (500 000 square meter surface) has been started.

7 CONCLUSION

The analysis of the logistic investments market in the ports of Baltic Sea Region leads to the conclusion that all of the regions (for example Leningrad Oblast, Mecklenburg-West Pomerania) and cities (for example Tallin, Vetspils, Lübeck) and ports (for example Stockholm, Copenhagen-Malmö , Gdynia, Rostock) are carring out activities to strengthen their position on the international market through the development of logistic functions. It is achieved through the infrastructural investments in the ports as well as in their surrounding.

REFERENCES

[1] *Baltic maritime transport*, Maritime, "Baltic Transport Journal", 1/2010
[2] *Baltic Port List 2009*, University of Turku, Centre for Maritime Studies, Turku 2010
[3] *Germany's Seaports 2011*, Association of German Seaport Operators, Germany Trade & Invest, Berlin 2011
[4] *Itella Annual Report 2010,* Helsiniki 2011.
[5] *Rail Services,* Port of Gothenburg, February 2011

---

[53] Grzybowski M. (2012). Port of Gdynia Turns 90, The Warsaw Voice, November 29, 2012.

Baltic Sea Logistic and Transportation Problems
STCW, Maritime Education and Training (MET), Human Resources and Crew Manning, Maritime Policy,
Logistics and Economic Matters – Marine Navigation and Safety of Sea Transportation – Weintrit & Neumann (Eds)

Redefining the Baltic Sea Maritime Transport Geography as a Result of a New Environmental Regulation for the Sulphur Emission Control Areas

M. Matczak
Gdynia Maritime University, Gdynia, Poland

ABSTRACT: The new Directive 2012/33/EU implements radical changes to shipping in the North and Baltic Seas (SECA zone). The severe limits concerning the sulphur content of marine fuels will definitely increase the costs of shipping. Utilization of clean marine fuels, installation of scrubbing devices or implementation of alternative propulsions (e.g. LNG) will simultaneously push up the freight rates. The price optimization of the Baltic system could lead to crucial changes in the spatial pattern of the cargo flow, as well as in the region. Finally, the derivate effect of the changes could result in a limitation of the competitiveness of the European economy.

1 THE BALTIC SEA AS A PART OF THE SULPHUR EMISSION CONTROL AREA (SECA)

1.1 New requirements defined in Directive 2012/33/EU

A balanced approach to the development of the world economy is equilibrium among economic, social and environmental benefits. Based on this idea, the International Maritime Organization (IMO) initiated a process of reducing the negative impact of shipping on the environment. The desire to reduce emissions of harmful substances (e.g. SO_X, NO_X, and particles), especially on areas heavily used by shipping, led to the creation of Emission Control Areas (ECA), and Sulphur Emission Control Areas (SECA)[54].

Table 1. Marine fuel sulphur limits.

Marine fuel Sulphur Content (% m/m[55])	Ratio Emission SO_2 (ppm) /CO_2 (% v/v[56])
3.50	151.7
1.50	65.0
1.00	43.3
0.50	21.7
0.10	4.3

Source: Directive 2012/33/EU.

The North Sea and the Baltic Sea are covered by SECA.

The regulations adopted by IMO have been confirmed and implemented by the European Union under Directive 2012/33/EU[57]. The Directive, in accordance with IMO, specifies the maximum sulphur content of marine fuel (optionally, the sulphur content of the exhaust's gases), that at the beginning of 2015 must be 0.1% by volume (Table 1)[58]. This means a radical change to the Baltic Sea transport system, including shipping and seaports. This applies important changes in both modal and spatial structures of the transport market.

1.2 Technological solutions concerning the new requirements of the eco-level

Currently, a few potential approaches concerning the fulfilment of the Directive's requirements have been defined. Key solutions have been considered in the following examples:

1 Paying penalties by carriers for failure to adapt to the new requirements,
2 Replacing traditional fuels (HFO, IFO) by low-sulphur fuels (MDO, MGO),

---

[54] *International Convention for the Prevention of Pollution from Ships* (MARPOL) 1973.
[55] m/m – mass per mass unite
[56] v/v – volume per volume unit

[57] *Directive 2012/33/EU of the European Parliament and the Council of 21 November 2012 amending Council Directive 1999/32/EC as regards the sulphur content of marine fuels.* OJ L 327/1, 27.11.2012.
[58] **MARPOL** *73/78,* **Annex VI** *Regulations for the* **Prevention** *of Air Pollution from Ships.* 01.07.2011.

3 Equipment of ships in reducing emissions to air devices (scrubbers),
4 Utilization of alternative fuels or their mixtures, among others LNG.

The first of the 'solutions' has a short-term nature, since the system of penalties must correspond to those specified in the Directive, which reads (article 11): *1) Member States shall determine the penalties applicable to breaches of the national provisions adopted pursuant to this Directive. 2) The penalties determined must be effective, proportionate and dissuasive and may include fines calculated in such way as to ensure that the fines at least deprive those responsible of the economic benefits derived from their infringement and that those fines gradually increase for repeated infringements.* It should be indicated that the amount of appropriate penalties is currently unknown, but it can be assumed that their level will have to be higher than the increase in costs resulting from the use of other solutions. In the case of irregular shipping connections, this solution could be cheaper than the use of complex systems. Such kind of practice is obviously unacceptable in the case of regular shipping lines operating in the Baltic Sea or on a Baltic - North Sea route.

The other way to fill the regulatory requirements of the Directive is to adopt low-sulphur marine fuel. In practice, this means that ships will have to use Marine Gas Oil (MGO) or Marine Diesel Oil (MDO). Unfortunately, these fuels are much more expensive than those that are currently used in emission control areas (see Table 2).

Table 2. Marine fuel prices (USD/mt, 24.01.2013)

	IFO 380	IFO 380 LS[59]	IFO 180	IFO 180 LS	MDO	MGO
Singapore	631		637		933	943
Genoa	636		668			1014
Rotterdam	613	640	639	668	-	950
Gothenburg	630	665	660	695	910	995

Source: BunkerIndex (www.bunkerindex.com), Bunkerworld (www.bunkerworld.com)

Taking into account the average fuel price calculated on the basis of the examples presented above, changing heavy low-sulphur fuel (IFO308LS/IFO180LS) for distillates fuels (MGO, MDO) causes an increase in the cost of bunkering from 35% to 49%. However, it should be stressed that prices can significantly increase as the effect of demand will grow on the carriers' side. On the one hand, the production of low-sulphur fuels must be based on sweet crude oil, which occurs only in very few deposits (among others: Daqing - China, Bonny Lt. - Nigeria, Brent – the United Kingdom). On the other hand, at the present time, a refinery has no

spare capacity and investment plans for the development of systems for the production of low-sulphur fuel (thermal cracking) or desulfurization fuels (VRDS-Vacuum Residue Desulfurization) refer to the period after 2020. What's more, the cost of development of such an installation is calculated between USD 0.5-1 billion, in one refinery[60]. High investment costs and the high risk of developing a future market has caused a lack of practical activities of the producers. A study on the impacts of the new IMO regulations on transportation costs prepared by the Ministry of Transport and Communications in Finland, based on the example of ships operating between Finland and other countries, indicating the share of the cost of fuel in the total cost, amounts to an average of about 30 (RoPax) - 54% (container)[61]. This differs depending upon the type, size and speed of the ship. It is estimated that from 2015 the share of the cost of fuel, as a result of the application of low-sulphur fuels, can increase to approximately 50-70%[62]. Consequently, the increase in fuel costs will lead to an increase in the level of freight rates in the SECA. Depending upon the price of MGO, increased freight rates range on average between 18.8%-22.9%[63]. The largest increase in rates, concerning the internal connections of the Baltic Sea, could reach a level of 40%. Such level of freight rates development is confirmed by the Finnish calculations (Table 3).

Table 3. Growth in freight rates as an effect of the implementation of the 0.1%, sulphur content of marine fuels.

Cargo type	Freight rates growth
Container	44-51%
Paper	35-40%
Truck	35-41%
Car	35-41%
Crude oil	28-32%
Bulk cargo	39-44%
Timber	35-40%
Steel	35-40%

Source: Sulphur content in ships bunker fuel in 2015 A study on the impacts of the new IMO regulations on transportation costs, Ministry of Transport and Communication in Finland, Helsinki, 2009.

Another solution, which allows complying with the requirements specified in the Directive is to apply desulphurization equipment, shipboard exhaust gas treatment - scrubbers. This technology is already widely used in land-based installations, however, shipping applications cannot be considered

---

[59] *Low sulphur (0.5%).*

[60] *Consequences of the Sulphur Directive.* SWECO Stockholm, October 2012.
[61] *Sulphur content in ships bunker fuel in 2015 A study on the impacts of the new IMO regulations on transportation costs*, Ministry of Transport and Communication in Finland, Helsinki, 2009.
[62] *Consequences of the IMO's new marine fuel sulphur regulations-Report*, Swedish Maritime Administration, Stockholm 2009.
[63] *Analysis of the Consequences of Low Sulphur Fuel Requirements*, ECSA, Antwerp, 2010.

commercially mature in a maritime environment. The technology can be based on dry and wet scrubbers. In the first case, the system uses dry chemicals as a scrubbing medium. Thus, due to the weight and high location of the scrubber, it is necessary to use additional stabilizing measures. On the other hand, this technology does not require additional equipment except for an electrical installation and measuring instruments. Handling of the waste from dry scrubbers in harbours is an unsolved question. Wet scrubbers use water as their scrubbing medium. There are two main types of wet scrubbers: fresh water scrubbers (the so-called closed loop scrubbers) and salt-water scrubbers (called open loop scrubbers). In the first case, the scrubbing medium is a caustic soda; in the second, salt water continuously cleans exhaust gases. The handling of caustic soda requires appropriate safety equipment and skills. In the second case, separated waste is dumped into the water. Therefore, common use of this method will depend upon the assessment of its impact on the environment. To date, scrubbers have been certified by two classification societies: *Germanischer Lloyd* and *Det Norske Veritas* (DNV) and have received a certificate of conformity for SECA. In December 2010, Wärtsilä received its first commercial order of scrubbers from the Finnish company Containerships Ltd Oy. It is estimated that the cost of installing a scrubber is about USD 5.84 million[64], of which:

– Scrubber machinery and equipment USD 2.6 m;
– Steel (150t) / pipe / electrical installation and modification USD 2.4 m;
– Design and classification costs USD 0.5 m;
– Off-hire (20 days) USD 0.34 m.

The cost is estimated as above, with return on investment (with reference to the change in fuel to MDO) about 4-6 years. The decision to install a scrubber is very complex; two issues are of particular interest: the difference in the cost of fuel (IFO – MDO) and the limited investment opportunities shipping lines have as they operate in large part on minimum profitability. Experts estimate that only 10% of the units operating in the zone will be equipped with this type of technology.

Another option for obtaining the parameters compatible with the Directive is the use of alternative fuels, including LNG. Liquefied Natural Gas is already considered the "fuel of the future" in shipping. It is estimated that most of the newly built vessels that will operate in the SECA zone will use LNG. LNG does not contain sulphur, therefore, it does not form during combustion of sulphur oxides, dust and does not arise as solid waste incineration slag, ashes, or soot. As a result, the combustion of LNG is about 85-90% less emissions of nitrogen

oxides and 15-25% less carbon dioxide emissions than burning conventional fuels. The use of LNG as a marine fuel can take place both in the case of operating vessels (retrofitting) and newly built vessels. In the first case, converting only the younger vessels is economically effective as the recovery period is about 2-4 years[65]. The cost of such reconstruction is estimated at USD 7.56 m. This would include:

– LNG machinery, tanks and equipment, main engine conversion USD 4.38 m;
– Steel (300t) USD 2.0 m;
– Design and classification cost USD 0.5 m;
– Off-hire (40 days) USD 0.68 m[66].

The return on investment in LNG retrofitting (based on the price of MDO) will in this case be from 3-9 years (with a price difference of 750 USD/mt).

Newly built vessels can be equipped with mono-fuel (LNG) or dual-fuel engines. The mono-fuel solution could be a good choice for vessels operating in areas with a well-developed network of LNG bunkering facilities (so far the Baltic does not offer such kinds of stations). Dual-fuel is much more flexible because the engine can be switched at any time without having to stop the engine. On the other hand, dual-fuel engines have a more complicated structure. LNG engines are offered by: Wärtsilä, Rolls-Royce, MAN Diesel, and Mitsubishi Heavy Industries.

The important challenge to using LNG is that much greater space is needed for tanks and their installation. The volume of LNG storage tanks must be approximately 1.8 times greater than the volume of MDO tanks. In addition, the application of the necessary isolation of LNG vessels, the required space is 2.3 times greater. There is also a need to use tanks of cylindrical shape, which in turn makes the required space on the vessel approx. 4 times more than conventional propulsion[67]. Because of this, LNG is best suited for use by vessels operating on short distances on a regular basis (e.g. container feedering, ferry and ro-ro). The cost of building ships propelled by LNG can be about 10-15% higher than the cost of conventional fuel-powered vessels. For a typical ro-ro vessel, the extra cost may amount to about EUR 3.2 m[68]. Similarly, DNV has calculated that operating costs driven by an LNG vessel may be about 35% less than the operating costs of a ship powered by MGO in terms of 10

[64] *Green ship of the future.* Green Ship Technology Conference, Copenhagen 2012.

[65] *North European LNG Infrastructure Project. A feasibility study for an LNG filling station infrastructure and test of recommendation.* Danish Maritime Authority, Copenhagen 2012.
[66] *Green ship of the future.* Green Ship Technology Conference, Copenhagen 2012.
[67] *Natural gas for ship propulsion in Denmark - Possibilities for using LNG and CNG on ferry and cargo routes,* Danish Ministry of the Environment, 2010.
[68] *Maritime Gas Fuel Logistics. Developing LNG as a clean fuel for ships in the Baltic and North Seas,* MAGALOG, December, 2008.

years. In a 20-year perspective, these costs can be reduced by up to 45%[69].

2 MARKET EFFECTS OF THE DIRECTIVE'S IMPLEMENTATION

There is no doubt that implementing the provisions of the Directive will increase the costs of maritime transport in the SECA. This situation will affect the competitive systems in the Baltic transport system in two dimensions:
- Inner-modal, what will be the changes to the spatial structure of the cargo flows on the Baltic Sea?
- Cross-modal, which refers to the change in modal split structure in transport (modal back-shift).

The main effects of implementing a new SOx regulation in the Baltic area, concerning the first issue, can be indicated as follows:
- Transferring shipping traffic to ports in the Baltic which are not regulated by the Directive, where particular importance will be given to Russian ports and to the ports of the states being covered by the Directive but located beyond the SECA (e.g. the seaport of Narvik, Norway).
- Relocating the shipping operation to the shortest maritime connections.

The most important effect of the cargo concentration could be a transfer of Russian foreign trade to ports in Russia. This concerns mainly the energy sources (oil, coal) and containers (Russia is the main box market in the BSR). Importantly, this trend is part of the investment activities of Russia which is intensely developing the Ust-Luga mega-port so it will be ready to handle both crude oil, coal and containers. For this reason transit traffic, so far served by other Baltic seaports, could be moved to Russian terminals.

At the same time, serving the mother container vessels in the Port of Gdańsk (in the case of these types of connections increased costs associated with the operation in the SECA zone will be marginal) can foster the acquisition of another cargo on the direct Asia – Baltic connections. It can result in a significant reduction in the feeder services provided by North Sea hubs. Similarly, it is a chance to introduce new services and urge other shipping operators to implement ocean-going vessels on the Baltic Sea.

The second effect should be a reduction in the average distance of carriage by sea. According to the abovementioned *Analysis of the Consequences of Low Sulphur Fuel Requirements*, the larger the distance traversed by sea transport, the greater the impact of using low-sulphur fuel on the cost of

transport is. Low-sulphur fuel costs will, therefore, have the greatest impact on the price of the carriage in the case of medium and long distance shipping. As a consequence, maritime transport could be used on shorter distances, and the rest of the route could be covered by road or rail transport. This would mean that the total quantity of cargo in shipping would remain unchanged, however, the cargo would be transported by sea on the shorter route. The ferry market, which is regarded as highly sensitive to such changes, could notice a considerable spatial redefinition. It is probable that Baltic ro-ro traffic could shift to the western part of the basin, which would mean a significant reduction in the turnover of northern and eastern ports in Sweden.

Another important alternative to west Baltic ferry traffic worth mentioning is the existing fixed links between Sweden and Denmark, such as the planned bridge on the Fehmarn Belt. What is more, the road/rail fixed connection will strongly be developed by intermodal transport, heavily supported by EU programs (e.g. Marco Polo).

The changes in spatial structure of shipping navigation will have a direct influence on the regional modal split. As has already been indicated, in order to reduce the average distance of sea transport on the Baltic, it is necessary to increase the distance overland. What's more, in continental relations (such as feeder traffic between the North Sea and the Baltic ports) intermodal land transport could completely eliminate the shipping connections. The good intentions of the regulator (decreasing emissions) could bring the opposite effect, especially as a modal shift to road haulage could occur. It should be noted that a rise in cargo volume on the road will not only have a negative impact on the environment (an increase in CO_2 emissions), but will also generate other externalities, such as: road fatalities, congestion, noise or vibration. In these cases the preferred alternative (from a social costs of transport point of view) should be rail, especially intermodal. Taking into regard the efficient and strong lobbying of European railway undertakings, it should be expected that such a development scenario could occur on the Baltic Sea and the North Sea-Baltic Sea routes. Seeking secondary effects it could be added that a reduction in the amount of cargo on the shipping lines, could contribute to a reduction in the number of employed ships which would decrease the frequency of maritime services. As was calculated, a loss of about 10-20% of cargo traffic could lead to a limitation in the number of vessels and frequency in a particular service. Higher fuel costs and the related operating expenses for this smaller frequency can lead to an even greater loss in the attractiveness and competitiveness of shipping. Consequently, such lines may no longer be profitable and, as a result, suspend their activity. In a study prepared for ECSA,

[69] *The age of LNG is here. Most cost efficient solution for ECAs*, DNV presentation, June 2010.

it was estimated that if the price reaches USD 1000 per ton of MGO, a vessel's capacity may drop by up to 40%. An illustration of the above-indicated scenarios may be found in the results of a survey conducted by the Swedish Maritime Administration. The authors of the study have clearly indicated the following, if a modal back-shift were to occur:
– a decline in maritime transport by 10%;
– an increase in rail traffic by 5%,
– an increase in the road haulage by 3%[70].

The Baltic ports, as a result of implementing the Directive can therefore become much less competitive in comparison to the ports located in the other regions of Europe. A change in the flow of goods may support European ports located outside the SECA zone. For example, part of the cargo transferred to Central Europe (e.g. the Czech Republic, Slovakia, Hungary, and Austria) may be in some part supported by Mediterranean or Black Sea ports. Finally, referring to the effects of the economic scale, implementation of Directive 2012/33/EU will lead to a reduction in the competitiveness of the European economy.

The changes in the North Sea and the Baltic market can cause a price increase in all other modes and kinds of transport. Higher demand for maritime diesel oil or maritime gas oil and limited supply performance will force up fuel prices (the so-called knock-on effect). Similarly, the process of internalizing the externalities, mainly in road transport (as a maritime alternative), will increase the costs of European production as well[71]. So, we can experience a new variation of the effect initially called *carbon leakage*. A new *sulphur leakage* will reduce the competitiveness of European industry while supporting the economies of the other countries.

3 CONCLUSIONS

The new market reality currently being created by the sulphur Directive will significantly change the structure of the Baltic transport market. Higher shipping costs, especially on medium and long distances, will cause a redefinition of cargo flows. It could be stated that the crucial effects are: a limitation on the average distance of a sea route especially on the ferry and ro-ro market, a higher concentration of traffic flow in Russian seaports (energy sources and containers), an increase in the attractiveness of direct container connections to the Baltic or a decrease in North Sea – Baltic feedering.

Similarly, a modal back-shift effect is very probable. In the end, a rise in transport costs will negatively influence the competitiveness of the European economy.

REFERENCES

1973. International Convention for the Prevention of Pollution from Ships (MARPOL).

2008. Maritime Gas Fuel Logistics. Developing LNG as a clean fuel for ships in the Baltic and North Seas, MAGALOG, December, 2008.

2009. Consequences of the IMO's new marine fuel sulphur regulations-Report, Swedish Maritime Administration, Stockholm 2009.

2009. Matczak M.: Maritime Safety in European Concept of the Internalization of External Costs of Transport. TransNav - International Journal on Marine Navigation and Safety of Sea Transportation, Vol. 3, No. 2, pp. 207-211

2009. Sulphur content in ships bunker fuel in 2015 A study on the impacts of the new IMO regulations on transportation costs, Ministry of Transport and Communication in Finland, Helsinki, 2009.

2010. Analysis of the Consequences of Low Sulphur Fuel Requirements, ECSA, Antwerp, 2010.

2010. Castanius M.: Consequences of the IMO's new marine fuel sulphur regulations. BPO Seminar, Copenhagen 2010.

2010. Natural gas for ship propulsion in Denmark - Possibilities for using LNG and CNG on ferry and cargo routes, Danish Ministry of the Environment, 2010.

2010. The age of LNG is here. Most cost efficient solution for ECAs, DNV presentation, June 2010.

2011. MARPOL 73/78, **Annex VI** Regulations for the **Prevention** of Air Pollution from Ships. 01.07.2011.

2012. Consequences of the Sulphur Directive. SWECO Stockholm, October 2012.

2012. Directive 2012/33/EU of the European Parliament and the Council of 21 November 2012 amending Council Directive 1999/32/EC as regards the sulphur content of marine fuels. OJ L 327/1, 27.11.2012.

2012. Green ship of the future. Green Ship Technology Conference, Copenhagen 2012.

2012. North European LNG Infrastructure Project. A feasibility study for an LNG filling station infrastructure and test of recommendation. Danish Maritime Authority, Copenhagen 2012.

Bunkerworld (www.bunkerworld.com).

BunkerIndex (www.bunkerindex.com).

---

[70] M. Castanius: *Consequences of the IMO's new marine fuel sulphur regulations.* BPO Seminar, Copenhagen 2010.

[71] See: M. Matczak: *Maritime safety in European concept of the internalization of external costs of transport.* 'Marine Navigation and Safety of Sea Transportation'. Edit. A. Weintrit. CRC Press Taylor & Francis Group 2009.

Baltic Sea Logistic and Transportation Problems
STCW, Maritime Education and Training (MET), Human Resources and Crew Manning, Maritime Policy,
Logistics and Economic Matters – Marine Navigation and Safety of Sea Transportation – Weintrit & Neumann (Eds)

Sustainable Transportation Development Prerequisites at the Example of the Polish Coastal Regions

A. Przybyłowski
Gdynia Maritime University, Gdynia, Poland

ABSTRACT: Sustainable transportation indicators are an important tool for better transportation planning. There is currently no standard set of sustainable transportation indicators. A variety of indicators are used, some of which are particularly appropriate and useful for planning and policy analysis. It is highly recommended to develop a research program concerning the collection, analysis and application of high quality, standardized transportation data in order to be able to provide a suitable framework for transportation planning and policy benchmarking. Based on the available strategic documents and statistical data 2004-2011 regarding transport indicators, the paper presents the sustainable transport development prerequisites at the example of selected regions representing three Polish coastal regions: Pomorskie, Zachodniopomorskie and Warmińsko-Mazurskie.

1 INTRODUCTION

More than ever, there is a need to combine more the processes of extending necessary transport infrastructure with the rule of balancing development by seeking selective and optimal solutions at the level of regional and at the local level. Also it is important to establish standardized sets of sustainable transportation indicators and to improve the collection of transportation statistics, expanding these efforts to reflect key economic, social and environmental impacts [Borys, 2008; Litman, 2008, p.11]. Indicators are important tools for making decisions and measuring progress.

The efficient and affordable transport systems are necessary for poverty alleviation and the need to mitigate adverse externalities to health and the environment. From an environmental and social point of view, the failure to unhitch growth in transport from growth in GDP is an extremely worrying tendency [Przybylowski 2011].

Based on the available strategic documents and chosen statistical data (2004-2011), the paper presents the sustainable transport development prerequisites in Poland at the example of three Polish coastal regions in the context of the sustainable development paradigm and sustainable transport policy guidelines in the EU policy. The analysis takes into consideration the following regions Pomorskie, Warmińsko-Mazurskie and Zachodniopomorskie.

2 SUSTAINABLE TRANSPORT PARADIGM AND INDICATORS

The idea of sustainable development contains within two key concepts [WCED, 1987, p. 43]:
- the concept of needs, in particular the essential needs of the world's poor, to which overriding priority should be given; and
- the idea of limitations imposed by the state of technology and social organization on the environment's ability to meet present and future needs.

It is possible to graphically represent (fig. 1) the achievement of sustainable development by the simultaneous coexistence of three capitals/capacities: environmental (natural), economic and social ones. There exist numerous other sustainability development models, taking into account four capitals: natural, social, economic and human ones [see more: Medhurst, 2003, p.4; Vivien, 2005, etc.].

According to the European Council of Ministers of Transport [ECMT, 2004] definition, a sustainable transport system:
- allows the basic access and development needs of individuals, companies and society to be met

safely and in a manner consistent with human and ecosystem health, and promotes equity within and between successive generations;
– Is affordable, operates fairly and efficiently, offers a choice of transport mode and supports a competitive economy, as well as balanced regional development;
– Limits emissions and waste within the planet's ability to absorb them, uses renewable resources at or below their rates of generation, and uses non-renewable resources at or below the rates of development of renewable substitutes, while minimizing the impact on the use of land and the generation of noise.

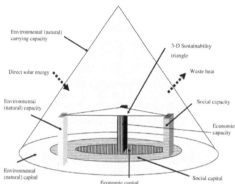

Figure 1. Sustainable Development triangle.
Source: Mauerhofer V., 2008, *3-D Sustainability: An approach for priority setting in situation of conflicting interests towards a Sustainable Development*, Ecological Economics, No 64/2008, p. 497-501.

There is a need for urgent action, ranging, inter alia, from the promotion of integrated transport policies and plans, the accelerated phase-out of leaded gasoline, the promotion of voluntary guidelines and the development of partnerships at the national level for strengthening transport infrastructure, promoting and supporting the use of non-motorised transport and developing innovative mass transit schemes. The international co-operation is required in order to ensure transport systems support sustainable development [www.un.org/esa/dsd/susdevtop, 2012].

In the EU, as a result of currently formed modal split in the transport sector, and as predicted realistically by 2020, there is no chance for any shift in it towards the more environmentally friendly modes of transport such as rail and inland waterways, reaching the set up transport policy's objective is thoroughly impossible. When this tendency is followed-up, sustainable mobility by still rapidly growing transport activity will even dash away [Grzelakowski, 2008]. For, sustainable mobility this means disconnecting mobility from its many harmful effects for the economy, society and environment.

Recently, the EU has proposed a new document: Europe 2020 Strategy [Europe 2020, 2010]. The Commission has identified three key drivers for growth, to be implemented through concrete actions at EU and national levels: smart growth (fostering knowledge, innovation, education and digital society), sustainable growth (making the production more resource efficient while boosting the competitiveness) and inclusive growth (raising participation in the labour market, the acquisition of skills and the fight against poverty).

In order to update the EU transport policy, the European Commission published a new document: WHITE PAPER 'Roadmap to a Single European Transport Area – towards a competitive and resource efficient transport system' [White Paper, 2011] where a set of very ambitious goals have been presented to be achieved by 2050 (tab. 1.).

Tab. 1. Ten goals for a competitive and resource efficient transport system: benchmarks for achieving the 60% GHG emission reduction target.

I. Developing and deploying new and sustainable fuels and propulsion systems	II. Optimising the performance of multimodal logistic chains, including by making greater use of more energy-efficient modes	III. Increasing the efficiency of transport and of infrastructure use with information systems and market-based incentives
Halve the use of 'conventionally-fuelled' cars in urban transport by 2030; phase them out in cities by 2050; achieve essentially CO$_2$-free city logistics in major urban centres by 2030	30% of road freight over 300 km should shift to other modes such as rail or waterborne transport by 2030, and more than 50% by 2050, facilitated by efficient and green freight corridors. To meet this goal will also require appropriate infrastructure to be developed	Deployment of the modernised air traffic management infrastructure (SESAR) in Europe by 2020 and completion of the European Common Aviation Area. Deployment of equivalent land and waterborne transport management systems (ERTMS, ITS, SSN and LRIT, RIS). Deployment of the European Global Navigation Satellite System (Galileo).
Low-carbon sustainable fuels in aviation to reach 40% by 2050; also by 2050 reduce EU CO$_2$ emissions from maritime bunker fuels by 40% (if feasible 50%).	By 2050, complete a European high-speed rail network. Triple the length of the existing high-speed rail network by 2030 and maintain a dense railway network in all Member States. By 2050 the majority of medium-distance passenger transport should go by rail.	By 2020, establish the framework for a European multimodal transport information, management and payment system.

256

| A fully functional and EU-wide multimodal TEN-T 'core network' by 2030, with a high quality and capacity network by 2050 and a corresponding set of information services. | By 2050, move close to zero fatalities in road transport. In line with this goal, the EU aims at halving road casualties by 2020. Make sure that the EU is a world leader in safety and security of transport in all modes of transport. |
| By 2050, connect all core network airports to the rail network, preferably high-speed; ensure that all core seaports are sufficiently connected to the rail freight and, where possible, inland waterway system. | Move towards full application of "user pays" and "polluter pays" principles and private sector engagement to eliminate distortions, including harmful subsidies, generate revenues and ensure financing for future transport investments. |

Source: WHITE PAPER Roadmap to a Single European Transport Area – Towards a competitive and resource efficient transport system, COM/2011/0144 final - eurlex.europa.eu.

As one may see, the EU is one of the most active promoters of this idea of sustainable development. Thus, in its transport policy the EU aims at changing the demand pattern through shifting potential demand from the road transport sector towards the rail, inland waterway and sea transport – short-distance shipping as well as promoting combined transport and collective public transport. Such solutions are more environmentally friendly, thus helping pursue sustainable development. The above presented goals will be very difficult to achieve, especially for such new EU members like Poland. It is necessary to adopt a high quality sustainable transport indicators monitoring system in order to better plan and manage the transport development [Borys, 2009, Nicolas, 2010].

Table 2. Sustainable Transportation Issues (Litman and Burwell, 2006).

Economic	Social	Environmental
Accessibility quality	Equity / fairness	Air pollution
Traffic congestion	Impacts on mobility disadvantaged	Climate change
Infrastructure costs	Affordability	Noise pollution
Consumer costs	Human health impacts	Water pollution
Mobility barriers	Community cohesion	Hydrologic impacts
Accident damages	Community livability	Habitat and ecological degradation
DNRR	Aesthetics	DNRR

For comprehensive and balanced analysis, indicator sets should include indicators from each of the major categories of issues, such as those listed in Table 2. For example, it is important to have indicators of transport cost efficiency (economic), equity and livability (social), and pollution emissions (environmental).

This table lists various impacts which should be reflected, as much as feasible, in sustainable transportation indicator sets. (DNRR=Depletion of Non-Renewable Resources).

Source: T. Litman and D. Burwell (2006), "Issues in Sustainable Transportation," International Journal of Global Environmental Issues, Vol. 6, No. 4, pp. 331-347; at www.vtpi.org/sus_iss.pdf.

These are examples of sustainable transportation issues, but the table is not intended to be comprehensive. Some indicators reflect multiple impact categories; for example, traffic accidents impose economic costs from damages and reduced productivity, and social costs from pain and reduced quality of life. Fuel consumption can be a useful indicator because it reflects energy consumption, pollution emissions, climate change, and total vehicle travel, and to a lesser extent mileage-related impacts such as congestion and crash rates. On the other hand it provides limited information about actual damage to the environment [Litman, 2008].

3 SUSTAINABLE TRANSPORT INDICATORS IN THE POLISH COASTAL REGIONS

Below, some available transport indicators regarding the selected Polish regions in the period 2004-2011 have been presented. The following indicators have been taken into consideration:

- Social: dead casualties in road accidents/100 thousand inhabitants;
- Economic: motorways in km, cargo traffic in the seaports, number of passengers in the seaports, inland waterway cargo transport in thousand tons;
- Environmental: railways electrified normal gauge in km, CO_2 emissions.

Figure 3 presents dead casualties in road accidents/100 th. inhabitants. Their number has been constantly decreasing in all the regions.

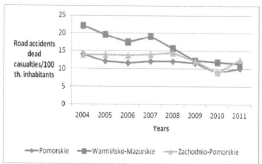

Figure 3. Dead casualties in road accidents/100 th. inhabitants in the selected Polish regions
Source: Own elaboration based on: Transport 2004-2010 - activities results, GUS, Warsaw.

Out of compared regions, in 2011 the 'safest' region was Pomorskie (10). The highest number of victims occurred in Zachodnipomorskie (12). Poland remains the country with one of the highest number of road accidents and casualties in the EU. This is due to the low quality of Polish road infrastructure and a high number of the drunk drivers.

The length of motorways has not increased in Zachodniopomorskie during the analized period of time. The construction took place only in Pomorskie where the first part of motorway A1 has been opened in 2008. In Warmińsko-Mazurskie there has not been any highways being constructed at all (fig. 4.).

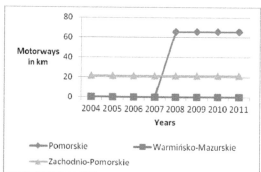

Figure 4. Motorways in km in the selected Polish regions
Source: Own elaboration based on: Transport 2004-2010 - activities results, GUS, Warsaw.

When it comes to cargo traffic in the seaports, one may see the domination of Pomorskie region, Zachodniopomorskie comes as second and Warmińsko-Mazurskie is placed far behind with minor transshipment results (fig. 5).

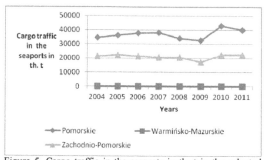

Figure 5. Cargo traffic in the seaports in th. t in the selected Polish regions
Source: Own elaboration based on: Transport 2004-2010 - activities results, GUS, Warsaw.

As far as the number of passengers entering and leaving the seaports is concerned, it has been increasing only in Pomorskie reaching roughly 1500000 passengers in 2008 (fig. 6).

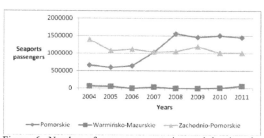

Figure 6. Number of passengers entering and leaving the seaports in the selected Polish regions
Source: Own elaboration based on: Transport 2004-2010 - activities results, GUS, Warsaw.

When it comes to the inland waterway cargo transport, the results prove that this form of transport remains neglected, except of Zachodniopomorskie region where there are signs of optimism after a growth in 2010 – a little bit more than 1500 thousand tons (fig. 7).

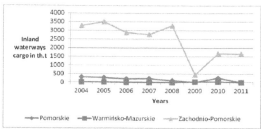

Figure 7. Inland waterway cargo transport in th. tons in the selected Polish regions
Source: Own elaboration based on: Transport 2004-2010 - activities results, GUS, Warsaw.

These results prove that the inland waterway transport plays unfortunately a minor role in reaching more sustainable development of the transport systems in the analyzed regions.

The length of railways electrified normal gauge (fig. 8) have remained at the same level which is an extremely worrying tendency, as well.

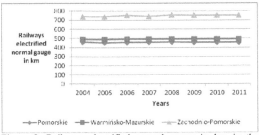

Figure 8. Railways electrified normal gauge in km in the selected Polish regions
Source: Own elaboration based on: Transport 2004-2010 - activities results, GUS, Warsaw.

This environmental-friendly form of transport could have contribute more to greater sustainability. The statistics prove that the railways stagnation concerns all the parts of Poland, reaching more than 700 km in Zachodniopomorskie and far less in Warmińsko-Mazurskie (around 500 km) and in Pomorskie (less than 500 only).

CO_2 emissions have been remaining more or less at the same level in Warmińsko-Mazurskie and their number has increased in the two other regions (fig. 9).

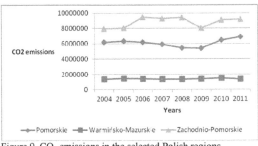

Figure 9. CO_2 emissions in the selected Polish regions
Source: Own elaboration based on: Transport 2004-2010 - activities results, GUS, Warsaw.

4 CONCLUSIONS

1 Transportation is expected to be the major driving force behind a growing world demand for energy. It is the largest end-use of energy in developed countries and the fastest growing one in most developing countries. Furthermore, adequate, efficient, and effective transport systems are important for access to markets, employment, education and basic services critical to poverty alleviation. Current patterns of transportation development are not sustainable and may compound both environmental and health problems.

2 Although transportation-related statistics are widely gathered, their quality is highly variable, and even the best data are often incompatible with those from other organizations and jurisdictions. It is highly recommended to develop a research program concerning the collection, analysis and application of high quality, standardized transportation data in order to be able to provide a suitable framework for transportation planning and policy benchmarking.

3 The statistical data analysis regarding transport indicators in 2004-2011 of the selected Polish regions confirm that the attainment of the sustainable transport development is a difficult an long-lasting process. There are some optimistic prerequisites like in the case of the decreasing number of dead casualties in road accidents. However, in the case of inland waterway and railways the indicators show that there is no big chances for shifting to environmental-friendly modes of transport.

4 Taking into consideration the analyzed sustainable transport indicators, the most eastern region – Warmińsko-Mazurskie - is lagging behind the others, with the exception of the number of CO_2 emissions being considerably the lowest when taking into account all the three coastal regions in Poland.

REFERENCES

Borys T. (2008), Analiza istniejących danych statystycznych pod kątem ich użyteczności dla określenia poziomu zrównoważonego transportu wraz z propozycją ich rozszerzenia, Raport z realizacji pracy badawczej, Ministerstwo Infrastruktury, Jelenia Góra–Warszawa.
ECMT (2004), Assessment and Decision Making for Sustainable Transport, European Conference of Ministers of Transportation, Organization of Economic Coordination and Development (www.oecd.org).
Europe 2020 Strategy (COM(2010)2020, final.
Grzelakowski A. S. G., (2008). European greener mobility, Baltic Transport Journal.
http://www.un.org/esa/dsd/susdevtopics/sdt_transport.shtml, (2010-07-10).
Litman T. [2008], Sustainable Transportation Indicators. A Recommended Research Program For Developing Sustainable Transportation Indicators and Data, http://www.vtpi.org/sustain/sti.pdf.
Litman T. and Burwell D. (2006), "Issues in Sustainable Transportation," International Journal of Global Environmental Issues, Vol. 6, No. 4, pp. 331-347; at www.vtpi.org/sus_iss.pdf.
Medhurst J. (2003), A Thematic Evaluation of the Contribution of Structural Funds to Sustainable Development: Methods and Lessons, REGIONET Network of Excellence Workshop, Manchester.
Nicolas J.-P., 2010, SIMBAD: un outil pour integrer le developpement durable dans les politiques publiques, [in :] Antoni J.-P. (red.), Modeliser la ville. Forme urbaine et

politiques de transport. Paris: Economica, coll. Methodes et Approches.

Przybylowski, A. (2011), Sustainable Transport Planning & Development in the EU at the Example of the Polish Coastal Region Pomorskie, [in:] Transport Systems and Processes, Marine Navigation and Safety of Sea Transportation, Edited by Weintrit, A. & Neumann, T., CRC Press, Taylor & Francis Group, London, UK, p. 101-110.

Transport 2004-2011 - activities results, GUS, Warsaw.

Vivien F.-D., (2005), Le developpement soutenable, La Decouverte, Paris.

WHITE PAPER Roadmap to a Single European Transport Area – Towards a competitive and resource efficient transport system, COM/2011/0144 final - eurlex.europa.eu.

World Commission on Environment and Development (WCED), Our common future, Oxford: Oxford University Press, (1987).

Baltic Sea Logistic and Transportation Problems
STCW, Maritime Education and Training (MET), Human Resources and Crew Manning, Maritime Policy,
Logistics and Economic Matters – Marine Navigation and Safety of Sea Transportation – Weintrit & Neumann (Eds)

Chosen Problems of Financing of the Logistic Centers in Polish Seaports

A. Salomon
Gdynia Maritime University, Gdynia, Poland

ABSTRACT: We should treat port logistic centers as intermodal knots, which by legislating permanent, regular connections of container terminals, are fastening together into the uniform logistic network. Intermodal logistic centers, with the public access to services offered on their area must to be equipped with technologies of handling, warehousing, standardized information exchanges and the like, opened to different clients, so that they don't create barriers on the road of the flow of goods and the information. From this point of view we should perceive logistic networks on the organizational, technical and functional levels.

The main purpose of the article is to show the chosen problems tied up with financing the logistic centers and chains of supplies in Polish Seaports, peculiarly based on the Public-Private Partnership.

1 INTRODUCTION

The evolution occurring in the international trade and the increase of attraction of Poland on the world markets are contributing to the development, on the territory of our country, main transport trails, get in touch with transport arteries in countries of Centre-Eastern Europe countries. The transport corridors, marked out by Polish territory, as well as the access to the Baltic Sea, are extorting changes in the approach towards current transit technologies and the transport policies. So that to live up these requirements and don't stay on edges of developing economic centers of the world, investors should take a lot of program action.

At present of increasing tendency in the production and the goods exchange, as well as in looking for the lowest costs, is turning up for modern solutions in the transport and warehousing. In order to cope with these challenges, the main pressure is being put on the building of the objects, in which, with the lowest financial outlays, a process storing and processing goods is being carried out. There are also handling cargo in intermodal systems, what increases the competitiveness of different branches of the transport and considerably is influencing over the environmental preservation. The perfect space for the location of such objects (the logistic centers) there are seaports. The distribution-logistic function, which developed in seaports, making the ports the chance aren't up till

now only place, in which the change the means of transport is coming, but became the logistic knot in the whole transport system.

Polish main seaports – Gdańsk, Gdynia and Szczecin with Świnoujście – are meeting requirements concerning construction in them of logistic centers. So that well compete with different ports located in the area of the Baltic Sea as well as become focal points of transport corridors on a map of Europe, is the most urgent need of construction of new logistic centers in Polish seaports.

2 KINDS OF CAPITAL FOR FINANCING LOGISTIC CENTRES

In the contemporary economy relations between individual elements of chains of supplies often surrender to the globalization. Managing the global chain of supplies consists in connecting action of all chain links of supplies in order to transform raw materials and half-finished products into finished goods and providing of them together with the service to hands of the customer [7]. Financing the construction of both the development of logistic centers and chains of supplies is a part of the realization of purposes of investments carried out by the Budget and the same sources are answering it. In financing construction and the development of logistic centers a really can take the participation:

public capital (budget), private capital and capital coming from measures of aid funds of the EU.

Public capital of construction and the development of logistic centers in Poland is being shaped by the National Plan of the Development (NPD) for years 2007-2013. It is a comprehensive program of the social-economic development of Poland in the financial perspective of union budget till 2013. An influence on the transport is one of NPD parts through:

1 the modernization and construction of the infrastructure of the transport;
2 improvement in the effectiveness of using of the infrastructure of the transport;
3 increase of the safety of the road traffic;
4 increase in the effectiveness of action of transport enterprises;
5 development of modern transport technologies;
6 development of the intermodal transport.

For years 2007-2013 budget took out over 134 bn euro. It consists of public national fund, European Union structural funds, the UE Fund of the Cohesion, the shared agrarian and fishing policies, domestic and international financial institutions and private enterprises. These funds were (and of minimal size still are) in the order of sector programs and regional programs. It is possible in them to search out sources of financing the development of logistic centers as well. The structure of these centers looks as follows [5]:

1 state funds and units of the local government – 30.8 bn euro;
2 UE capital – 80.6 bn euro;
3 private funds – 23 bn euro.

The very important instrument of supporting projects from the scope of the transport, the freight forwarding and the logistics is a possibility of giving the guarantees by the State Treasury. The bank of the Domestic Household (Bank Gospodarstwa Krajowego) is authorised to give of the guarantee the bonds and credits taken out by participating subjects in financing the development of logistic centers.

Important place in financing the development of logistic centers, especially in seaports, is taking the Strategy of the Regional Development, which considers the regional politics and policies of the regional development of provinces.

3 ROLE OF THE PUBLIC-PRIVATE PARTNERSHIP IN FINANCING THE DEVELOPMENT OF LOGISTIC CENTERS

In Poland there are limited funds of private and public capital, so the foreign aid budget and the public-private partnership (PPP) are being a chance for financing the development of logistic centers [2].

Public-private partnership (PPP) is a form of the long-term cooperation of sectors public and private at the completion of investment. According to the PPP conception an achievement of shared benefits in the field of the commercial and social measurement is a purpose of the cooperation. Getting bigger value and the higher quality for the lower price than in the case is a result of the cooperation of financing investment in the traditional way [1].

In the case of the PPP formula both sectors can together carry the investment undertaking out, even though each of these sectors can have different purposes inducing it for the cooperation. The public sector can have the following purposes in mind:

1 carrying out setting about the wider social measurement in the shorter time;
2 lowering costs of the completion of investment and making smaller investing own funds;
3 assuring optimal financing investment in the longer temporary compartment;
4 achieving social benefits into forms of the inflow of new investments and new places of employment, stimulating the growth in the economy.

The private sector can be interested in the realization of completely different purposes [3]:

1 achieving notable benefits from the completion of investment in forms of the turn from invested capital;
2 possibility of generating additional gross receipts from operational activity;
3 lowering costs of investment and operating costs;
4 limiting the risk by moving the that kind of risk which this sector better is managing to the public sector.

The undertaking carried out according to the PPP formula is enabling the better identification of the risk, his allocation and managing it, than in the case of the undertaking carried out in the traditional way [4]. But we have to be careful. A series of studies of large transport projects found that PPP projects costs are systematically under-estimated, and benefits often under-estimated. For example: a study of 258 transport projects found that, on average, actual costs were 28 percent higher than planned costs – and 65 percent higher on average for projects outside Europe and North America. On the other hand a study of 25 rail projects found traffic was heavily overestimated, at over twice actual traffic, on average. The accuracy of traffic forecasts for 183 road projects was also found to be highly variable, but without a tendency to over-estimate on average [6].

The following categories of risk are common to many PPPs:

1 Site – risks associated with the availability and quality of the project site, such as the cost and timing of acquiring the site, needed permits and the cost of meeting environmental standards;

2 Design, construction and commissioning – risk that construction takes longer or costs more than expected;
3 Operation – risks to successful operations, including the risk of interruption in service or asset availability;
4 Demand, and other commercial risk – the risk that usage of the service or revenues is different than was expected;
5 Regulatory or Political – risk of regulatory or political decisions, or changes in the sector regulatory framework, that adversely affect the project;
6 Change in legal framework – the risk that a change in general law or regulation adversely affects the project;
7 Sponsor, or default – the risk that the private party to the PPP contract turns out not to be financially or technically capable to implement the project;
8 Economic or financial – risk that changes in interest rates, exchange rates or inflation adversely affect the project outcomes;
9 Force Majeure – risk that external events beyond the control of the parties to the contract, such as natural disasters, war or civil disturbance;
10 Asset ownership – risks associated with ownership of the assets, including the risk that the technology becomes obsolete or that the value of the assets at the end of the contract is different than was expected.

Between the scale of the risk and the waited payback from investment a distinct relation is appearing. Private investor taking the great risk of the failure into account he will expect the tall base of the turn, in addition the assessment of the risk can completely discourage it for the completion of investment. Taking the risk to itself the public sector in exchange will be offering few benefit considerably to the private investor. Finding the appropriate proportion between the risk and expected benefits is one of two crucial elements of the success of realization of the public-private partnership. Making the size of the risk smaller is second, for example by raising financial funds from UE programs which will make smaller the size of invested private capital in investment as well as can induce the bank of the part for financing remained of investments with credit given on better conditions [10].

4 BENEFITS OF THE COOPERATION BETWEEN THE PUBLIC INVESTORS AND PRIVATE

Benefits of the cooperation between the public and private investors are possible to put in order as follows [1]:

1 main benefit of the public sector is a possibility of long-term investments about the commercial and social dimension, burdened the considerable investment risk;
2 agreements most often concern the investment and exploitation phase, what is correcting the effectiveness of using financial funds and of the wealth making the non-cash refill of partners;
3 commercial partners are contributing experiencing into managing with investment project and are interested in the optimization of costs in the entire cycle of the realization and the exploitation of the object being an object of investment;
4 the sides of agreement better dividing the risk between themselves, what managing them is more effective and in the smaller rank than in the case of the traditional collaboration can to discourage partners at the cooperation;
5 the model of financing investment permits for reducing costs of the realization, thanks of the possibility of non-commercial financial funds and including in the schedule financing next investment tasks of these gross receipts which are turning up after the realization of determined phases of investment before the whole project will be finished.

The public-private partnership is becoming the main formula for the completion of investment into logistic centers. In order to functioning that formula in Poland with full benefits, it needs legal acts, which, from one side will distance the suspicions of the sale of the public business for commercial subjects, on the other side – will remove barriers which are accumulating ahead of investment consortiums, which they are supposed to order effectively with capital and risks accompanying investment undertakings [9].

Very important factor, which at the moment is impeding the development of PPP in Poland, is legal acts making uncomfortable and stopping the public sector and the private access of subjects for public funds and the completion of reserved tasks for the public sector.

Obstacles which they are preventing PPP from entering it into the transportation contracts, concern [1]:

1 the procedures of choice of the private investors as the partner in the completion of investment, which public money is involved in;
2 dragging obligations by public investors in frames of the PPP agreement;
3 legal limitations of rewarding private investors from public money;
4 limitations of the possibility of making described public objectives by private investors.

In the case of very expensive projects, where the private sector is need to switch on for the co-realization, is also a possibility of mix financing: in the PPP formula with co-financing using UE funds (hybrid model). Such a cooperation can take place at applying different model solutions. In the classic hybrid form the whole of the responsibility for financing, construction and the exploitation of the infrastructure is incurring the private investors, which at the same time is applying for EU funds and independently he is accounting for it.

This method is enabling the amount of expenses incurred by the public sector to lower and, thanks to including of funds the EU, reducing payments for the private investors. PPP formula is making the possible to use full advantages of the financing method for private investor.

A division of the project into two sub-projects is an alternative method to the classic hybrid form. First part is co-realized in the PPP formula and second – using EU funds. After finishing the construction stage, the private investors are responsible for exploitations of the project. The other possibility is less complicated, however it requires from the public partner to guarantee the invest money on the stage of investment.

But independently from obstacles, the present state and perspectives of local authorities finances are pointing, around in the most recent years administrative districts, second levels of local government administration in Poland and regions will more and more often reach for Public-Private Partnership. The private sector capital investment into providing the public service has a future.

In practice, the development of logistic centers is a domain of the private sector, however the role of the state in this area is not for overrating. It is responsible, above all, for the development of the infrastructure enabling the get the logistic center. As main places, in which the development would be more rapid, thanks to the more modern train and road infrastructure, it is possible to name the main Polish seaports [8].

An interaction of private and public capital is also possible inside logistic centers. It is often met for example exactly in seaports, which in the case of Poland, are making the national property. Private capital, with reference to new logistic centers, is playing not only a role of financing the source, but it should be also the guarantor of accepting market premises to the uprising center. The location of logistic center must be result of real needs of the market, rather than only administrative decision. The role of public capital, besides with fact that there is this source of financing, is also a word of the transport policy of the state. By supporting the development of infrastructure and transport technologies, it is possible to promote the defined kind of the transport, for example – for the more environmental.

5 ENDING

PPP projects are a reply to the limited possibility of financing of infrastructure investments from public funds. They are actively supported by international financial institutions for example: European Bank for Reconstruction and Development – EBRD and European Investment Bank – EIB. They are supporting projects which are contributing for increasing European integration. The EIB participation cannot exceed the 50% of investment costs, whereas the minimum support direct in the form of loans must to take the 25 m. euro out. Apart the financial aspects, EIB is operational on also advisory activity directed to authorities of membership states and of the institution the EU, in particular in the scope of the development and using instruments in the PPP formula.

REFERENCES

[1] Fechner I., Centra logistyczne. Cel – Realizacja – Przyszłość, Instytut Logistyki i Magazynowania, Poznań 2004 (in Polish)
[2] Fechner I., Centra logistyczne i ich rola w procesach przepływu ładunków w systemie logistycznym Polski, http://www.it.pw.edu.pl/prace-naukowe/z76/fechner.pdf (in Polish)
[3] http://www.partnerstwopublicznoprywatne.info (in Polish)
[4] Korbus B. (ed.), Partnerstwo Publiczno-Prywatne. Poradnik, Warszawa 2010, http://ppp.parp.gov.pl/uploads/files/CK_PPP_Partnerstwo_Publiczno_Prywatne_UZP_2010_PL.pdf (in Polish)
[5] Mindur M., Logistyka. Infrastruktura techniczna na świecie. Zarys teorii i praktyki, Instytut Technologii Eksploatacji – RIB, Radom 2008 (in Polish)
[6] Public-Private Partnerships. Reference Guide, World Bank Institute, Washington 2012, http://www.ppiaf.org/sites/ppiaf.org/files/publication/Public-Private-Partnerships-Reference-Guide.pdf
[7] Wiśniewska A., Rola operatorów logistycznych w łańcuchach dostaw, http://www.akademor.webd.pl/download/lt_8.pdf (in Polish)
[8] Wołek M., PPP w tworzeniu usług logistycznych w obrocie portowo-morskim, Instytut Morski, Gdańsk 2007, http://www.logvas.com/fileadmin/Logvas/Intermediate_Conference/City_Hall_Gdynia.pdf (in Polish)
[9] Wskazówki dla podmiotów publicznych zainteresowanych podejmowaniem partnerstwa publiczno-prawnego, Ministerstwo Gospodarki, Warszawa 2007, http://www.mg.gov.pl/NR/rdonlyres/12B5A328-3485-4858-AC39-88C2976E830B/29863/wskazowkiPPP.pdf (in Polish)
[10] Wytyczne dotyczące udanego partnerstwa publiczno-prawnego, Komisja Europejska, Bruksela 2003, http://www.ppp.gov.pl/Poradnik_inwestora/AktyPrawne/Documents/Wytyczne_Komisji_PPP_190111.pdf (in Polish)

Baltic Sea Logistic and Transportation Problems
STCW, Maritime Education and Training (MET), Human Resources and Crew Manning, Maritime Policy,
Logistics and Economic Matters – Marine Navigation and Safety of Sea Transportation – Weintrit & Neumann (Eds)

Trans-shipping Terminal in Polish Logistic System

Z. Łukasik & A. Kuśmińska-Fijałkowska
University of Technologies and Humanities, Radom, Poland

ABSTRACT: Real object is the subject of the authors' research- inland trans-shipping terminal functioning in the logistic system in Poland.

1 INTRODUCTION

The proper functioning of the intermodal transport chain depends largely on the proper functioning of inland terminals, and above all on their location, the ability to carry cargo, handling infrastructure, cost, scope of services, quality and reliability (Kuśmińska – Fijałkowska & Łukasik. 2011). Land and sea terminals in our country create a network of intermodal transport infrastructure.

Eastern border passage open for containers

Existing terminals PKP CARGO S. A.

Planned terminals of PKP CARGO S. A.

Private terminal, public

Service in container terminals

Containers

Semitrailer

Replaceable bodies

Possibility of connecting cooling sets to the admission

International railway lines- AGTC

Different railway lines

SUW 2000

Figure 1. Loading terminals in Poland (www.pkpcargo.pl)

It is expected that Poland will increase European trade turnover as a result of a combination of factors such as:

- natural increase - the increase in trade between the EU and the neighboring countries of Central and Eastern Europe,
- increase in the limit - strengthening environmental factors that are conducive for the development of intermodal transport,
- increase in the efficiency - the development of the EU policy of charging for transport users, resulting in competition between different modes of transport and promotion of intermodal transport,
- EU enlargement to new countries - should create significant opportunities for further development of the sea - land intermodal transport chains, and this applies to, among others, the Baltic Sea

region, especially Poland, where a significant increase in turnover with the EU is particularly important from the point of view of Polish maritime position in the region.

Having regarded the above trends, the terminals in Poland must be prepared to take more incoming JTI (Intermodal Transport Units). Poland, in terms of number of terminals, does not differ significantly from European countries except for: Czech Republic (65), France (83), Germany (119) and Italy (104). (Kuśmińska–Fijałkowska & Łukasik, 2012)

2 INFRASTRUCTURE AND TECHNICAL POSSIBILITIES OF THE EXAMINED TRANS-SHIPPING TERMINAL

Road vehicles arrive at terminal on land (Fig. 2) to deliver the units for a given set of cars. Most goods arrive much earlier than the cars to be loaded. These observations are checked by this research. The change in the type of road and railway transportation is the aim of applying intermodal transportation in the transportation network and their integration. (Kuśmińska-Fijałkowska & Łukasik, 2011)

Figure 2. Units entring and leaving land terminal (Kuśmińska–Fijałkowska & Łukasik, 2011)

Infrastructure and technical capabilities of the terminal handling are shown in Table 1.

Table 1. Technical capabilities of Terminal A (own study)

Terminal A - Technical features	
Trans-shipping capabilities	41 tons
number and the length of railway tracks	$2 \times 700 = 1400$
surface of the terminal	76 000 m^2
area of storing	53 800 m^2
possibility of storing	6000 TEU, 3 coset
kind of JTI reloaded	containers, swap bodies, semi-trailers

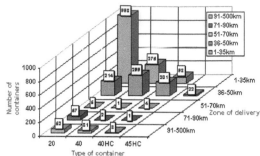

Figure 3. Number of the definite type of containers sent depending on the zone of delivery by road transportation to the customer in years 2011 (own study the land terminal)

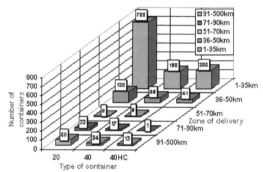

Figure 4. Number of the definite type of containers sent depending on the zone of delivery by road transportation to the customer in years 2012 (own study of land terminal)

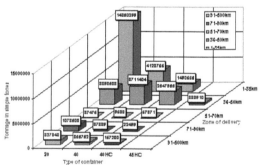

Figure 5. Tonnage of JTI sent from land terminal by road transportation depending on the zone of delivery in years 2011 (own study of land terminal)

266

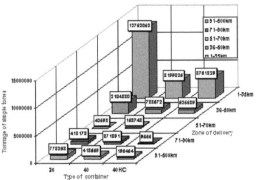

Figure 6. Tonnage of JTI sent from land terminal by road transportation depending on the zone of delivery in years 2012 (own study the land terminal)

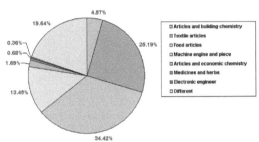

Figure 7. Proportion of goods sent from land terminal by road transportation depending on distance (1-35 kilometers) in years 2011-2012 (own study of land terminal)

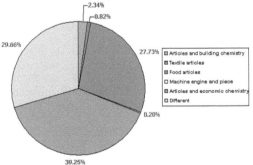

Figure 8. Proportion of goods sent from land terminal by road transportation depending on distance (36-50 kilometers) in years 2011-2012 (own study of land terminal)

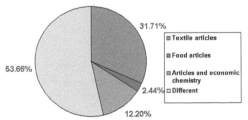

Figure 9. Proportion of goods sent from land terminal by road transportation depending on distance (51-70 kilometers) in years 2011-2012 (own study of land terminal)

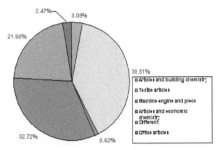

Figure 10. Proportion of goods sent from land terminal by road transportation depending on distance (71-90 kilometers) in years 2011-2012 (own study of land terminal)

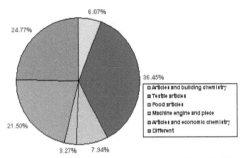

Figure 11. Proportion of goods sent from land terminal by road transportation depending on distance (91-500 kilometers) in years 2011-2012 (own study of land terminal)

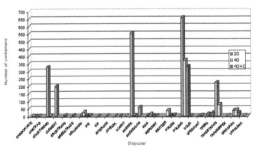

Figure 12. Number and type of containers sent from land terminal by road transportation depending on the dispatcher and on distance (1-35 kilometers) in years 2011-2012 (own study the land terminal)

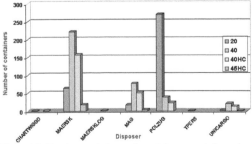

Figure 13 Number and type of containers sent from land terminal by road transportation depending on the dispatcher and on distance (61-50 kilometers) in years 2011-2012 (own study of land terminal)

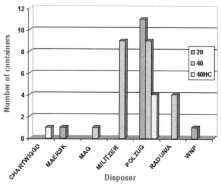

Figure 14. Number and type of containers sent from land terminal by road transportation depending on the dispatcher and on distance (51-70 kilometers) in years 2011-2012 (own study of land terminal)

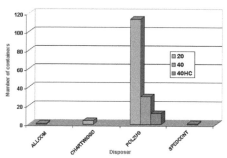

Figure 15. Number and type of containers sent from land terminal by road transportation terminal depending on the dispatcher and on distance (71-90 kilometers) in years 2011-2012 (own study of land terminal)

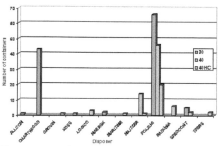

Figure 16. Number and type of containers sent from land terminal by road transportation terminal depending on the dispatcher and on distance (91-500 kilometers) in years 2011-2012 (own study of land terminal)

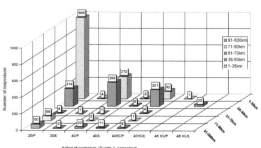

Figure 17. Number of new units on road vehicle terminal depending on the zone of delivery in years 2011 (own study of land terminal)

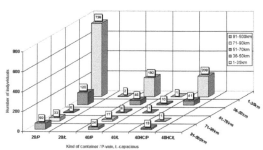

Figure 18. Number of new units on road vehicle terminal depending on the zone of the delivery in years 2012 (own study of land terminal)

The analysis of the data shows that road vehicles come to the land terminal to deliver units for the train, considerable long time before the train. These observations are confirmed by research carried out. The situation with the real object was "very poor" case corresponding to option from the chart in Figure 19 (storage of JTI). However, the aim is to ensure that JTI arrive at such a time to the terminal to undergo a direct transshipment - Figure 20. If in our national conditions, the intermodal transport is to be faster than road transport, the processes of JTI arriving by road vehicles should be improved, as well as the flow of JTI in the terminal, which is the link of the road and railway. [1] (Kuśmińska–Fijałkowska & Łukasik, 2011).

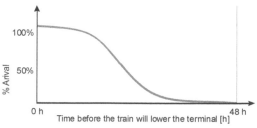

Figure 19. Proportion of road vehicles arrival with units in the function of time (poor option of the arrival) (S.C.2.70 Deliverable 3. 1999)

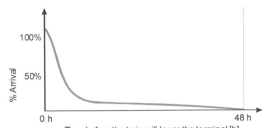

Figure 20. Proportion of road vehicles arrival with units in the function of the time (best option of the arrival) (S.C.2.70 Deliverable 3. 1999)

3 CONCLUSIONS

As a result of the statistical analysis of the data obtained in the land terminal it can be seen that in the number of units sent to customers by trucks, within a radius of 1-35km, fully laden 20-foot containers (Figure 3, Figure 4) are most numerous. It results from tonnage and percentage comparison of cleared goods (Fig. 5 Fig. 11). Type of cleared goods of JTI results from the location of the terminal in the Polish logistics system. However, the greatest number and type of units leaving the terminal, according to statistics, has a dispatcher from Polzug

at different distances, and Maersk, Chartwig, Tradetrans, as shown in the Figures (Figure 12-16). Considering the different flow direction of JTI, namely arriving at terminals on road vehicles, to be later transported by railway it can be easily seen that these units are empty. The largest number of JTI in 2011-2012 came from zones 1-35km and 36-50km within the inland terminal (Figure 17, Figure 18) which implies that the tested terminal has a good location in the Polish logistics system.

REFERENCES

Kuśmińska A., Łukasik Z.: „Identification of the individuals of intermodal transpotion and technical solutions in the trans-shipping terminal" 14th International Scientific and Technical Conference on Marine Traffic Engineering 12-14 October, Świnoujście 2011, p. 281-291, ISBN 978-83-89901-63-7

Kuśmińska A., Łukasik Z.: „Identification in the process of flow individuals intermodal transportation in the trans-shipping terminal" Akademia Morska w Szczecinie, Zeszyty Naukowe nr 29(101), Szczecin 2012, pp. 101-108, ISSN 1733-8670

S.C.2.70 Deliverable 3: the „Design of of Platforms Simulation environmentt" Lugano 31.01.99

www.pkpcargo.pl

Chapter 7

Financial Indices, Freight Markets and other Economic Matters

Financial Indices, Freight Markets and other Economic Matters
STCW, Maritime Education and Training (MET), Human Resources and Crew Manning, Maritime Policy,
Logistics and Economic Matters – Marine Navigation and Safety of Sea Transportation – Weintrit & Neumann (Eds)

The Impact of Freight Markets and International Regulatory Mechanism on Global Maritime Transport Sector

A.S. Grzelakowski
Gdynia Maritime University, Gdynia, Poland

ABSTRACT: The main purpose of the paper is to analyze the forms of direct and indirect impact of two global regulatory systems on the world shipping industry. The author intends to characterize the freight market mechanism as well as its relations with global commodity markets in creating growth potential for the global maritime transport sector. Subsequently, the existing international regulatory scheme introduced by IMO and other international organizations (UE, EMSA, ILO, etc.) will be analyzed and viewed in terms of its efficiency and impact on the shipping industry. At the end, the results of functioning of both regulatory subsystems will be assessed with the aim to indicate how the existing conflicting areas as well as inconsistencies existing between them might deteriorate the system itself and hamper the development of the global shipping industry, and subsequently, the global merchandise trade and the whole economy.

1 REGULATORY SCHEME OF THE GLOBAL MARITIME TRANSPORT SECTOR

1.1 The main systems and mechanisms of regulation of the global maritime transport

Maritime transport sector just like many other forms and areas of international commercial and economic activity has been regulated for ages. Nowadays, real activity of that global sector is subject to regulation of the complex system which consists of two independent, however different in their nature, regulatory subsystems. Each of them has its own specific characteristics. It is a public subsystem, the essence of which consists in international as well as autonomous subsystem based on freight market regulatory processes. The first one is external to the transport sector, while the second, is both external and internal, largely because one of the markets, namely freight market, is an integral element of the sector itself – it is one of its subsystems. Each of them, along with the set of regulatory tools typical for itself, affects to some degree the efficiency of the operational activity of maritime transport and its relationship with global environment. The mutual relationship between the regulatory scheme of the maritime transport sector and its subsystems, and the maritime transport real sector, is presented, in a schematic form, in fig.1.

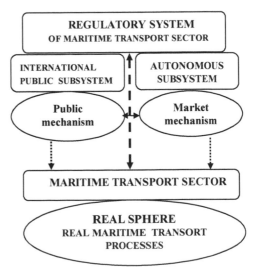

Figure 1. The regulatory system of the international maritime transport and its components
Source: A.S. Grzelakowski, Maritime Transport Development in the Global Scale – the Main Chances, Threats and Challenges. *International Journal on Marine Navigation and Safety of Sea Transportation.* Vol. 3, No. 2, 2009, p.199

The presented scheme shows that international maritime transport is subject to the regulation of the two, running in parallel, although to a large extent independently from each other, and hence, not fully coordinated with each other, regulatory subsystems. Each of them has its own independent regulatory mechanism influencing directly the real activity of the maritime transport sector and indirectly its various relationships with other sectors of the global economy. The primary regulatory subsystem remains as before the market subsystem. On a short-, mid- and long-term basis it plays a steering role in regulating the real activity of the international, widely globalized maritime transport sector.

The global maritime transport sector performs its duties within well-developed typical international market framework, still substantially fragmented. It is characterized currently in general by well-advanced degree of liberalization. However, in some of its segments, mainly regional ones, relatively strong state regulatory measures and simultaneously, protectionist practices are maintained. While characterizing market environment of the maritime transport sector, one should note that it consists not only of the global freight market but also of other types of transport markets as well as commodity, capital and labor markets. Each of them has a significant impact on the maritime transport sector. All these markets through their regulatory mechanisms shape real processes carried out in the maritime transport sector, i.e. its real activity and eventually determine the form of its links with the global environment which is a global logistics system (logistics mega-system).[72] The position and function as well as the relations between the maritime sector and other components of the logistics mega-system, i.e. the logistics supply chains and networks are presented in figure 2.

It is obvious that the efficiency and effectiveness of the autonomous regulation of the real activity of maritime transport depends on the degree of coherence of various types of markets and, in fact, it is specified by the level of liberalization of each of them (fig.2). The more it is aligned, the more efficiency and productivity is generated by market regulation regarded as a complex autonomous subsystem. Such subsystem, along with its equally liberalized market mechanisms, able to produce synergy effects, is fit enough only to act as an instrument of optimization of the real activity of the international maritime transport. One should note that such a complex market optimization of the short-, mid-, and long-term real processes carried out in the international shipping sector, applies not only to the sector itself but also to the whole environment it is closely linked with, i.e. to logistics mega-system (fig.2).

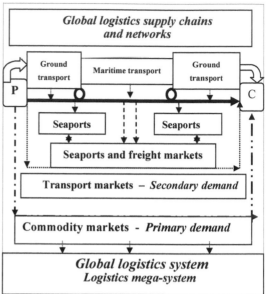

Figure 2. International maritime transport in the global logistics system.
Source: As in fig. 1, p. 203,

1.2 *Diffusion of regulatory processes within the global maritime transport regulatory scheme*

The mechanism of market regulation of the global maritime transport sector which determines its position and function in the global logistics supply chains and networks constituting logistics mega-system is, in fact, autonomous. However, as such, it has long been subject to regulation. There are many reasons for this, but unlike other transport markets (road, rail, etc.), which have not been fully developed, and immature markets (infant markets) even at the stage of global deregulation (WTO) are nowadays still operating in many regions as generally imperfect ones, the freight markets do not need any form of de-monopolization of their supply side or support in terms of construction a fully-developed, competitive and transparent regulatory pricing mechanism.[73] In the maritime transport sector, which is one of the most internationalized areas of global economic activity, the reason for public intervention in market activity is primarily to: 1/ provide adequate conditions and safety standards in the carriage of people and goods (in terms of safety and security), as well as ensure security of the supply chain (supply chain security) along with ecological and social security in this area of

[72] A. S. Grzelakowski., Internalization of External Costs in the EU Transport Sector as an Instrument of Rationalization of the Logistics Supply Chains. *Logistics and Transport*., No. 2, 2011, p. A. S. Grzelakowski, Mechanizm regulacji sektora transportu morskiego w skali globalnej i w UE Międzynarodowa regulacja versus rynek frachtowy „Przegląd Komunikacyjny" 2012, Nr 12, p. 7 – 8

[73] Grzelakowski A. S., Transportation Markets as the Instruments Transportation Systems Regulation and Optimization. Methodological Aspects. [in:] Contemporary Transportation Systems. Selected Theoretical And Practical Problems. The Development of Transportation Systems. Edited by: R. Janecki, G. Sierpiński. Wyd. Politechniki Śląskiej. Gliwice 2010

transport, 2/ establish uniform rules for access to the profession in the shipping sector and entry to its market as well as perform, in a proper way, all regulatory functions by maritime administration of coastal states, 3/ harmonize legal, organizational and technical standards (standardization of loading units, documents, etc.) in international transport by sea and land, in order to ensure better integration of intermodal transport and logistics operations within the supply chains and networks in the global logistics system. The market mechanism is not able to pursue automatically all these tasks and goals in the international dimension. Attempts taken to regulate freight markets are generally focused on the supply side, both on potential and effective one.

Public intervention, which from the very beginning has been perceived as a form of public control in the maritime transport sector, and as such, is usually implemented indirectly through the market mechanisms, has been applied by two subsystems: These are: 1. national regulation subsystem and 2. international one (fig.1). The first one contains general standards as well as legal and administrative arrangements set in each maritime country related to this sector of transport - its operational sphere of activity. The second one, however, comprises the package of international regulations related directly and indirectly to the maritime transport activity such as: conventions, resolutions and recommendations of international organizations, governmental and non-governmental organizations, as well as various trade and economic associations and communities. Nowadays, international public regulation of the world shipping sector has been carried out by many specialized international organizations such as: IMO, EU, WTO, ILO, UNCTAD and other regional branch associations e.g. ECSA. The second public regulation system since the very beginning of globalization has played the dominant role and in many areas has replaced the national subsystems the role of which is being gradually diminished. [74] This process of growing internationalization of the public regulatory system in the shipping sector is irreversible on a world scale. It is the most advanced in the countries being members of free trade zones and especially in those ones integrated in form of economic communities, like the EU. The EU member states have worked out a common shipping policy which is oriented at substantial international (IMO) goals and instruments, especially as far as safety and security standards are concerned.

Therefore, the real activity of the global maritime transport sector is formed by very complex regulatory system based on two mechanisms which in fact do not have complementary character. At that time when the leading shipping countries played the dominant role in the world maritime transport sector their national shipping policies strongly affected the international public regulation subsystem. As the freight market regulation subsystem was at that time barely in its initial phase and in fact unable to assure free and competitive market regulation, such public regulatory mechanism tried to substitute to some extent the market mechanism. Therefore, public regulatory subsystem replaced the market one in some areas through protectionist practices and, as a result, deformed market relations in the global shipping sector. Only as a result of ongoing processes of globalization of the maritime transport sector, growing integration and liberalization of cargo and freight markets and, consequently, the development of logistics and global supply chains, those subsystems and their mechanisms have begun to lose their typically substituting character bringing gradually closer each other. It is expressed in the form of still closer and closer adaptation of the area of international public regulation (international shipping policy) to the requirements of the global freight markets and is being affected by the shrinkage of the national regulation in favor of the international one which has already been performed according to the much more liberalized principle: not instead of the market, but through the (freight) market. It means that the main area of its impact is becoming the freight market itself and predominantly, its supply side.

Such a strategy has been effectively implemented by IMO and the EC for more than 15 years. It has gradually increased the scope of international cooperation in the area of the global freight market regulation. The IMO has played the leading role, but in recent years the driving force, especially as far as the environmental standard in the shipping industry is concerned, was taken by the EC, becoming the trigger of initiatives and real changes in this field.

2 SHIPPING MARKET AS REGULATORY MECHANISM AND ITS IMPACT ON THE GLOBAL MARITIME TRANSPORT SECTOR

2.1 Commodity and freight markets as autonomous regulatory mechanism in the international maritime transport

Commodity markets as primary ones determine the dynamics and the efficiency of freight markets. As a result, the shipping markets, belonging to the group of secondary ones, in a natural way have to reflect the main tendencies and processes occurring on the primary markets. It means that all freight market components and primarily its demand side, is influenced to a great extent by numerous fluctuations of effective supply and demand for

[74] G. Steinerts, Effectiveness of the European Maritime Policy Instruments. *International Journal on Marine Navigation and Safety oft Sea Transportation* .Vol. 6, No. 2, 2012, p. 274-275

commodities transported by sea. Such a direct form of correlation existing between particular segments of adequate commodity markets and freight markets creates a special pattern of economic relations and ties. They are expressed in a traditional way in market terms by parameters of price elasticity of demand for shipping services. Therefore, the indices of price elasticity of demand for this kind of services (E_{DMT}) are calculated as: [75]

$$E_{DMT} = E_{DCM} \times K_{MT} / P_{CC}, \text{ where:}$$

E_{DCM} - price elasticity for commodities transported be sea,
K_{MT} - maritime transport costs related to those commodities,
P_{CC} - final price of commodities in the point of destination (consumption center).

The method of calculating price elasticity of demand for services provided by shipping companies clearly indicates that:
1 demand for maritime transport services has derivative character and is induced by demand for commodities carried by sea,
2 commodity markets play vital role in shaping and structuring the freight markets, simultaneously forming their dynamics as well as the effectiveness of meeting the effective demand for maritime transport services.

Taking the above into account it must be concluded that commodity markets constitute, alongside the international maritime transport policy, an important direct tool of freight market regulation. It means that these markets, together with other markets such as: transport, capital or labor ones, influence indirectly the real sphere of activity of the maritime transport sector, determining its efficiency as well as the virtue of freight market mechanism. As a result, their mechanism, autonomous in its nature, is subject to relatively strong impact of other autonomous regulatory forces, induced, in fact, by the market-oriented environment of the global maritime transport sector. This means, de facto, that the autonomous mechanism in this transport sector is fairly comprehensive in its character. However, as such, due to regulatory weaknesses, accounts for the global public regulatory mechanism. The complexity of the nowadays existing autonomous regulatory system, along with all interactions occurring among all its components and especially between demand and supply on commodity and freight markets, are briefly presented below.

The international seaborne trade as an absolutely dominant component of world merchandise trade and world maritime transport that remains the backbone of international trade and global economy, has supported strongly not only the ongoing processes of globalization but also the rapid development of the global supply chains and networks. Both are fueled by the world economic growth, i.e. increasing world production and consumption stimulating GDP on a global scale and consequently, merchandise trade. The last one, due to highly advanced processes of deregulation of economies and liberalization of all types of markets on a global scale, has grown very fast in recent decades, leaving far behind other macro parameters reflecting those tendencies.

In 2011, the volume of international seaborne trade reached 8.72 billion tons. It means that almost 80 per cent of world merchandise trade by volume (tones) and more than 90 per cent in ton-miles have been carried by sea.[76] In 2011 world seaborne trade grew by 4 per cent, whereas the global economy, measured in terms of the world GDP, expanded only by 2.7 per cent (compared with 4.1 per cent in 2010). Such a significant slowdown was caused among others by decelerated industrial production in the OECD countries which grew at modest 2.1 per cent (down from 8.5 per cent in 2010) and sharp drop in the world merchandise trade. Growth in the world merchandise trade by volume expanded at an annual rate (2011/2010) of 5.9 per cent as compared with 13.9 per cent in the previous year.[77] Referring to WTO rough estimates, deceleration in the global trade continued in 2012 and the world merchandise trade volumes expanded presumably only by 2.5 per cent, rate far below the 6 per cent average recorded over the period of 1990-2011.[78]

The recent growth in trading commodity volumes transported by sea, 4.0 per cent increase year-on-year, was not very much impressive, despite the still existing economic downturn, higher than the one recorded in the last decades. Indeed, since 1970 the annual average growth rate of the world seaborne trade has been estimated at 3.2 per cent.[79] In 2012, the volume of cargoes transported by sea was 2.4 times higher than in 1990. After all, its growth dynamics was still more than two times lower than

[75] J. G. Jansson, *Transport System Optimization and Pricing*. The Economic Research Institute. Stockholm School of Economics. Stockholm 1980, p. 399 and A. Grzelakowski, *Natural equilibrium as optimum condition of port services market*. "International Journal of Transport Economics" 1985, no. 3, p. 49 Stockholm

[76] Ton-miles is a unit which offers a measure of real demand for shipping services and tonnage involved in their manufacturing as it takes into account distance, which determines true ship availability. Compare: 2012. *Review of Maritime Transport 2012*. Report by the UNCTAD secretariat. UNTAD/RMT/2012, New York and Geneva 2008, p. 4-17.

[77] The merchandise trade in value terms (based on export) increased however, due to the growing commodity prices in 2011, by 19 per cent and reached $ 18.2 trillion. In 2010 it average annual growth rate accounted for 22 per cent. *Ibidem*, p. 19-21

[78] *Ibidem*, p. 19, 21

[7] Insight& Analysis. *World Trade Service Brochure* Global Insight 2011, p. 2

[8] A. Stachniol, The expected overall impact on trade from a maritime *Market Based- Mechanism* (MBM), March 2011, s. 2

the world merchandise export. Its annual average growth rate accounted for 8.7 per cent over the period of 2000-2012.

Taking into account that:

1 global seaborne shipments are growing in tandem with the world merchandise trade (measured on export base) as well as the world GDP and

2 currently existing barriers to development of the main economies being the trigger of global growth will have to be smoothly overcome in the next two years and the world production and trade will enter their previous path of growth;

one may assume that in 2020 the seaborne trade is likely to increase by 36-40 per cent as compared to 2010, reaching 12.0-12.5 billion tones.[80] By such dynamics, the volume of goods carried by sea can exceed in 2031 twice the level reached in 2010.

Therefore, the international forecasts regarding the development of seaborne trade and freight markets on a global scale in the next decades, despite the envisaged economic turbulences, are as previously assumed optimistic. They are based on assumption, as it was mentioned earlier, that despite growing global risks and uncertainties, the seaborne trade will increase in tandem with the world economy measured in the growing world GDP and production as well as global merchandise trade. It is assumed that the correlations between all these factors remain in the predicted period more or less at the same level as they were over the period of 1990-2011. At that time (1990=100) the global merchandise export grew 3.1 times what corresponded with the dynamics of seaborne trade (2.72 times), whereas the world GDP was enlarged 1.78 times and industrial production only 1.41 times.[81]

It is obvious that these relations are determined by many other economic factors inherited in other segments of the global economy which influence significantly the world maritime transport, too. They originate from other type of markets, e.g. capital or energy ones as well as primary markets themselves. The main factors here include: rising energy prices (bunker) with their potential implications for transport costs and trade, soaring non-oil commodity prices, the uncertainties on global financial markets, etc. All of them, as relatively strong regulatory instruments, will influence the global freight markets, both their demand and supply side, determining the scope and the structure of the global maritime transport real sphere of activity. Despite the growing role of non-autonomous measures, their

regulatory driving force in this sector seems to be still dominant

2.2 Freight market mechanism as regulatory instrument of global maritime transport sector

Maritime transport sector operating on a global scale has constantly undergone pressure of numerous market forces, especially these stemming from the freight market itself. Therefore, freight market mechanism as a powerful regulatory tool of the real sphere of activity of the maritime transport sector, able to fulfill its traditional autonomous regulatory functions in the global shipping industry, becomes the primary driving force of any changes in this area. Closely connected with other markets, it steadily affects the decision making processes of demand and supply side representatives concerning both their short-term operational activity and long-term investment engagements (forms of capital allocation). In such a way, freight markets not only inform all sides involved in the maritime transport operation about the current fluctuations of effective demand and potential supply, determining to some extend the future market pattern too, but also stimulate them to cause appropriate reaction and ensure pro-market oriented behavior. How efficiently it works as regulatory tool of this sector able to adjust the supply side, being less sensitive in adopting itself to demand fluctuations, to dynamic changes occurring on the other market side, may be viewed on the data base concerning predominantly the last five years. It was a very turbulent time, rich in many extraordinary and spectacular events, the effects of which significantly affected the global shipping industry. Hence, it can be used to examine the functioning of the freight market mechanism as well its evaluation as a regulatory instrument.

As a result of the growing world economy and subsequently, international merchandise trade, the world merchant fleet, representing the potential supply of maritime transport sector, due to deliveries of new buildings expanded by almost 10 per cent during 2011.[82] The potential global supply in maritime transport reached in 2012 1.6 billion dwt. It means that the world tonnage grew 2.5 times faster than the word merchandise trade carried by sea in 2011.

The merchant fleet recorded an impressive increase of over 45 per cent in just five years. However, it should be noted that since the economic crisis of 2008 and 2009, far fewer new orders have been placed by global ship operators than tonnage delivered by the world shipyards. As a result, the existing order book of the global shipyard sector has been significantly reduced. Nevertheless, in 2010 and 2011 more tonnage was added to the existing

[81] WTO 2011. Trade growth to ease in 2011 but despite 2010 record surge, crisis hangover persists. World Trade 2010. Prospects for 2011, *Press Releases*/628 7, April 2011, p. 5

[82] 2012. *Review of Maritime Transport 2012*.Op. cit., p. 51-62,

world fleet than in any previous year, which in fact resulted from orders placed prior to the economic crisis.

This general tendency, however, concerns in varying degree particular groups of world tonnage. Looking at this trend at the supply side in long term period, one must conclude that growth dynamics of each group of tonnage in recent 30 years was in line with the needs determined by the demand side of the freight markets, reflecting the quantitative and structural changes observed on global commodity markets. For example, since 1980 the general cargo fleet has declined by 7 percent, whereas the rest of the tonnage categories grew by more than 150 per cent; oil tankers grew by more than 164 per cent, dry bulk by almost 335 per cent and containers grew 18 times in deadweight tonnage.

Such tendency observed at the supply side of the global maritime transport freight market as compared with the demand fluctuations (its deceleration), has caused several consequences for both shipowners and shippers. On the one hand, the world tonnage has been renewed and its average age per dwt slightly diminished (11.5 years). Thanks to that, its potential earning capability is being increased. However, it does not mean that operational productivity of the world fleet has to follow the said capability (tab.1). On the other hand, the threat of speeding up the already ongoing tonnage overcapacity (oversupply) was getting much more realistic with all market consequences, e.g. for freight and charter rate distortions. In fact, all these negative economic and financial consequences were unfold on the freight markets at that time, painfully hitting not only ship operators but also other participants of the global logistics supply chain.[83] However, due to the existing freight market regulatory power, expressed in the efficient reactions generated by the ship operators and ship owners, the global freight market as well as its major segments have great viability to survive under turbulent time.

Such elasticity and adaptability of the supply side of the freight market to the demand side, which obviously differs accordingly to the type of market: liner or tramp one, has its roots in free mechanism of ship registration (shipping assets distribution). In fact, any ship owner has generally free choice of national (ownership) or foreign flag, indicating the country where their ship is registered. The use of international open registers as well as the so-called second registers (DIS, FIS, NIS, etc) generates to ship-owners still more economic and fiscal privileges than flying the national (country owned) flag. Due to that, the share of the foreign –flagged fleet is gradually growing in the global total tonnage and currently accounts for more than 72 per cent. As open registers are in use, and many countries offering all ship-owners such administrative and economic instrument are nowadays competing with each other, the process of tonnage concentration in this "secondary shipping nations" is relatively high. The top 35 open and international registers run by foreign countries, the so-called *flag of convenience* have concentrated almost 92 per cent of flagged-out fleet. Ten major open registers concentrate 57 per cent of tonnage registered under foreign flags. This tendency intensifies since vast majority of new-buildings is registered under foreign flag.[84] As a result, the growth of most of the major flags of registration is getting higher than the growth of the total global fleet with all consequences for ship-owning countries.

Due to that free scheme of ship registration, ownership of the fleet which belongs to substantial maritime assets in this sector not necessarily implies that the ship-owning country effectively operates and controls the shipping operation). The ship owning countries can not fully benefit from possessing their maritime assets (taxes and other incomes) and try to take economic and administrative measures to stimulate the process of reflagging ships using open international register. However, the process seems to be irreversible.

This group of countries, quite different from those offering open registers, is characterized by very high degree of world tonnage concentration. As of January 2012, only 35 countries and territories with the largest owned fleet controlled 95.5 of the global fleet. Their market share is estimated to grow further. What is more, almost 50 per cent of the world tonnage is owned by shipping companies from just four countries.

As a result, tonnage concentration in a relatively small group of countries as far as vessels ownership is concerned, is getting some characteristics of the contemporary global freight market and may influence the primary markets as well as global logistics costs and final prices of commodities transported by sea. All external factors unequivocally indicate that such tendency will gradually go ahead in the next years, partially as a result of still growing international competition and already achieved position of the main shipping countries on the commodities and freight markets (economies of scale).

[83] A. S. Grzelakowski, Transport kontenerów drogą morską w gospodarce globalnej. Podstawowe wyzwania na przełomie I i II dekady XXI wieku. „*Biuletyn Polskiej Izby Spedycji i Logistyki*" No. 06-08, 2012, A. S. Grzelakowski, Maritime Transport Development in the Global Scale – the Main Chances, Threats and Challenges. *International Journal on Marine Navigation and Safety of Sea Transportation.* Vol. 3, No. 2, 2009, p.203

[84] Among the tonnage delivered in 2011 almost 83 per cent was registered abroad, i.e. in countries running open international registers. *2012 Review*, op. cit., p. 62

The above indicated tendencies observed in the international maritime transport on its supply and demand side as well as in its contemporary existing regulatory mechanism, especially relating to merchant fleet distribution on the basis of tonnage ownership (real control) and vessels registration (fleet management), have great impact on the world fleet operational productivity and its effectiveness. As maritime transport constitutes a very important link in the global supply chains, servicing primary markets (see fig.1), such trends and tendencies have to influence significantly the efficiency and elasticity of the logistics supply chains and the international seaborne trade. To examine the scope and intensity of their impact on secondary and primary market use by global supply chains operators, the indices of the world fleet operational productivity need to be analyzed.

The main indexes of this kind are defined in tons and ton-miles per deadweight ton (dwt). They show the still changing relations between the growth in the supply of tonnage and the growth in the total seaborne trade as well as in ton-miles performed by the world fleet, which corresponds with a distance one ton was carried over. Consequently, as the growth in the supply of the fleet outstrips the growth in the total seaborne trade, which on the highly competitive global freight market has been a standard relation in recent years, the tons of cargo carried per deadweight ton (dwt) decreases. In 2007 the global average of tons of cargo carried per dwt of cargo carrying capacity was 7.7 and in 2011 only 5.7 (see tab.1). In other words, it may be interpreted, that the average ship was fully loaded 7.7 and 5.7 times respectively during those years. Since 2007, due to the growing imbalances between the demand and supply side on the global freight market, the operational productivity has decreased significantly (see tab.1).

Table 1. Operational productivity of the total world fleet in the period of 1970 – 2011 (selected years)

Year	Tons carried per dwt	Thousands of ton-miles performed per dwt
1970	7.9	32.7
1980	5.4	24.6
1990	6.1	26.0
2000	7.5	29.7
2007	7.7	31.6
2008	7.4	35.1
2009	6.6	31.0
2010	6.0	29.3
2011	5.7	27.9

Source: Calculations on Lloyd's Register-Fairplay, Fernleys, Review and Review of Maritime Transport 2008, 2010, 2012.

During the same period, the ton-miles performed per deadweight dropped from 31.6 to 27.9 (tab.1). These indices inform all market representatives that the average dwt of cargo carrying capacity of the

world fleet transported one ton of cargo over a distance of 31,600 nautical miles in 2007, i.e. 87 miles per day and, due to the growing overcapacity of the word fleet and slightly shrinking distances (growing fuel prices), its productivity decreased in 2011 as few as 27.900 miles.

The indices of operational productivity of the world fleet presented in tab.1 indicate that it varies significantly on year-to-year basis. On the one hand, it is a result of the high demand fluctuations, and on the other, it reflects the level of overcapacity generated by shipping operators accomplishing on the highly competitive freight markets the strategy of flexible and efficient demand meeting. The level of world tonnage overcapacity (tonnage oversupply in all groups and categories) is presented in tab.2.

Table 2. Tonnage regarded as idle (tonnage oversupply) as percentage of world fleet in selected years (1990 - 2011).

Year	1990	2000	2005	2008	2010	2011
	9.7	2.3	1.0	2.2	1.4	0,9

Source : Elaborated on data presented by Lloyd's Register – Fairplay and Lloyd's Shipping Economics as well Review of Maritime Transport 2008, 2010 and 2012 p. 68

As regards the operational productivity of the world tonnage, one should mention that in response to still rising oil prices, ship operators usually run a slow stemming strategy. They are interested in reducing even by 20 percent the service speed of their vessels with the aim to save fuel. Such a strategy has been widely used by shipping operators since 2007, especially in liner shipping. However, with lower service speeds, more vessels are required on a given route. On the one hand, it helps to reduce overcapacity, leading on the other, at the same time to significant decrease in operational productivity of the world fleet. The capacity constraints and congestion at ports also have negative impact on the fleet productivity, as ship capacity is tied up while queuing. All these factors stemming from commodity and freight markets (fig.1), their relations and existing imbalances, have influenced not only the degree of the idle world merchant fleet and its operational productivity, but also the level of maritime freight costs as a percentage of the value of imported and exported goods in the global merchandise trade.

Maritime freight costs still remain an important component of the price of goods purchased by final consumers.[85] High maritime transport costs for imported goods may impact significantly the price level of the basket of consumers' goods and in the end can stimulate inflation. On the other hand, excessive freight rates for exported goods affect sometimes painfully the trade competitiveness of the products of a country on the global markets. Hence,

---

[85] *Review of Maritime Transport 2012, op. cit., p.89-91*

each country, their economic groupings and regions have to be interested in working out proper approaches to reduce both inbound and outbound maritime transport costs in their trade relations with partners. Their activity in this area is focused on commodity and freight markets and is expressed in the form of maritime transport policy, regarded as a regulatory measure able to affect the maritime transport sector and its freight markets.

The share of global freight payments in the value of world imports has reached on average 6.7 per cent as compared with the recent five years.[86] In 2000 it amounted to 6.26 per cent in the developed countries and 8.63 in the group of developing countries from Asia and America and increased ten years later in the first group of countries to 6.52 percent, concurrently decreasing in the second group to 7.57 per cent.[87] These indices reflecting maritime transport costs as a percentage of the total value of imported goods as well as their volatility over time simultaneously indicate in the long term a tendency towards a lower ratio between freight costs and value of goods occurred among all groups of countries. Furthermore, the freight rates share of developing countries which is relatively high as a result of the existing trade imbalances (in value and volume terms of imports and exports), tend to converge to these developed countries. It is a positive sign indicating that the freight market price mechanism is efficiently reducing the economic differences generated by existing world trade pattern created by the developed countries. Moreover, this tendency created by the freight market mechanism, not rarely to the detriment of the shipping sector which does not belong in the economic terms to the beneficiaries of its activity, does not hamper the globalization processes. The more integration between global freight markets and commodity markets, the more effects for shippers and ship operators as well as in the end for the final consumers of goods transported by sea.

3 INTERNATIONAL MARITIME TRANSPORT POLICY AND ITS REGULATORY IMPACT ON THE GLOBAL SHIPPING INDUSTRY

The international maritime transport policy, created directly or indirectly by international organizations (i.e. IMO, ILO, HELCOM, and EMSA) and international (regional) groupings such as EU, NAFTA, BSSC, etc., constitutes in the contemporary world a very important and powerful regulatory mechanism of the whole shipping sector. It completes the still functioning, typical for this open, international transport sector, autonomous regulatory mechanism.

The said group includes primarily IMO, which plays the most important role in composing such regulatory subsystem in the world scale. The majority of conventions adopted under the auspices of IMO or for which this organization is otherwise responsible, fall into three main categories. The first group concerns maritime safety; the second the prevention of marine pollution; and the third the liability and compensation, especially in relation to damage caused by pollution. Outside these major groupings there are a number of other conventions dealing with facilitation, tonnage measurement, unlawful acts against shipping and salvage, etc. Taking into account the number of the IMO regulatory instruments existing in the form of conventions and protocols amending the first ones, as well as a number of contracting parties (countries) and the percentage of world tonnage covered by each of those legal instruments, it may be claimed that this organization creates real global shipping policy constituting the backbone of the world maritime transport regulatory mechanism.

In addition to IMO, ILO also participates in formatting the widely understood economic, social, technical and environmental order in the world shipping industry. ILO prepares conventions and recommendations concerning regulation of social standards in maritime sector. The organization has set out many minimum requirements for decent work in the maritime industry. Recently, in 2006, ILO adopted a new consolidated Convention (C 186) that provides comprehensive labor charter for the world's 1.2 million or even more seafarers, addressing the evolving realities and needs of the sector that handles 90 per cent of the world's trade. The Convention sets minimum requirements for seafarers to work on a ship and contains provisions on the terms of employment, hours of work and rest, accommodation, recreational facilities, food and catering, health protection, medical care, welfare and social security protection.

Upon discussing the international maritime transport policy, one should notice that the EU is strongly committed to setting up such regulatory mechanism and not only within the Community. The European Commission's transport policy aims at harmonious performance of the European maritime transport system as a whole. It performed two strategic goals, at once. Over the years, the Commission built quite comprehensive regulatory framework encouraging the efficiency of ports and maritime transport services, *inter alia* reinforcing market position of the EU fleet flying member states' flags and strengthening competitive advantages for the EU shipowners for the benefit of all other economic sectors and of the final consumers on the one hand and safety and security

---

[86] See: *Review of Maritime Transport 2008, 2010 and 2012, op. cit.,* p. 91.

[87] Ibidem, p. 91

Financial Indices, Freight Markets and other Economic Matters
STCW, Maritime Education and Training (MET), Human Resources and Crew Manning, Maritime Policy,
Logistics and Economic Matters – Marine Navigation and Safety of Sea Transportation – Weintrit & Neumann (Eds)

Forecasting Financial Indices: The Baltic Dry Indices

E. Thalassinos
University of Piraeus, Department of Maritime Studies, Piraeus, Greece

M.P. Hanias
TEI of Chalkis, General Dept. of Applied Sciences, Psachna Euvoias, Greece

P.G. Curtis & J.E. Thalassinos
TEI of Chalkis, Department of Accounting, Psachna Euvoias, Greece

ABSTRACT: The main aim of this paper is to use chaos methodology in an attempt to predict the Baltic Dry Indices (BDI, BCI, BPI) using the invariant parameters of the reconstructed strange attractor that governs the system's evolution. This is the result of the new emerging field in econo-physics which mainly consists of autonomous physic-mathematical models that have been already applied to financial analysis.

The proposed methodology is estimating the optimal delay time and the minimum embedding dimension with the method of False Nearest Neighbors (FNN). Monitoring the trajectories of the corresponding strange attractor we achieved a 30, 60, 90 and 120 time steps out of sample prediction.

1 INTRODUCTION

Predicting Baltic Dry Indices is possible applying algorithms used in physical sciences. This article combines finance and non linear methods from chaos theory to examine the predictability of the Baltic Dry Indices. The proposed methodology applies non linear time series analysis for the BCI, BDI and BPI indices covering the period from 04-01-2000 until 04-01-2008. In particular the method of the False Near Neighbor (FNN) is used in the first step to evaluate the invariant parameter of the system as the minimum embedding dimension. In a second stage using the reconstructed state space the article achieves an out of sample multi step time series prediction. The methodology is very dynamic compared to traditional ones (Thalassinos et al., 2009). The most important property of the proposed methodology is the fact that there is no need to know from trial to error some constants to fit and extrapolate the time series. The methodology calculates the invariant parameters of the system itself. The fact that shipping indices are sensitive in irregular shocks and crises which are innate elements in chaotic systems, their predictability is much easier using chaotic methodology than any other forecasting method. Other benefits as they are pointed out in Thalassinos et al., (2009) are:

1 The possibility to extract information about a complex dynamic system, which generates several observed time series by using only one of them.

2 To analyze the image-system with the same topology that preserves the main characteristics of the genuine system.

The article combines maritime time series daily data for the period of 04-01-2000 to 04-01-2008 (total number of observations 2000) with chaos methodology to examine the predictability of the three Baltic Dry Indices as the most characteristic indices in the maritime industry namely the BCI, BDI and BPI.

2 BCI TIME SERIES

According to the theory of observed chaotic data (Abarbanel, 1996; Sprott, 2003; Hanias et al., 2007a, 2007b; Thalassinos et al., 2009) any non-linear time series can be presented as a set of signals $x=x(t)$ as it shown at Figure 1. The sampling rate is $\Delta t=1$ day and the number of data N=2000. As it is presented in Figure 1 the corresponding index had a significant number of structural shifts in the period under study with a max of 16,000 and a min of 1,000 points.

From the time series plots of the BCI in Figure 1, 1700 data points has been selected as the "training data set", in other words the data that are used for the state space reconstruction and the other 300 data points as the "test data set" for out of sample period prediction.

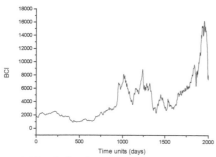

Figure 1. Time Series plots of the BCI

3 STATE SPACE RECONSTRUCTION

3.1 *Time delay τ*

From the given data a vector $\vec{X}_i$, i=1, 2, ... N, where N=1700, in the m order dimensional phase space given by the following relation (Kantz et al., 1997; Takens, 1981; Hanias et al., 2007a, 2007b; Thalassinos et al., 2009) is constructed as shown in equation (1):

$$\vec{X}_i = \{x_i, x_{i-\tau}, x_{i-2\tau}, \ldots, x_{i+(m-1)\tau}\} \qquad (1)$$

where $\vec{X}_i$ represents a point in the m dimensional phase space in which the attractor is embedded each time, where τ is the time delay $\tau = i\Delta t$. The element x_i represents a value of the examined scalar time series in time, corresponding to the *i*-th component of the time series. By using this method the phase space reconstruction to the problem of proper determining suitable values of values of m and τ is constructed. The next step is to find the time delay *(τ)* and the embedding dimension *(m)* without using any other information apart from the historical values of the indices.

Equation (2) is used to calculate the time delay by using the time-delayed mutual information proposed by Fraser et al., 1986 and Abarbanel, 1996.

$$I(\tau) = \sum_{x_i, x_{i+\tau}} p(x_i, x_{i+\tau}) \log_2 \left(\frac{p(x_i, x_{i+\tau})}{p(x_i)p(x_{i+\tau})} \right) \qquad (2)$$

In equation (2), $p(x_i)$ is the probability of value x_i and $p(x_i, x_{i+\tau})$ denotes joint probability. $I(\tau)$ shows the information (in bits) being extracted from the value in time xi about the value in time $x_{i+\tau}\tau$. The function I can be thought of as a nonlinear generalization of the autocorrelation function. For a random process x(t) the I(t), is a measure of the information about x(t+τ) contained in x(t). The first nadir of I(τ) gives the delay, tau0, such that x(t+tau0) adds maximal information to that already known from x(t). This tau0 is returned as an estimate of the proper time lag for a delay embedding of the given time series. The time delay is calculated as the first minimum of the mutual information (Kantz et al., 1997, Fraser et al., 1986, Casdagli et al., 1991). Mutual information against the time delays (with a minimum at 54 time steps) for our time series is presented in Fig 2.

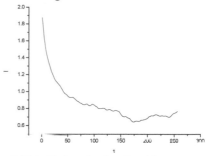

Figure 2. Mutual Information I vs. time delay τ

3.2 *Embedding dimension m*

After obtaining the satisfactory value of τ, the embedding dimension *m* is to be determined in order to finish the phase space reconstruction. For this purpose the method of False Nearest Neighbors (Kantz. H., et al., 1997, Kennel M.B., et al., 1992) is used. More specifically, the method is based on a fact that when embedding dimension is too low, the trajectory in the phase space will cross itself. If we are able to detect these crossings, we may decide whether the used *m* is large enough for correct reconstruction of the original phase space, when no intersections occur or not. When intersections are present for a given *m*, the embedding dimension is too low and we have to increase it at least by one. Then, we test the eventual presence of self-crossings again (Kennel M.B., et al., 1992, Abarbanel 1996). The practical realization of the described method is based on testing of the neighboring points in the *m*-dimensional phase space. Typically, we take a certain amount of points in the phase space and find the nearest neighbor to each of them. Then we compute distances for all these pairs and also their distances in the (*m*+1) dimensional phase space. The rate of these distances is given by equation (3) as follows:

$$P = \frac{\left\| y_i(m+1) - y_{n(i)}(m+1) \right\|}{\left\| y_i(m) - y_{n(i)}(m) \right\|} \qquad (3)$$

where $y_i(m)$ represents the reconstructed vector, belonging to the *i*-th point in the *m*-dimensional phase space and index the $n(i)$ denotes the nearest neighbour to the *i*-th point. If P is greater than some value P_{max}, we call this pair of points False Nearest Neighbors (i.e. neighbors, which arise from trajectory self-intersection and not from the closeness in the original phase space). In the ideal case, when the number of false neighbors falls to zero, then the value of m is found. For this purpose we compute the rate of false nearest neighbours in the reconstructed phase space using the formula in equation (4):

$$\left| x_{i+m\tau} - x_{n(i)+m\tau} \right| \geq R_A \qquad (4)$$

where R_A is the radius of the attractor,

$$R_A = \frac{1}{N} \sum_{i=1}^{N} \left| x_i - \bar{x} \right| \qquad (5)$$

and

$$\bar{x} = \frac{1}{N} \sum_{i=1}^{N} x_i \qquad (6)$$

is the average value of time series.

When the following criterion

$$P \geq P_{max}, \qquad (7)$$

is satisfied then it can be used to distinguish between true and false neighbours (Kennel M.B., et al., 1992). The dimension m is found when the percent of false nearest neighbors decreases below some limit, (Kugiumtzis D., et al., 1994) so we choose $P_{max}=10$ as used typically. Before we apply the above procedure we must determine the Theiler window for out time series and exclude all pairs of points in the original series which are temporally correlated and are closer than this value of Theiler window. For this purpose we produce space time sepatarion plots (Kantz H., et. Al., 1997) as shown in Fig 3. (a)

a)

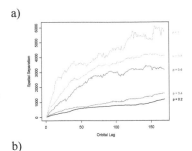

b)

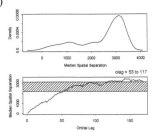

Figure 3. Space time separation plot (a). Density function estimate of the median contour (upper graph) in addition to a suggested range of suitable orbital lags. The most populous values of the median contour are highlighted by a cross-hatched area that covers a plot of the median curve (lower) (b).

The space time separation plots where produced with a desired embedding dimension 5, τ=54, and the probability associated with the first contour equal to 0.2.

In Figure 3 above we had plot a summary of the space-time contours including a density function estimate of the median contour in addition to a suggested range of suitable orbital lags. In the latter case, the most populous values of the median contour are highlighted by a cross-hatched area that covers a plot of the median curve. The suggested range for a suitable orbital lag is based on the range of values that first escape this cross-hatched region, so we can choose a Theiler window from 53 to 117. In our case we choose the Theiler window to be 65. Then we used matlab code to calculate the quantity P.

The rate of false neighbors that is under the above limit $P_{max}=10$ is achieved for $m = 4$, thus this value should be suitable for the purpose of phase space reconstruction. This is shown at Figure 4 for our time series.

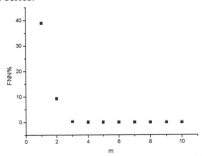

Figure 4. Percent of false nearest neighbors number FNN vs. m

4 TIME SERIES PREDICTION OF BCI

The next step is to predict the evolution of BCI, by computing weighted average of evolution of close neighbors of the predicted state in the reconstructed phase space (Miksovsky J. et al., 2007, Stam C. J. et al., 1998). The reconstructed m-dimensional signal projected into the state space can exhibit a range of trajectories, some of which have structures or patterns that can be used for system prediction and modeling. Essentially, in order to predict k steps into the future from the last m-dimensional vector point $\{x_N^m\}$, we have to find all the nearest neighbors $\{x_{NN}^m\}$ in the ε-neighborhood of this point. To be more specific, let's set $B_\varepsilon(x_N^m)$ to be the set of points within ε of $\{x_N^m\}$ (i.e. the ε-ball). Thus any point in $B_\varepsilon(x_N^m)$ is closer to the $\{x_N^m\}$ than ε. All these points $\{x_{NN}^m\}$ come from the previous trajectories of the system and hence we can follow their evolution k-steps into the future $\{x_{NN+k}^m\}$. The final prediction for the point $\{x_N^m\}$ is obtained by averaging over all neighbors' projections k-steps into the future. The algorithm can be written as in equation (8) as follows:

$$\{x^m_{N+k}\} = \frac{1}{\left|B_\in(x^m_{NN})\right|} \sum_{X^m_{NN} \in B_\in(x^m_{NN})} x^m_{NN+k} \quad (8)$$

where $\left|B_\in(x^m_{NN})\right|$ denotes the number of nearest neighbors in the neighborhood of the point $\{x^m_N\}$ (Kantz H., et al., 1997). As an example we suppose that we want to predict k=2 steps ahead. The basic principle of the prediction model is visualized in Figure 5. The blue dot $\{x^m_N\}$ represents the last known sample, from which we want to predict one and two steps into the future. The blue circles represent ε-neighborhoods in which three nearest neighbors were found.

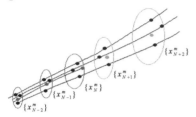

Figure 5. Basic prediction principle of the simple deterministic model

The next step in the algorithm is to check that the projections, one and two steps into the past, of the points in $\{x^m_{NN}\}$ are also nearest neighbors of the two previous readings $\{x^m_{N-1}\}$ and $\{x^m_{N-2}\}$ respectively. This criterion excludes unrelated trajectories that enter and leave the ε-neighborhood of $\{x^m_N\}$ but do not 'track back' to ε- neighborhoods of $\{x^m_{N-1}\}$ and $\{x^m_{N-2}\}$, thus making them unsuitable for prediction. Assuming that any nearest neighbors have been found and checked using the criterion detailed previously, we project their trajectories into the future and average them to get results for $\{x^m_{N+1}\}$ and $\{x^m_{N+2}\}$. We used the values of τ and m from our previous analysis so the appropriate time delay τ was chosen to be τ=54. We use as embedding dimension the 2*m = 8 (Sprott J. C) and for the optimum number of nearest neighbors we used the value of embedding dimension m= 4. These values of embedding dimension and number of nearest neighbors gave the better results for k=30 time steps ahead. We apply the procedure for in sample forecasting until data point 1700 as shown at Fig 6 then we applied the procedure for out of sample prediction from data point 1700 to data point 1730 as shown at Figure 7.

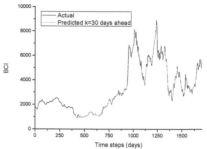

Figure 6. Actual and in sample predicted time series for k=30 days ahead

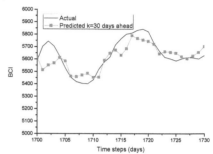

Figure 7. Actual and out of sample predicted time series for k=30 days ahead

Actual and predicted time series for k=60, 90, 120 time steps ahead are presented at figs 8, 9, 10 respectively for out of sample period.

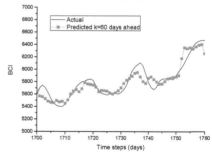

Figure 8. Actual and out of sample predicted time series for k=60 days ahead

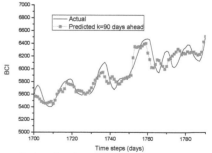

Figure 9. Actual and out of sample predicted time series for k=90 time steps ahead

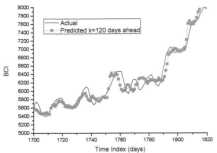

Figure 10. Actual and out of sample predicted time series for k=120 time steps ahead

The prediction error for establishing the quality of the fit was chosen to be the classical root mean square error (RMSE) and found to be 8.99×10^{-2}, 9.54×10^{-2}, 2.17×10^{-1}, 2.26×10^{-1} for k=30,60,90,120 respectively.

5 TIME SERIES PREDICTION OF BDI

The BDI time series is presented as a signal $x=x(t)$ as it shown at Figure 11. It covers data from 04-01-2000 to 04-01-2008. The sampling rate was $\Delta t=1$ day and the number of data are N=2000.

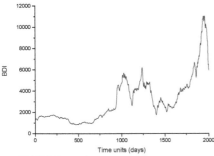

Figure 11. BDI time series

From this time series we have choose 1700 data as the "training data set", in other words the data that we used for the state space reconstruction and the other 300 data as the "test data set" for our out of sample prediction. We used the same procedure as before and we had estimated the delay time, Theiler window and embedding dimension. Mutual information against the time delays (with a minimum at 55 time steps) for BDI time series is presented in Figure 12.

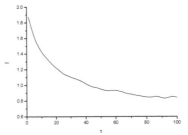

Figure 12. Mutual Information I vs. time delay τ

The Space time separation plot and the density function estimation plots are shown in Figure 13. From this plot we estimate the Theiler window to be 100.

a)

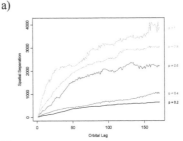

b)

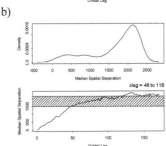

Figure 13. Space time separation plot (a). Density function estimate of the median contour (upper graph) in addition to a suggested range of suitable orbital lags. The most populous values of the median contour are highlighted by a cross-hatched area that covers a plot of the median curve (lower) (b).

The percent of false nearest neighbors number FNN vs. m is shown at Figure 14. From this it is clear that the embedding dimension is 4.

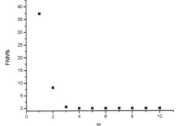

Figure 14. Percent of false nearest neighbors number FNN vs. m

The next step is to predict 30, 60, 90 and 120 time steps ahead. Figures 15, 16, 17, 18 and 19 are shown the in sample and out of sample period predictions.

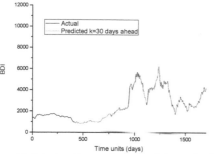

Figure 15. Actual and in sample predicted time series for k=30 days ahead

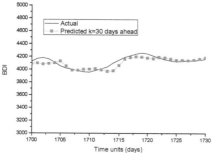

Figure 16. Actual and out of sample predicted time series for k=30 days ahead

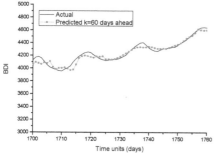

Figure 17. Actual and out of sample predicted time series for k=60 days ahead

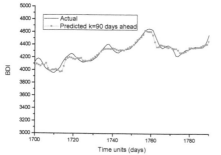

Figure 18. Actual and out of sample predicted time series for k=90 time steps ahead

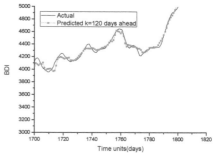

Figure 19. Actual and out of sample predicted time series for k=120 time steps ahead

The prediction error for establishing the quality of the fit was chosen to be the classical root mean square error (RMSE) and found to be 5.68×10^{-2}, 5.99×10^{-2}, 6.53×10^{-2}, 6.80×10^{-2} for k=30, 60, 90 and 120 respectively.

6 TIME SERIES PREDICTION OF BPI

The BPI time series is presented as a signal x=x(t) as it shown at Figure 20. It covers data from 04-01-2000 to 04-01-2008. The sampling rate was $\Delta t=1$ day and the number of data are N=2000.

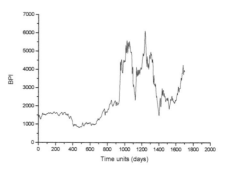

Figure 20. BPI time series

From this time series we have choose 1700 data as the "training data set", in other words the data that we used for the state space reconstruction and the other 300 data as the "test data set" for our out of sample prediction.

We used the same procedure as before and we have estimated the delay time, Theiler window and embedding dimension. Mutual information against the time delays (with a minimum at 55 time steps) for BDI time series is presented in Figure 21.

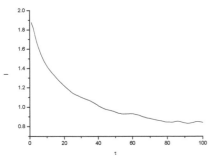

Figure 21. Mutual Information I vs. time delay τ

The space time separation plot and density function estimation plots are shown in Figure 22. From this plot we estimate the Theiler window to be 100.

a)

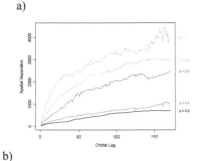

b)

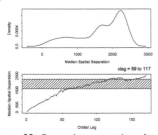

Figure 22. Space time separation plot (a). Density function estimate of the median contour (upper graph) in addition to a suggested range of suitable orbital lags. The most populous values of the median contour are highlighted by a cross-hatched area that covers a plot of the median curve (lower) (b).

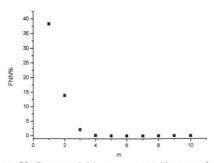

Figure 23. Percent of false nearest neighbors number FNN vs. m

The percent of false nearest neighbors number FNN vs. m is shown in Figure 23. From this fig it's clear that the embedding dimension is 4.

The next step is to predict 30, 60, 90 and 120 time steps ahead. Figures 24, 25, 26, 27 and 28 are shown the in sample and out of sample period predictions.

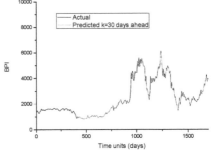

Figure 24. Actual and in sample predicted time series for k=30 days ahead

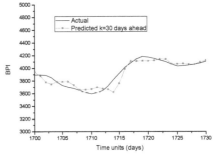

Figure 25. Actual and out of sample predicted time series for k=30 days ahead

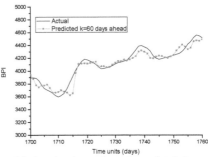

Figure 26. Actual and out of sample predicted time series for k=60 days ahead

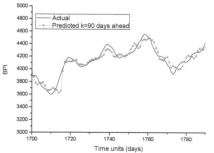

Figure 27. Actual and out of sample predicted time series for k=90 days ahead

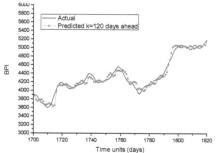

Figure 28. Actual and out of sample predicted time series for k=120 days ahead.

The prediction error for establishing the quality of the fit was chosen to be the classical root mean square error (RMSE) and found to be 7.10×10^{-2}, 7.56×10^{-2}, 1.24×10^{-1}, 2.36×10^{-1} for k=30, 60, 90 and 120 respectively.

7 CONCLUSION

In this paper chaotic analysis has been used to predict Baltic Dry Indices time series. After estimating the minimum embedding dimension, the proposed methodology has pointed out that the system is characterized as a high dimension chaotic. From reconstruction of the systems' strange attractors, it has been achieved a 30, 60, 90 and 120 out of sample time steps prediction. In Table 1 the RMS values of the corresponding prediction are shown.

Table 1: RMS values

	RMS			
	30 DAYS	60 DAYS	90 DAYS	120 DAYS
BDI	5.68×10^{-2}	5.99×10^{-2}	6.53×10^{-2}	6.80×10^{-2}
BCI	8.99×10^{-2}	9.54×10^{-2}	2.17×10^{-1}	2.26×10^{-1}
BPI	7.10×10^{-2}	7.56×10^{-2}	1.24×10^{-1}	2.36×10^{-1}

It is clear that the BDI has the minimum error and gives the best prediction compared to the other two indices. The prediction power of the suggested

method is limited by the properties of the original system and the series alone. Because the series represents the only source of information for the system it is important to be as long as possible. Too short time series will not worsen the prediction only but it may also make the proper reconstruction of the phase space impossible. Another important limitation factor is the eventual contamination by noise – evidently, the more noise is included in the series, the less accurate result can be gained. However the prediction horizon of 120 days would be useful for economical analysis as it can reveal the trend in a very good manner.

REFERENCES

[1] Abarbanel H.D.I., (1996), Analysis of Observed Chaotic Data, Springer, New York.
[2] Casdagli M, Eubank S., Farmer J.D., and Gibson J., (1991), "State Space Reconstruction in the Presence of Noise", Physica D, 51, 52- 98.
[3] Fraser A.M., Swinney H.L., (1986), "Independent coordinates for strange attractors from mutual information", Physics Review A, 33, 1134.
[4] Hanias M.P., Curtis, G.P., and Thalassinos, E.J., (2007), "Non-Linear Dynamics and Chaos: The Case of the Price Indicator at the Athens Stock Exchange", International Research Journal of Finance and Economics, Issue 11, pp. 154-163.
[5] Hanias M.P., Curtis, G.P., and Thalassinos, E.J., (2007), "Prediction with Neural Networks: The Athens Stock Exchange Price Indicator", European Journal of Economics, Finance And Administrative Sciences, Issue 9.
[6] Hanias M. P., Karras, A.D., (2007), "Efficient Non Linear Time Series Prediction Using Non Linear Signal Analysis and Neural Networks in Chaotic Diode Resonator Circuits", Springer Berlin / Heidelberg Lecture Notes in Computer Science, Advances in Data Mining. Theoretical Aspects and Applications, Volume 4597, pp. 329-338.
[7] Kantz H., Schreiber, T., (1997), Nonlinear Time Series Analysis, Cambridge University Press, Cambridge.
[8] Kennel M.B., Brown R., Abarbanel H.D.I., (1992). "Determining embedding dimension for phase-space reconstruction using a geometrical construction", Physics Review A, 45, 3403.
[9] Kugiumtzis D., Lillekjendlie B., Christophersen N., (1994) "Chaotic Time Series", Modelling Identification and Control, 15, 205.
[10] Miksovsky J., Raidl A., (2007), "On Some Nonlinear Methods of Meteorological Time Series Analysis", Proceedings of WDS Conference.
[11] Stam C. J., Pijn J. P. N., Pritchard W. S., (1998), "Reliable Detection of Nonlinearity in Experimental Time Series with Strong Periodic Component", Physica D, vol. 112.
[12] Sprott J. C., (2003), "Chaos and Time series Analysis", Oxford University Press.
[13] Takens F., (1981), "Detecting Strange Attractors in Turbulence", Lecture notes in Mathematics, vol. 898, Springer, New York.
[14] Thalassinos. E., Hanias M.P., Curtis, G.P., and Thalassinos, E.J., (2009), "Chaos Theory: Forecasting the Freight Rate of an Oil Tanker", International Journal of Computational Economics and Econometrics, vol. X, issue 1.

Financial Indices, Freight Markets and other Economic Matters
STCW, Maritime Education and Training (MET), Human Resources and Crew Manning, Maritime Policy,
Logistics and Economic Matters – Marine Navigation and Safety of Sea Transportation – Weintrit & Neumann (Eds)

Analysis of the Toll Collection System in Poland in the Context of Economy

J. Mikulski

Silesian University of Technology, Katowice, Poland

ABSTRACT: The aim of the paper is to examine the impact of the tolling system implementation on various types of roads at the expense of transport companies on the Polish example. It is clear that in this approach a wide range of other costs incurred by transport companies is neglected.

1 INTRODUCTION

Nowadays, transport plays one of the most important roles in everyday life. The safety of various modes of transport and the rate of movement of cargo can be discussed. Each type of transport has its advantages and disadvantages, which more or less impact on comfort, time of reaching the place or costs.

Transport, especially the road transport (the share of road transport in the carriage of goods in recent years has significantly increased) is treated as an important link in the supply chain of goods. The dominant role of road transport – mainly bus – in passenger transport cannot be forgotten.

The task of transport companies is to provide the goods at the right time to the destination at an acceptable cost.

Transport always involves additional costs, among other things, a fare, or manipulation charges.

The aim is to examine the impact of the tolling system implementation on various types of roads at the expense of transport companies on the Polish example. It is clear that in this approach a wide range of other costs incurred by transport companies is neglected.

2 OLD VIGNETTE SYSTEM AND VIATOLL

Road pricing is a general term which may be used for any system where the driver pays directly for use of a particular roadway or road network.

To date, several different systems of collecting fees for the use of motorways, expressways and national roads were used in Poland. The easiest tolling way is a gantry system, which consists in counting fees "for the travelled gateway" or "for the distance travelled." Another system is a system of vignettes, regulating the passage on all roads.

Figure 1. Example of vignette cards

A vignette tolling system was in force in Poland generally till the 1st of July 2011. Charge rates were dependent on the travel time on roads, maximum vehicle weight and emissions standards. The new viaToll system introduced on that date is obligatory for trucks, buses and minibuses (the carriage of at least 9 people), if the total permissible weight exceeds 3.5 tonnes (it is necessary to mention here, for example, personal cars with trailers, where the total mass exceeds this value).

An example of vignette cards (daily charge) for vehicles other than buses with a maximum weight of 12 tonnes (number of axles max. 3) satisfying requirements of the standard Euro 4 is show in Fig. 1.

Figure 2. Toll collection system (German example)

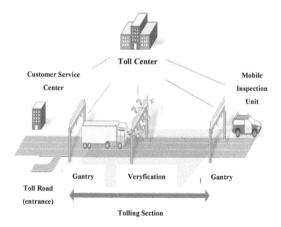

Figure 3. Toll collection system in Poland

Fig. 2. and 3 show how an automated toll collection system operates in Germany and ViaToll in Poland (there are variations - a satellite system in Germany and the DSRC system in Poland. Satellite system is based on the GPS and web application for booking "truck routes". DSRC – Dedicated Short Range Communications – are oneway or twoway short range wireless communication channels specifically designed for automotive use, including road pricing. Operation of the electronic tolling system in Poland is based on such modern microwave technology, which has absolutely no negative impacts on the environment).

Gantries, equipped with an antenna, are installed on roads (section charge). The antennas enable communication between the "relay" and ViaBOX-installed in the vehicle. Every time a vehicle (with a viaBOX) passes under a gantry it will be charged for crossing a particular section of the toll road. The truck is notified, but the fees calculation process is fully automated without the need to reduce the vehicle speed or to stop it.

ViaToll system in Poland (Fig. 4) is implemented in stages, and initially fees are charged on approx. 600 km of motorways, approx. 550 km of expressways and approx. 450km of national roads. The electronic viaToll system does not include toll roads, managed by private concessionaires.

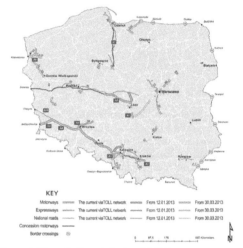

Figure 4. Map of toll roads in the viaTOLL system

The use of the toll roads network in Poland will require an on-board electronic device called via-BOX (Fig. 5). Its installation is simple and can be done according to instructions in the manual that comes with the device. Refundable deposit for the viaBOX device is approx. 40 €. It is a device, in size similar to a mobile phone, which sends the information about the vehicle encoded in it to "Relays" installed on the gantries. ViaBOX transmits the vehicle data based on which the fee is charged. To get a viaBOX device it is necessary to register for viaTOLL.

Vehicles that are equipped with windows metallised on the entire surface should use a via-BOX-2 device. It is equipped with an additional external antenna. However, during the registration process users must ask for a special viaBOX-2 on-board equipment. Security deposit for this device is the same as in the case of via-BOX for all other vehicles.

Figure 5. Via-Box device

3 COMPARISON OF THE TWO SYSTEMS

New toll rates depend on the permissible vehicle weight and emissions standards. The fees are set out in Tables 1 and 2.

Table 1. The toll amounts of on motorways and expressways in Poland

Emission standards	Fees for vehicles weighing	
	3.5 t – 12 t	exceeding 12 t
Euro 2	0.09 € /km	0.12 € /km
Euro 3	0.08 € /km	0.10 € /km
Euro 4	0.06 € /km	0.08 € /km
Euro 5	0.05 € /km	0.06 € /km

Note: exchange rate on 12.10.2012

For transport companies the start of viaToll system means a considerable increase in fees. Until now, passing one km cost them about 0.02 €. The new rate for the most popular trucks in Poland (with engines satisfying Euro 3 standard) is as high as 0.10 € / km. It is therefore a large increase, which significantly increases the cost of transport.

Table 2. The toll amounts on national roads in Poland

Emission standards	Fees for vehicles weighing	
	3.5 t – 12 t	exceeding 12 t
Euro 2	0.07 € /km	0.10 € /km
Euro 3	0.06 € /km	0.08 € /km
Euro 4	0.05 € /km	0.07 € /km
Euro 5	0.04 € /km	0.05 € /km

A comparative analysis of cost changes was made by adopting a vehicle with a maximum weight exceeding 12 tones (truck and trailer). Two main routes from Western Europe to Eastern Europe and from Northern Europe to Southern Europe. lead through Polish territory Therefore corridors considered as an example: Berlin – (Warsaw) – Minsk (route I) and Lund – (Szczecin) – Legnica – Prague (route II).
Route I specification is shown in Table 3, and route II specification is shown in Table 6.

The cost of travel on route I is provided by Table 4 (system of vignettes - daily fee) and Table 5 (electronic toll collection system).

Table 3. Data of route I

Total length of the route	1250 km
Length of the route on the Polish territory	730 km
Number of paid km (viaToll)	100 km
Number of gates on the commercial motorway	3 (corresponds to 43 €)
Journey time	11.5 hours

The cost of travel on route II is provided by Table 7 (system of vignettes - daily fee) and Table 8 (electronic toll collection system).

Table 4. The cost of travel on route I - system of vignettes

Emissions standards	Euro 2	Euro 3	Euro 4	Euro 5
Cost of purchasing vignettes		19.5 €		
Costs of commercial highway		0		
Total cost		19.5 €		
Cost of travel per 1 km		0.03 €		

Table 5. The cost of travel on route I - viaToll system

Emission standards	Euro 2	Euro 3	Euro 4	Euro 5
Cost in the system	12 €	10.40 €	8,30 €	6.07 €
Costs of commercial motorway		43 €		
Total cost	55 €	53.30 €	51.30 €	49.07 €
Cost of travel per 1 km	0.07 €	0.07 €	0.07 €	0.07 €

Table 6. Data of route II

Total length of the route	592 km
Length of the route on the Polish territory	420 km
Number of paid km (viaToll)	133 km
Number of gates on the commercial motorway	0
Journey time	7 hours

Table 7. The cost of travel on route II - system of vignettes

Emission standards	Euro 2	Euro 3	Euro 4	Euro 5
Cost of purchasing vignettes		9.75 €		
Costs of commercial motorway		0		
Total cost		9.75 €		
Cost of travel 1 km		0.02 €		

Table 8. The cost of travel on route II - system viaToll

Emission standards	Euro 2	Euro 3	Euro 4	Euro 5
Cost in the system	16 €	13.90 €	11.20 €	8.15 €
Costs of commercial motorway		32.86 €		
Total cost	48.86 €	46.76 €	44.06 €	41.01 €
Cost of travel 1 km	0.10 €	0.10 €	0.09 €	0.08 €

Difference in the increase in fees (also in percentage) are presented in Table 9.

Table 9. Changes in the cost of fees

Emission standards	Euro 2	Euro 3	Euro 4	Euro 5
Route I	36.30 €	35.09 €	31.71 €	29.46 €
	275 %	267 %	258 %	250 %
Route II	39.10 €	36.98 €	34.27 €	31.25 €
	489 %	467 %	444 %	411 %

The presented comparative analysis of changes in the toll collection system in Poland shows that the owners of transport companies will pay much more than in the case of vignettes. This will have a significant impact on overall costs of the transport services, and thus on the rates of carriers. It should be remembered, of course, about varying fees increase depending on the class of vehicle emissions. This should result in the need for change and modernization of the fleet held by such companies.

The results of both comparisons show that the introduction of a new toll collection system (although this is an expensive investment) will bring considerable revenue to the state budget.

On 1 June 2012 a wider charging system on motorways was started in Poland and the fees on this section cover also the vehicles with a maximum permissible weight less than or equal to 3.5 t. The toll rates per 1 km is approx. 0.01 € for motorcycles and approx. 0.02 € for cars.

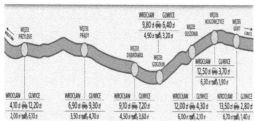

Figure 6. Map and toll stations

A description of the charging on the A4 motorway: section Gliwice (junction Sośnica) – Wrocław (junction Wrocław Bielany) will be shown as an example (Fig. 6). The A4 motorway is the primary road connecting Germany with mining - industrial district of the southern Poland. This is one of the most important Polish roads, part of a Pan-European Corridor that runs from West to East, from Dresden in Germany to Kiev in Ukraine, through Wrocław, Katowice and Krakow. The journey throughout the more than 160 kilometer section of the A4 motorway between Sośnica (near Gliwice) and Wroclaw Bielany passenger car drivers have to pay 3,5 € and motorcyclists 1,8 €.

Section from Wroclaw to the border with Germany is free of charge, since on the whole length there is no an emergency lanes and does not fulfill standards of toll motorway.

Free is also a highway in the Silesian agglomeration.

On the other hand traveling by car between Katowice and Krakow (about 70 miles) costs 4,1 €, a motorcycles 2,05 €.

Two toll collection plazas operate at the beginning and end of the section, and also toll collection points on the exits from the motorway – on the slip roads.

Manual payments can be made by bank cards (also in proximity system without entering a PIN), fuel cards and in cash. A viaAUTO equipment dedicated for cars may be also used. It will operate under the viaTOLL that charges e-toll on trucks. A viaAUTO device (Fig. 7) placed at the top of the windshield (slightly larger than a matchbox) will enable automatic registration of a gate and then charging a fee at the exit.

Figure 7. ViaAUTO device

A survey among road users would be very helpful to determine whether for road users the changes are clear, readable and fair. 100 drivers were interviewed in Silesia. The respondents every day are working as international lorry drivers in various companies.

The questions were as follows:

Question 1: Which way of charging in Poland is more cost-effective, old system– vignette (A) or new – viaTOLL (B) ?

The answer was as follows: (A) – 60 %, (B) – 40%.

Question 2: Which way of charging in Poland is more convenient to use ?

The answer was as follows: (A) – 70 %, (B) – 30%.

Question 3: Will the increased rates improve the condition of roads in Poland?
   (A)– yes, (B) - no

The answer was as follows: (A) – 20 %, (B) – 80%.

Question 4: Is the access to viaTOLL and points operating this system easy or difficult?

(A) – easily accessible
(B) – difficult to access

The answer was as follows: (A) – 50 %, (B) – 50%.

4 CONCLUSION

Both, the question and the previous calculations show that, despite the recognition that the introduction of automated toll collection system in Poland is right and despite the fact that the system had to be introduced, so far it is more expensive and less friendly.

But everybody hopes that it is certainly a "childhood" disease.

REFERENCES

Mindur L.: Technologie transportowe XXI wieku. Instytut Technologii Eksploatacji, Radom 2008.

Młyńczak J. Analysis of Intelligent Transport Systems (ITS) in transport of Upper Silesia. in: Modern transport telematics, Selected papers. Mikulski J.[ed], Berlin : Springer, 2011, pp. 164-171 Communications in Computer and Information Science ; vol. 239

Nowacki G. [ed] :Telematyka transportu drogowego. Wydawnictwo Instytutu Transportu Samochodowego, Warszawa 2008.

Rydzkowski W.: Transport. PWN, Warszawa 2009

Rozporzadzenie Ministra Infrastruktury z dnia 5 czerwca 2009 w sprawie opłat za przejazd po drogach krajowych, Dziennik Ustaw z 8 czerwca 2009 poz. 721 http://www.infor.pl/dziennik-ustaw,rok,2009,nr,86/poz,721,rozporzadzenie-ministra-infrastruktury-w-sprawie-oplat-za-przejazd-po-drogach.html 12.09.2012

http://www.dkv.pl/images/schemat-duzy.jpg 14.09.2012

http://viabox.pl/jak-dziala-system-viatoll/ 1.09.2012

http://viabox.pl/viabox/ 11.09.2012

http://viabox.pl/viaauto/ 1.09.2012

http://pliki.nto.pl.s3.amazonaws.com/pdf/op%C5%8 2ata-na-autostradzie-a4.jpg 15.09.2012